Biomolecular Structure, Function and Interactions

Biomolecular Structure, Function and Interactions

Guest Editor

Ivo Crnolatac

Basel • Beijing • Wuhan • Barcelona • Belgrade • Novi Sad • Cluj • Manchester

Guest Editor
Ivo Crnolatac
Organic Chemistry and
Biochemistry
Ruđer Bošković Institute
Zagreb
Croatia

Editorial Office
MDPI AG
Grosspeteranlage 5
4052 Basel, Switzerland

This is a reprint of the Special Issue, published open access by the journal *International Journal of Molecular Sciences* (ISSN 1422-0067), freely accessible at: https://www.mdpi.com/journal/ijms/special_issues/PK7CUKJJM8.

For citation purposes, cite each article independently as indicated on the article page online and as indicated below:

Lastname, A.A.; Lastname, B.B. Article Title. *Journal Name* **Year**, *Volume Number*, Page Range.

ISBN 978-3-7258-6340-2 (Hbk)
ISBN 978-3-7258-6341-9 (PDF)
https://doi.org/10.3390/books978-3-7258-6341-9

Cover image courtesy of Ivo Crnolatac

Contents

About the Editor

Ivo Crnolatac

Ivo Crnolatac is a Senior Scientist in the Division of Chemistry and Biochemistry, Ruđer Bošković Institute in Zagreb, Croatia. He is dedicated to comprehending the intricacies of biomolecular interactions, with a particular focus on thermodynamics of the binding processes. In addition to his position at the Ruđer Bošković Institute, he has also held senior positions in the context of industrial pharmaceutical research and at the Faculty of Pharmaceutical Sciences at KU Leuven, contributing to both fundamental and applied research in molecular biophysics and chemical biology. His work integrates experimental biophysical techniques with pharmacology and bioimaging. He has expertise in a multitude of biophysical and spectroscopic techniques. His great passion is science popularization, advocating science to students, pupils at different levels of education, and the public.

Preface

Biophysical chemistry is a truly interdisciplinary research area combining the principles of physics, biology, and chemistry to explore the processes in biological systems and their underlying physical and chemical properties. The focus of the researchers in this field is on understanding how biological molecules, such as proteins, nucleic acids, and lipids, interact with each other and with their environment at the molecular level. Thus, biomolecular interactions are a fundamental part of biophysical chemistry. The structure and function of biomacromolecules are also topics that biophysical chemists are occupied with. Furthermore, the ambition is to explain how these structures, functions, and interactions are related to processes in the cells, tissues and organisms. This Reprint contains eleven papers published in the Special issue of the *International Journal of Molecular Sciences* entitled "Biomolecular Structure, Function and Interactions". These papers bring together examples of various experimental, computational, and machine-learning methodologies.

Ivo Crnolatac
Guest Editor

International Journal of
Molecular Sciences

MDPI

Editorial

Special Issue "Biomolecular Structure, Function and Interactions"

Ivo Crnolatac

Division of Organic Chemistry and Biochemistry, Ruđer Bošković Institute, Bijenička 54, 10000 Zagreb, Croatia; icrnolat@irb.hr

Welcome to the Special Issue of *International Journal of Molecular Sciences*, titled "Biomolecular Structure, Function, and Interactions." The ten original research contributions and one review gathered here reflect a rapidly evolving field in which experimental, computational, and machine learning methodologies converge to provide deeper insights into how the sequence, structure and dynamics of biomolecules give rise to function, and how those functions shape interactions in biological systems.

The classical idea in molecular life sciences states that the linear amino acid sequence dictates the folded three-dimensional structure, which then determines the biological function of the protein. This sequence–structure–function system is a foundation; however, we now have two parallel highways that are reshaping the way we examine this axis. Advances in structural biology (X-ray crystallography, NMR, cryo-EM, single-molecule spectroscopy) continue to provide ever higher-resolution snapshots of biomolecular machines, their complexes, and conformational assemblies. On the other hand, the explosion of sequence data (e.g., UniProt now lists hundreds of millions of sequences) together with algorithmic and hardware advances in machine learning and artificial intelligence (AI) have rendered powerful data-driven models of structure, dynamics, and interaction [1,2]. Structural modelling has been transformed by AI (e.g., the success of deep neural-networks in structure prediction), and models of interaction (protein–protein, protein–small molecule) have followed suit. In this context, the contributions in this Special Issue illustrate three major trends:

Integration of experimental and computational methods—structures that were derived empirically provide scaffolds for modelling, while computational predictions guide experiment.

Machine-learning-driven abstraction—the embedding of molecular features (sequence, structure, dynamic descriptors) into predictive frameworks.

Function—an emergent property of interactions and dynamics, not a static endpoint of structure alone.

1. Highlights of This Issue

López-Cortés et al. in Contribution 1 curate chemical descriptors (for ligands), protein sequence/structure features (for odorant/pheromone-binding proteins), and functional assay data (ligand-binding responses) to then build regression models. This work appears to be the first to build a quantitative bridge across these three data types in the insect olfaction context.

The authors Gálvez-Ramírez et al. (Contribution 2) combine biochemical assays with in silico structural modelling to elucidate the molecular basis of three class A G6PD variants—a more classical structure–function inquiry still enriched with computational tools.

Shih and Hsu (Contribution 3) demonstrate how an engineered peptide can stabilize a protein of biomedical relevance, illustrating an interface of structure, function, and therapeutic potential.

Filić et al. (Contribution 4) use site-directed mutagenesis, enzymatic assays, and computational modelling to detail how conserved residues mediate multifunctionality in an industrially relevant enzyme, showing the enduring importance of classic biochemistry when accompanied with in silico insight.

A deep learning approach for understanding structure and consequently the function of *Chlorella vulgaris* hydrogenase is presented by Botticelli et al (Contribution 5).

Desai et al. (Contribution 6) used the Deep Learning Model for Transmembrane Topology Prediction and Classification (Deep TMHMM) [3] combined with UniProt to tackle the experimentally challenging field of membrane proteins.

The review by Bui and Inaba (Contribution 7) offers a comprehensive view of transporter structure–mechanism–function relationships, highlighting how conformational transitions, metal binding, and topology intersect in homeostasis.

Additional contributions (8 and 9) (e.g., "Genome-wide analysis of AUX/IAA gene families, structural insights into PBP4 from M. tuberculosis", "Cortical and Striatal Astrocytes of Neonatal Rats Display Distinct Molecular and Pharmacological Characteristics of Dopamine Uptake") round out a rich mixture of methodological and thematic diversity. These works demonstrate not just isolated advances but underline an emergent methodological system: high-throughput sequencing, machine-learning embedding or regression, structure prediction/docking, experimental validation, and ultimately function or interaction assay.

2. Outlook and Future Directions

Looking ahead, several themes merit emphasis for our research area:

From static to dynamic: While structure prediction (via AI) has achieved remarkable fidelity, capturing dynamics, ensembles and transient interactions remain a challenge. The integration of time-resolved experimental data (e.g., single-molecule, cryo-EM of flexible complexes) with machine learning will become increasingly important [4].

Interpretable machine learning (ML) and mechanistic insight: ML models can provide strong predictive performance, but mechanistic interpretability is often limited. We seek meaningful mechanistic explanations and insights. Prediction alone, no matter how precise, is not enough. The newly emerged area of explainable artificial intelligence (XAI) tries to cover this gap [5]. Hybrid approaches that embed physical constraints or known biochemical mechanisms (rather than treating ML as a "black box") will likely dominate the next wave of impactful contributions.

Deep integration of ligand/chemical space, protein sequence/structure, and functional output: The work by López-Cortés et al. (Contribution 1) in this Special Issue is a promising leap in that direction; we anticipate more studies that integrate all three modalities.

Expanding the interaction universe: Beyond protein–protein and protein–ligand interactions, interactions with nucleic acids, membranes, small metabolites, and the cell-microenvironment will be increasingly modelled using integrative pipelines. The recent "AI meets biomolecular interaction prediction" review [6] attests this trend.

Of course, the future will not only involve decoding natural biomolecular phenomena but also designing novel biomolecules (enzymes, binders, sensors) and translating structure–function knowledge into therapeutic or biotechnological applications.

3. Concluding Remarks

This Special Issue is timely because it brings together examples of how experimental rigour, computational sophistication, and machine learning innovation can be interwoven into coherent investigative threads. The mesh of these approaches represents not a sum of parts, but rather a synergistic methodology now accessible to many researchers.

I invite you to take a system-level view while reading these contributions: consider how sequence, structure, ligand, descriptor, model, and assay all interconnect, and how the methodologies themselves are evolving. I hope this Special Issue not only showcases current achievements but also stimulates new collaborations across disciplines (structural biology, data science, computational chemistry, bioengineering) and encourages the development of integrative workflows you may adopt or adapt in your own research.

I thank all authors for their excellent contributions, the reviewers for their insightful feedback, and the editorial office of *IJMS* for their support. I believe this Special Issue will serve as a valuable resource for scientists interested in the frontiers of biomolecular structure–function–interaction, and that it stimulates further advances in this rich and expanding area of research, with best wishes for your explorations of the structure, function, and interaction of biomacromolecules.

Conflicts of Interest: The authors declare no conflict of interest.

List of Contributions:

1. López-Cortés, X.A.; Lara, G.; Fernández, N.; Manríquez-Troncoso, J.M.; Venthur, H. Insight into the Relationships Between Chemical, Protein and Functional Variables in the PBP/GOBP Family in Moths Based on Machine Learning. *Int. J. Mol. Sci.* **2025**, *26*, 2302. https://doi.org/10.3390/ijms26052302.
2. Gálvez-Ramírez, A.; González-Valdez, A.; Hernández-Ochoa, B.; Canseco-Ávila, L.M.; López-Roblero, A.; Arreguin-Espinosa, R.; de la Cruz, V.P.; Hernández-Urzua, E.; Cárdenas-Rodríguez, N.; Enríquez-Flores, S.; et al. Evaluation of Three Mutations in Codon 385 of Glucose-6-Phosphate Dehydrogenase via Biochemical and In Silico Analysis. *Int. J. Mol. Sci.* **2024**, *25*, 12556. https://doi.org/10.3390/ijms252312556.
3. Shih, J.-Y.; Hsu, Y.-H.H. Peptide Activator Stabilizes DJ-1 Structure and Enhances Its Activity. *Int. J. Mol. Sci.* **2024**, *25*, 11075. https://doi.org/10.3390/ijms252011075.
4. Filić, Ž.; Bielen, A.; Šarić, E.; Ćehić, M.; Crnolatac, I.; Tomić, S.; Vujaklija, D.; Abramić, M. Evaluation of the Structure–Function Relationship of SGNH Lipase from Streptomyces rimosus by Site-Directed Mutagenesis and Computational Approach. *Int. J. Mol. Sci.* **2024**, *25*, 595. https://doi.org/10.3390/ijms25010595.
5. Botticelli, S.; La Penna, G.; Minicozzi, V.; Stellato, F.; Morante, S.; Rossi, G.; Faraloni, C. Predicting the Structure of Enzymes with Metal Cofactors: The Example of [FeFe] Hydrogenases. *Int. J. Mol. Sci.* **2024**, *25*, 3663. https://doi.org/10.3390/ijms25073663.
6. Desai, M.; Singh, A.; Pham, D.; Chowdhury, S.R.; Sun, B. Discovery and Visualization of the Hidden Relationships among N-Glycosylation, Disulfide Bonds, and Membrane Topology. *Int. J. Mol. Sci.* **2023**, *24*, 16182. https://doi.org/10.3390/ijms242216182.
7. Bui, H.B.; Inaba, K. Structures, Mechanisms, and Physiological Functions of Zinc Transporters in Different Biological Kingdoms. *Int. J. Mol. Sci.* **2024**, *25*, 3045. https://doi.org/10.3390/ijms25053045.
8. Gao, X.; Liu, X.; Zhang, H.; Cheng, L.; Wang, X.; Zhen, C.; Du, H.; Chen, Y.; Yu, H.; Zhu, B.; et al. Genome-Wide Identification, Expression, and Interaction Analysis of the Auxin Response Factor and AUX/IAA Gene Families in Vaccinium bracteatum. *Int. J. Mol. Sci.* **2024**, *25*, 8385. https://doi.org/10.3390/ijms25158385.
9. Sočan, V.; Dolinar, K.; Kržan, M. Cortical and Striatal Astrocytes of Neonatal Rats Display Distinct Molecular and Pharmacological Characteristics of Dopamine Uptake. *Int. J. Mol. Sci.* **2024**, *25*, 911. https://doi.org/10.3390/ijms25020911.

References

1. Zhang, Y.; Li, S.; Meng, K.; Sun, S. Machine Learning for Sequence and Structure-Based Protein–Ligand Interaction Prediction. *J. Chem. Inf. Model.* **2024**, *64*, 1456–1472. [CrossRef] [PubMed]
2. Lin, B.; Luo, X.; Liu, Y.; Jin, X. A comprehensive review and comparison of existing computational methods for protein function prediction. *Brief. Bioinform.* **2024**, *25*, 289. [CrossRef] [PubMed]
3. DTU. 2025. Available online: https://dtu.biolib.com/DeepTMHMM (accessed on 24 November 2025).
4. Franke, L.; Peter, C. Clustering and Analyzing Ensembles of Residue Interaction Networks from Molecular Dynamics Simulations. *J. Chem. Inf. Model.* **2025**, *65*, 11203–11214. [CrossRef] [PubMed]
5. Lavecchia, A. Explainable Artificial Intelligence in Drug Discovery: Bridging Predictive Power and Mechanistic Insight. *WIREs Comput. Mol. Sci.* **2025**, *15*, e70049. [CrossRef]
6. Wang, H.; Meng, X.; Zhang, Y. Biomolecular Interaction Prediction: The Era of AI. *Adv. Sci.* **2025**, *12*, e09501. [CrossRef] [PubMed]

International Journal of *Molecular Sciences*

MDPI

Article

Insight into the Relationships Between Chemical, Protein and Functional Variables in the PBP/GOBP Family in Moths Based on Machine Learning

Xaviera A. López-Cortés [1,2,*], Gabriel Lara [2,†], Nicolás Fernández [2,†], José M. Manríquez-Troncoso [2,†] and Herbert Venthur [3,4,*]

1 Department of Computer Sciences and Industries, Universidad Católica del Maule, Talca 3466706, Chile
2 Centro de Innovación en Ingeniería Aplicada (CIIA), Universidad Católica del Maule, Talca 3466706, Chile; gabriel.lara@alu.ucm.cl (G.L.); nicolas.fernandez@alu.ucm.cl (N.F.); manriquez.josematias@gmail.com (J.M.M.-T.)
3 Laboratorio de Química Ecológica, Departamento de Ciencias Químicas y Recursos Naturales, Facultad de Ingenieria y Ciencias, Universidad de La Frontera, Temuco 4811230, Chile
4 Centro de Investigación Biotecnológica Aplicada al Medio Ambiente (CIBAMA), Universidad de La Frontera, Temuco 4811230, Chile
* Correspondence: xlopez@ucm.cl (X.A.L.-C.); herbert.venthur@ufrontera.cl (H.V.)
† These authors contributed equally to this work.

Abstract: During their lives, insects must cope with a plethora of chemicals, of which a few will have an impact at the behavioral level. To detect these chemicals, insects use several protein families located in their main olfactory organs, the antennae. Inside the antennae, odorant-binding proteins (OBPs), as the most studied protein family, bind volatile chemicals to transport them. Pheromone-binding proteins (PBPs) and general-odorant-binding proteins (GOPBs) are two subclasses of OBPs and have evolved in moths with a putative olfactory role. Predictions for OBP–chemical interactions have remained limited, and functional data collected over the years unused. In this study, chemical, protein and functional data were curated, and related datasets were created with descriptors. Regression algorithms were implemented and their performance evaluated. Our results indicate that XGBoostRegressor exhibits the best performance (R^2 of 0.76, RMSE of 0.28 and MAE of 0.20), followed by GradientBoostingRegressor and LightGBMRegressor. To the best of our knowledge, this is the first study showing a correlation among chemical, protein and functional data, particularly in the context of the PBP/GOBP family of proteins in moths.

Keywords: chemical ecology; lepidoptera; odorant-binding proteins; artificial intelligence; ligand binding; regression algorithm

1. Introduction

Insects play an important role in ecosystems. However, as a result of globalization, invasive species have spread quickly and are now a problem in many countries [1,2]. These insect pests have an extraordinary sense of olfaction, adapting to new regions and climates using plants as hosts for feeding or oviposition [3]. In particular, moths have become serious pests throughout the world, where the cotton leaf worm *Spodoptera littoralis*, spongy moth *Lymantria dispar*, codling moth *Cydia pomonella*, oriental fruit moth *Grapholita molesta*, Indian meal moth *Plodia interpunctella*, and grapevine moth *Lobesia botrana* are few examples of highly invasive widespread polyphagous species [4].

Olfaction-driven behaviors in moths have proven to be key for the development of traps baited with odorants, either sex pheromones or attractants (i.e., semiochemicals) [5]. Traditionally, these chemicals have been identified by time-consuming methods using live insects, volatile trapping in polymers, chromatographic analysis, and, ultimately, behavioral assays in both the laboratory and field. In addition, pheromone identification has remained elusive in some species following this traditional approach [6]. More worryingly, new insects are being introduced in new countries, and pest management strategies must be implemented, involving time and expenses for local governments. Normally, insecticides are the primary and cheapest resource for insect control, and are sometimes complemented by odorant-baited traps taking advantage of the well-tuned olfactory system of moths [7]. In that sense, key odorants that elicit behavioral responses in these species must be identified [8].

In insects, the primary olfactory organs are sensilla on the antennae, inside which some protein families, called chemosensory proteins, play pivotal roles in detecting odorants (i.e., volatile organic compounds or VOCs) [9–12]. The odorants that bind to these proteins are highly specific, and in moths, it has been shown that behavioral responses can be elicited at very low concentrations (i.e., micro- or nanomolar) [13]. The first chemosensory protein was discovered in 1981 from antennae of the giant moth *Antheraea polyphemus*, and called the odorant-binding protein (OBP) [14]. More than forty years later, OBPs have become the most studied chemosensory protein and the target of choice for studying insect chemosensation, evidenced by multiple review articles [8,10,12,15–18]. Briefly, OBPs transport odorants from olfactory pores located in hair-like structures called sensilla, which are distributed across antennae. Afterwards, OBPs deliver odorants to ORs for olfactory transduction, and, ultimately, insect behavioral responses are unleashed. In moths, an evolutionarily conserved clade of OBPs are significantly expressed in antennae rather than other tissues, showing high binding affinities (i.e., K_i) to odorants with semiochemical functions [19–21].

Nowadays, genomic and transcriptomic approaches allow the identification of dozens to hundreds of insect moth OBPs, with 15–45 OBPs usually identified in each lepidopteran species [16,22]. OBPs, from olfaction-derived data, are considered the first filter of odorants in the antennae of insects, and are extensively studied [22]. To date, approximately 28,700 amino acid sequences for OBPs have been deposited in UniProt database. This is 2.3 times the number of sequences for ORs, the other olfactory proteins that recognize odorants in insects [23]. Although recent evidence suggests that OBPs appear to have both chemosensory and non-chemosensory functions, OBPs still are crucial for insect olfaction [12].

Furthermore, OBPs have become the target of choice for odorant discovery due to their inherent binding affinities [10,16]. The use of OBPs for in vitro functional evidence can overcome factors related to live insects, such as life cycle, size, abundance, and colony ñrearing. Furthermore, ample sets of chemicals can be used, accelerating the identification of odorants with behavioral effects. For instance, the fluorescence binding assay (an in vitro assay) has become frequent method to test the binding affinity of VOCs to insect OBPs, resulting in inhibition constant (K_i) values measured in the nano- or micro-molar (nM or μM, respectively) range [16,24,25]. Currently, 215 functional studies combining OBPs and VOCs have been reported, and subsequently, 622 VOCs have been counted with quantifiable data through initiatives such as iOBPdb [26], a centralized database that reunites OBP and VOC information along with their binding affinities. In moths, a particular evolutionary clade occurs with OBPs named general-odorant-binding proteins (GOBPs) and pheromone-binding proteins (PBPs), which are highly conserved among lepidopterans [27,28]. Although increasing evidence suggests that OBPs might play other chemosensory and non-chemosensory roles, research has shown that PBPs and GOBPs are still crucial in transporting sex pheromone components or both plant volatiles and sex pheromones [12,28–31]. Currently, it is known that moths have three types of PBPs, whereas butterflies have two, probably because of the nocturnal habits in moths and, therefore, their odorant-guided behaviors [28]. Likewise, it is common to identify two GOBPs in moths [16,32].

Considering the above, current initiatives have focused on comprehensively studying the main olfactory organ of insects (i.e., antennae), and related proteins that bind VOCs, providing an opportunity to identify novel behaviorally active chemicals and, consequently, use them in pest management. Hitherto, molecular and bioinformatics approaches have addressed the above with some success, where OBPs have played a role as targets [16]. However, new cheaper, reproducible and scalable methodologies are needed to identify odorants with the potential to be implemented in pest management. In this sense, advances in computer science have resulted in software capable of learning, helping in visual perception, translation between languages, speech recognition and decision-making tasks, i.e., artificial intelligence (AI) [33]. Applications of AI to biological problems are becoming powerful methods of solving biological problems at different scales [34–38]. One type of AI enables a computer to learn on its own, and this is called machine learning (ML). ML can identify patterns from databases and make predictions [39–42]. Nowadays, most of the research that integrates ML and pest management seems to be focused on identification and monitoring rather than control [43]. Thus, ML has been applied to monitoring insects' flight based on their dependence on abiotic factors, such as temperature, wind, humidity, etc. Interestingly, a neural network method with four layers (a type of ML model) was developed to track the flight of the grapevine moth L. botrana, which is highly influenced by temperature [44]. Likewise, a 79% accuracy in predicting thrip and squamous and black weevil incidence using a supervised ML algorithm in the form of logistic regression and vector machine has been reported [45]. One step further was reported for the noctuid moth *S. littoralis*, whereby the authors found the chemical space of volatiles that could elicit behavioral activity (attraction or repellency) based on the activation of odorant receptors (ORs) (i.e., SlitOR25) from a panel of 3 million compounds and using a QSAR model (as a supervised ML model) [46]. Despite the suitability of using ML models for the identification of behaviorally active odorants, current reports are from a chemical perspective. Also, sequence-based predictions have been investigated, and to the best of our knowledge, function-based predictions through ML have not been considered for the previously mentioned purposes. The closest approach to this has been through AlphaFold2 (a deep learning-based tool), predicting the 3D structure of olfactory proteins and identifying odorants with putative biological activity [47–49]. However, these studies have focused on binding characterization from a structural perspective only, and final applications (e.g., traps baited with new odorants) are still lacking. Therefore, the objective of this study was to evaluate and select suitable ML models that can integrate not only chemical and sequence descriptors (i.e., odorants and OBPs, respectively), but also functional data in the form of the inhibition constant K_{is}, which represents the binding affinity between proteins and ligands, but focusing on PBPs and GOBPs from moths. Here, a proposed methodology consisted in six main steps with collection and filtering of data, descriptor search, dataset creation, preprocessing of datasets, model selection and optimization and evaluation (Figure 1).

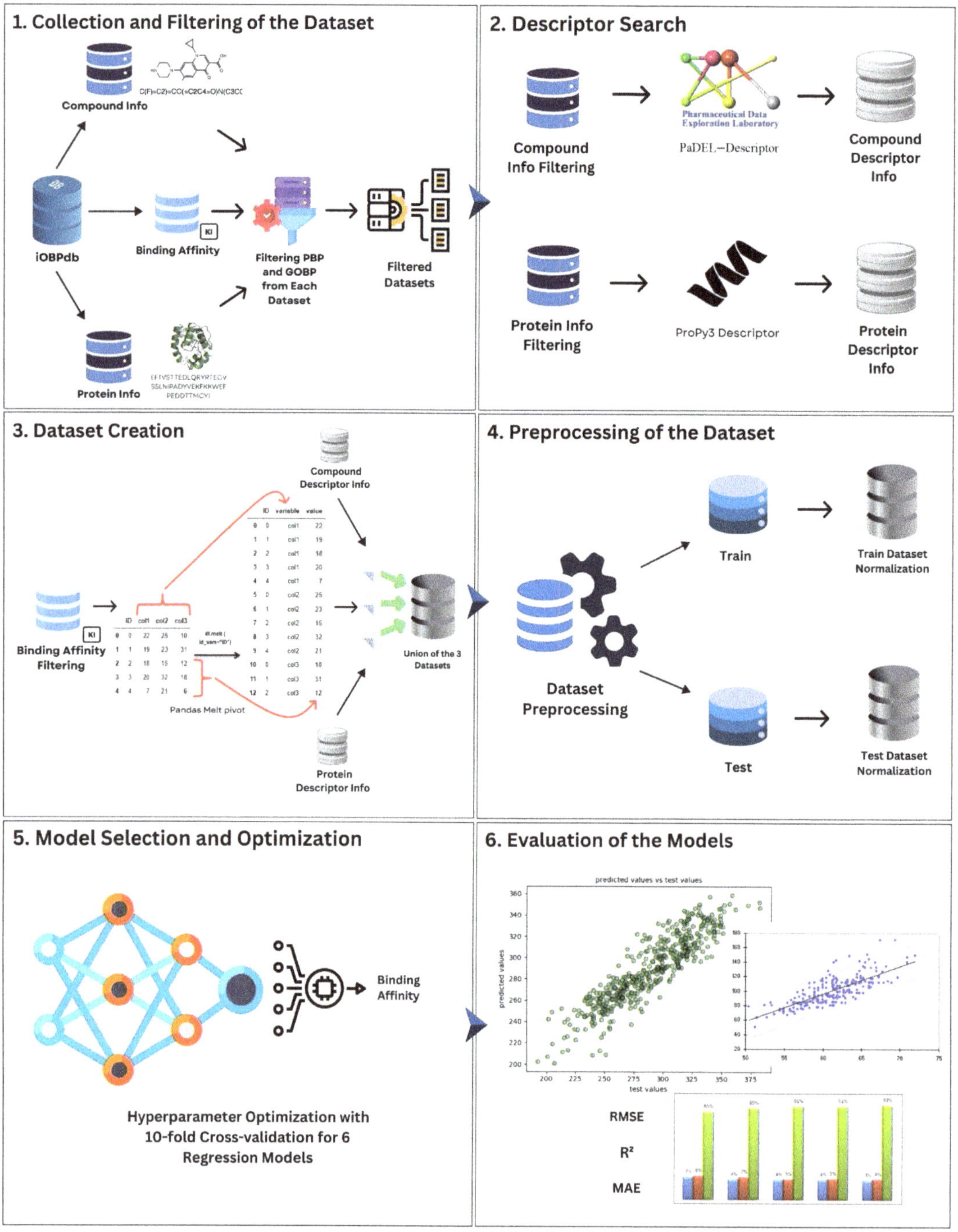

Figure 1. Scheme of proposed methodology. (**1**) Collection and filtering of PBP and GOBP related dataset. (**2**) Descriptor search based on amino acid sequences and SMILES of compounds. (**3**) Dataset creation of protein and chemical descriptors along with their respective binding affinity.

(**4**) Cluster of data divided into 80% training and 20% testing, and normalization of StandardScaler characteristics followed by transformation of binding affinities for each cluster. (**5**) Implementation of 6 ML regression models: For training, optimization for hyperparameters with ten-step cross-validation for each ML was used. (**6**) For testing, training with best parameters for 6 ML regression models was used in order to evaluate the results on plots as shown in the circles in the figure and compare performance according to Root-Mean-Square Error (RMSE), Coefficient of Determination (R^2) and Mean Absolute Error (MAE).

2. Results

The performance of the optimized models was evaluated using a cluster test (20% of the data), with the best parameters found through the Root-Mean-Square Error (RMSE), Coefficient of Determination (R^2) and Mean Absolute Error (MAE) [50]. The results from the metrics of each model are presented in Table 1 and Figure 2.

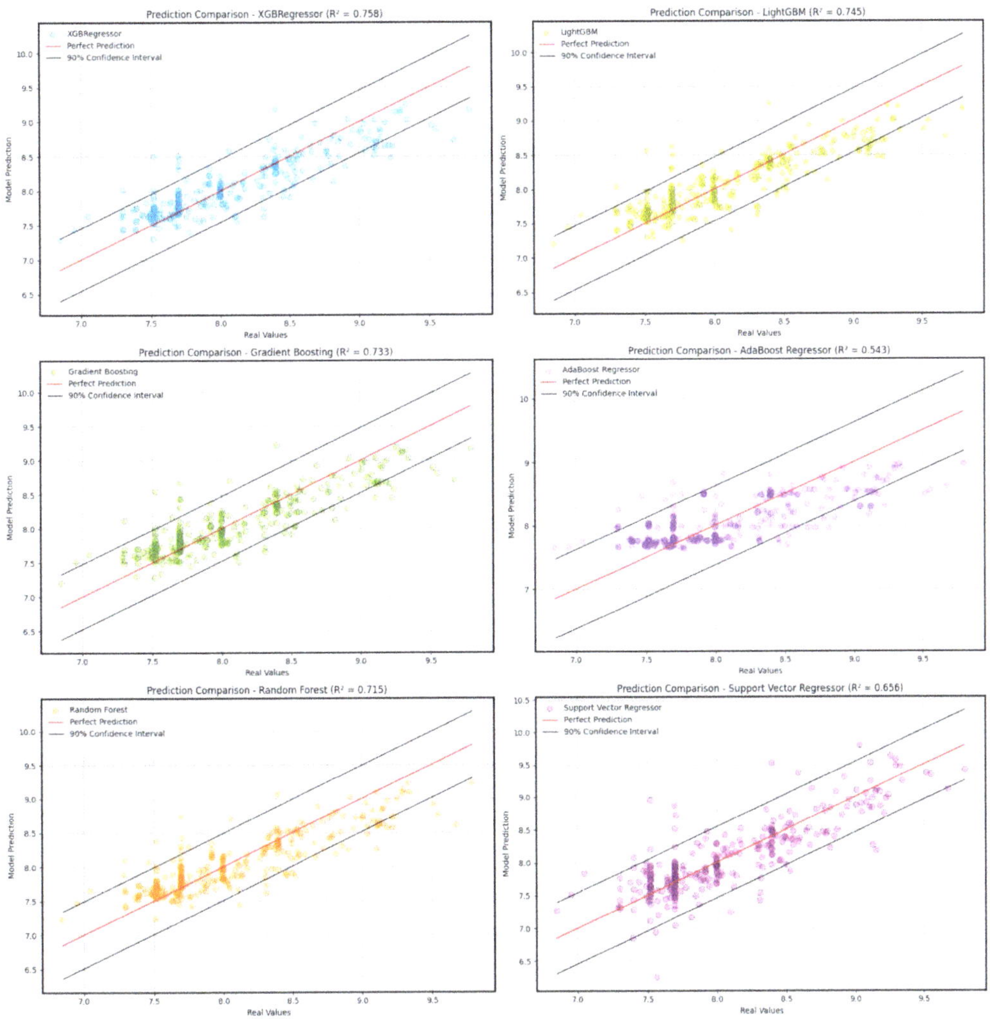

Figure 2. A Comparison of the predictions with the 20% test data using different regression models. The actual values used are from the iOBPdb database [26].

Table 1. Evaluation of predictions with 20% test.

Models	RMSE	R^2	MAE
XGBoostRegressor	**0.276**	**0.758**	**0.202**
LightGBMRegressor	0.284	0.745	0.208
GradientBoostingRegressor	0.290	0.733	0.216
AdaBoostRegressor	0.380	0.543	0.292
RandomForestRegressor	0.300	0.715	0.222
SupportVectorRegressor	0.329	0.656	0.236

In terms of accuracy, the XGBoostRegressor model showed the best performance, reaching an R^2 of 0.758, with an RMSE of 0.276 and an MAE of 0.202. This model presented the best predictions in comparison with the real values, which positions it as a suitable curated dataset. It was closely followed by GradientBoostingRegressor and LightGBM-Regressor, which showed an R^2 of 0.733 and 0.745, respectively. Both models presented similar RMSE and MAE values when compared with XGBRegressor, although with a slight decrease in accuracy.

The RandomForestRegressor also showed a satisfactory yield, with an R^2 of 0.715 and an RMSE of 0.300, which suggests a robust capacity for predictions, although inferior to other boosting methods such as XGBoostRegressor, GradientBoostingRegressor or Light-GBMRegressor. On the other hand, Support Vector Regressor (SVR) obtained an R^2 of 0.656, indicating a lower adjustment capacity in comparison with previous models.

Finally, AdaBoost Regressor presented the lowest performance of all the evaluated models in this study, with an R^2 of 0.543 and higher values of RMSE and MAE, suggesting a limited capacity to capture the relationships among variables (chemical, protein and functional).

The scatter plots (Figure 2) compare the actual values with the values predicted by each model. The 90% confidence intervals are included to assess the accuracy of the predictions. XGBoost and LightGBM show strong alignment with the perfect prediction line (red line), confirming their high predictive power. Gradient Boosting and Random Forest also present good fits, albeit with a slight dispersion. AdaBoost and SVR exhibit higher errors and higher variability in their predictions, which is reflected in their lower R^2 values.

Table 1 and Figure 2 illustrate the superiority of XGBoostRegressor in predicting affinities with fewer errors and closer to a perfect prediction line, followed by LightGBM-Regressor and GradientBoostingRegressor. On the other hand, AdaBoostRegressor is not recommended due to its high error rate and low fit. These findings suggest that decision tree-based methods with boosting are the most effective for this problem.

In order to evaluate the feature contribution, the model with the best performance for K_i prediction was used. Specifically, the XGBoost model was used to conduct Shapley value analysis. Figure 3 shows the Shapley values of the 20 features (descriptors obtained with PaDEL or Propy3) with the highest average contribution for K_i prediction. Specifically, the features TIC5, TIC4, ZMIC1, SdsCH and TIC3 correspond to molecular descriptors calculated using the PaDEL-Descriptor tool. Specifically, TIC5, TIC4 and TIC3 refer to the Total Information Content Index—Neighborhood Symmetry of the n-th order. These descriptors quantify the total structural information based on the symmetry of atoms within the molecule at different neighborhood levels (n-th order). Since the SHAP analysis shows that TIC5, TIC4 and TIC3 are the most influential features, this suggests that the structural symmetry of the molecule at different neighborhood levels is key for interaction with OBP/PBP proteins. Similarly, ZMIC1 (Z-modified Information Content Index of the first order) is a modified variant of the Information Content Index (TIC), adjusted with the Z constant, and captures the first-order neighborhood symmetry of the molecule.

Finally, the SdsCH descriptor is an electrotopological descriptor that belongs to the atom-type descriptors of the electrotopological state (E-State). These descriptors are used in cheminformatics to quantify the electronic and topological properties of specific atoms within a molecule. In this case, SdsCH specifically refers to the sum of the electrotopological state values for =CH-type atoms. The fact that these features appear in the top five SHAP values could indicate a strong correlation with the prediction of the K_i value.

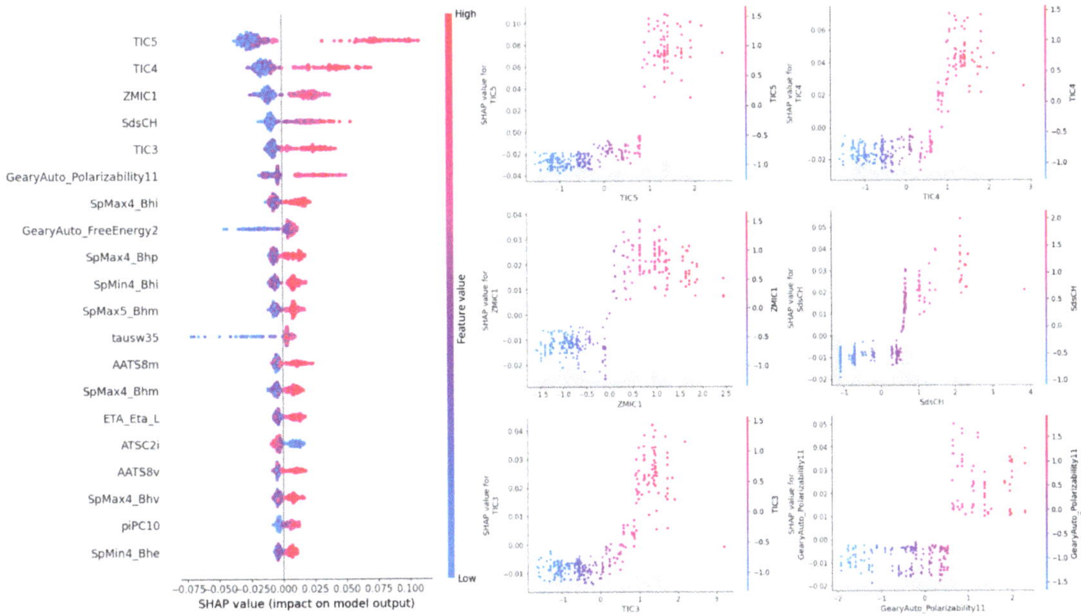

Figure 3. A SHAP (Shapley Additive exPlanations) summary plot and individual feature impact plots for the XGBoostRegressor model. The left panel shows the global importance of the features ranked by their impact on the model's prediction of binding affinity (K_i) between OBPs (PBP/GOBP family) and VOCs. Each dot represents a SHAP value for a specific interaction, with colors indicating the feature value (red: high; blue: low). The right panel presents detailed SHAP dependency plots for the five most impactful features (TIC5, TIC4, ZMIC1, SdsCH and TIC3), highlighting how variations in these descriptors influence the predicted binding affinity. The gray histograms in the dependency plots represent the distribution of feature values in the dataset.

3. Discussion

Olfaction plays a crucial role in the life cycle of an insect. Over time, research has provided insights into how odorants are recognized by a well-tuned olfactory system, especially in lepidopterans, where it all started. In 1959, the first sex pheromone was identified from the silk moth *Bombyx mori* [50,51]. To date, more than 6500 compounds have been identified and deposited in the Pherobase database [52] that mediate interactions among insects, whether sex pheromones (intraspecific) or allelochemicals (interspecific), both classed as semiochemicals [53]. Some of these chemical compounds, usually volatiles, are currently used in field traps for monitoring and control. However, their discovery is highly demanding in terms of time and laboratory expenses and dependent on insect availability. Therefore, alternative strategies that can overcome these difficulties and, at the same time, use data from insect olfaction are necessary.

Hitherto, research has focused on insect olfaction following two pathways. The first is studying chemical information from odorants with or without semiochemical function. For example, a supervised ML model through QSAR was used to screen a panel of 3 million compounds that could elicit behavioral activity (attraction or repellency) in the moth *S. littoralis* based on the activation of OR25 [54]. Similarly, a novel set of antagonistic volatile compounds were reported for Orco (a conserved insect OR co-receptor) from the fruit fly *Drosophila melanogaster* using ML models, such as a Naïve Bayesian classifier and Extended Connectivity Fingerprints. Their results suggested 2-tert-butil-6-metilfenol was the best at inhibiting behavioral responses in larvae of *D. melanogaster* [55]. In this sense, chemical, structural and functional descriptors appear to be key in decoding the odorant–protein relationship. Thus, SMILES notations have been used for finding odorant–smell relationships through deep learning approaches, such as deep neural networks (DNN) and convolutional neural networks (CNN) [56]. Likewise, the use of chemical descriptors has provided predictions of odor perception based on chemical structure [57]. ML has also been applied for OBP sequence classification through the Regularized Least-Squares Classifier (RLSC) [58]. Hence, chemicals and proteins have been analyzed independently, and therefore, the odorant–OBP relationship has not been decoded by ML algorithms.

A third pathway could be the implementation of functional properties derived from the binding of a given odorant to an OBP (i.e., K_i) and evaluated through ML algorithms. To the best of our knowledge, this approach has remained unexplored. In this study, three datasets were unified and implemented under supervised ML. Thus, XGBoostRegressor resulted in an R^2 of 0.758, and a prediction of the binding between OBPs and ligands by combining, for the first time, both chemical and protein descriptors along with functional data in the form of K_i. These findings suggest that the prediction of binding affinities in the context of OBPs is feasible. The functional properties from OBPs in moths, particularly PBPs and GOBPs, have been obtained through fluorescence-based assays. Here, chemical compounds (such as odorants) displace *N*-phenyl-1-napthylamine (1-NPN, also called a fluorescent probe) from the unique binding site present in OBPs, acting as competitors. Although a comparative study reported that K_i could change depending on the fluorescent probe, this competitive fluorescent assay is still the most widely used technique for insect OBPs [16,59]. Consequently, initiatives such as iOBPdb, as a database that reunites VOCs (or odorants as ligands), OBPs and the resulting binding affinities (i.e., K_{is}), have emerged [26]. Thanks to this database, it is possible to download current deposited data related to OBPs, VOCs and Kis separately. Furthermore, researchers can contribute their own data, helping to constantly update iOBPdb [60]. Notably, outside the insect olfaction field, other studies have used unified functional and chemical data for ML algorithms. An example is the identification of two lactones as potential inhibitors of acetylcholinesterase (AChE), an important target of research in Alzheimer's disease, found based on 7032 molecules with IC_{50} and another 8593 secondary metabolites through classification models [61]. Similarly, datasets of cytochrome P450 inhibitors and IC_{50} were implemented for Random Forest and SVM, resulting in over 80% accuracy [62].

With respect to the models implemented, it is possible to say that the best performance in predicting K_i corresponded to the XGBoostRegressor model, which outperformed LightGBMRegressor, GradientBoostingRegressor, AdaBoostRegressor, RandomForestRegressor and SupportVectorRegressor. The Shapley value analysis (3) revealed that most of the features with high predictive contributions were derived from the ligand, specifically the PaDEL-Descriptors. TIC descriptors measure the topological structural complexity of a molecule, while ZMIC evaluates its connectivity based on Zagreb indices. Among the top five contributors to the prediction of K_i, it is worth noting that TIC5, TIC4, ZMIC1 and TIC3 represent ligand descriptors. These indices capture key aspects of molecular structural diversity, including electrical, geometric, symmetric and topological properties. This finding underscores the critical role of ligand structural features in influencing binding affinity predictions. Moreover, the SdsCH descriptor, which pertains to protein-specific characteristics, particularly hydrophobicity, further highlights the importance of protein–ligand interactions in the predictive model. The integration of both ligand-centric and protein-specific features suggests a multifaceted approach to modeling, where the roles of the structural and physicochemical properties of both entities are essential for accurate predictions. These observations provide valuable insights into the mechanisms driving protein–ligand interactions, particularly the binding of PBPs and GOBPs to VOCs, and pave the way for refining future computational models.

Although the implemented models yielded promising results, due to the predictive power obtained through R^2, there are several avenues for enhancing the proposed approach. First, increasing both the volume and the diversity of the data could significantly improve the generalizability of the model [63]. Likewise, expanding the dataset by incorporating additional OBP and VOC protein data from various databases would extend the analysis to encompass a broader range of species and chemical compounds. Moreover, performance evaluation could be beneficial for the inclusion of alternative metrics, such as the Mean Absolute Percentage Error (MAPE), Concordance Index (CI) or domain-specific metrics like Receive Operating Characteristic (ROC) and Regression Error Characteristic Curves (REC), providing deeper insights into critical protein–ligand interactions.

Another potential enhancement involves adopting periodic retraining techniques to account for new data or leveraging transfer learning models to capitalize on prior knowledge from related problems. Furthermore, exploring advanced neural network architectures, such as DeepDTA or Affinity2Vec [64,65], could better capture the intricate complexities of protein–chemical interactions. Finally, employing advanced feature engineering strategies, such as dimensionality reduction tailored to domain-specific relationships or custom feature creation, can optimize data representation and improve predictive accuracy [66,67].

These advances will not only enhance the robustness of the model, but also broaden its applicability in future studies on ecological chemistry and integrated pest management.

4. Materials and Methods

4.1. Data and Preproccessing

4.1.1. Data

The dataset was extracted from the iOBPdb database [26], a free bioinformatics resource containing information on odorant-binding proteins (OBPs), volatile organic compounds (VOCs) and the interaction affinity between them. In particular, three specific subsets of data were used:

- Odorant-binding proteins

This database contains information on 436 OBPs such as their names, species, cystine counts, protein types, and amino acid sequences with and without signal peptides, among other characteristics.

- Volatile organic compounds

This database contains information on 621 VOCs such as their names and properties like molecular formulas, SMILES and functional groups to which they belong.

- Binding affinity (K_i)

This database contains a 621 × 436 matrix that records the binding affinity (K_i) between VOCs and OBPs.

4.1.2. Preprocessing

The process of filtering and transforming the iOBPdb data began by selecting OBPs only from species in the order Lepidoptera that fit the PBP and GOBP subcategories, applying a taxonomic filter to exclude non-Lepidopteran species using the Python Pandas library [68,69]. Next, volatile organic compounds (VOCs) related to these species were identified, ensuring that only VOCs directly linked to Lepidoptera remained. The binding affinity values for each protein–ligand pair were then extracted from iOBPdb (https://www.iobpdb.com (accessed on 17 April 2024)) [26,60], providing key interaction data between odorant-binding proteins (PBPs and GOBPs) and VOCs. To eliminate redundancy, duplicate SMILES structures and repeated amino acid sequences were identified and removed, ensuring the uniqueness of each protein–ligand pair. This resulted in three new sets of 110 OBPs, 254 VOCs and a 254 × 110 affinity matrix reflecting the binding affinity (K_i) of each protein–ligand combination (Figure 4).

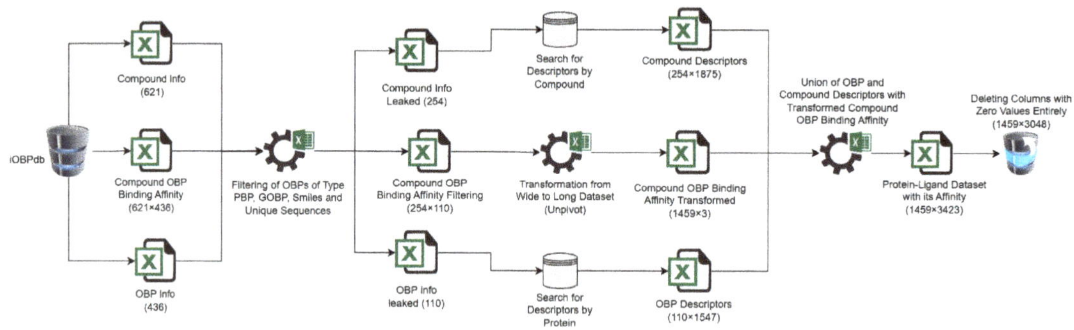

Figure 4. Flow chart scheme of proposed methodology (available on the Supplementary Material).

4.1.3. Extraction of Descriptors

For the 254 VOCs, using the PaDEL-Descriptor library [70] (through PaDEL-Py, its Python 3.9 implementation [71]), 1875 descriptors were generated (431 3D features and 1444 2D and 1D features) per compound, including physicochemical, topological, geometrical and other characteristics. For the 110 OBPs generated through Python's PyPro3 [72] library, 1547 descriptors were calculated based on the amino acid sequence without signal peptides, such as amino acid composition, physicochemical properties, hydrophobicity and other relevant characteristics.

4.1.4. Dataset Creation

The affinity matrix was transformed into a long format, where each row represents a single VOC-OBP interaction with its binding affinity value (K_i), yielding a dataset of 1459 compound–protein interactions. Two merges were performed to enrich the dataset with descriptors of VOCs and OBPs with the Merge function of the Python Pandas library [68,71]. Features with zero were then removed in their entirety, generating a final set of 1459 interactions and 3048 descriptor features.

4.1.5. Machine Learning Models

- Training and testing data.

To assess the generalizability of the models, the dataset was split into 80% (1167 × 3048) for training and 20% (292 × 3048) for testing with a seed of 41 to ensure reproducibility. The training subset was used to fit and optimize the models, while the test set allowed performance to be assessed on unseen data, ensuring a robust measure of their accuracy.

- Dataset preprocessing

Considering variation in data scales and ranges, normalization was applied to improve the performance and stability of the models. In particular, the StandardScaler method of the Scikit-Learn library [73] was used, which transforms each characteristic according to the mean and standard deviation of its values, allowing a normalized distribution to be obtained with a mean of 0 and a standard deviation of 1:

$$X_{scaled} = \frac{X - \mu}{\sigma} \quad (1)$$

where the definitions are as follows:

- X is the original value of the feature;
- μ is the mean of the feature in the dataset;
- σ is the standard deviation of the feature.

In addition, the binding affinity (K_i) values were transformed into a logarithmic scale (pK_i), similar to the SimBoost, DeepDTA and Affinity2Vec [54,65–73] methods, by applying the following equation:

$$pKi = -\log 1\left(\frac{K_i}{1 \times 10^9}\right) \quad (2)$$

- K_i represents the value of the inhibition constant in units of molarity (M).
- The factor 1×10^9 (or 10^9 nM) is used to convert Ki to nanomolarity (nM) so that the resulting logarithmic values are on a comparable scale.

This transformation converts the value of K_i into a scale that is easier to interpret. These values in the dataset range from 6.30 to 9.80, where high values of pK_i indicate a strong binding affinity (i.e., lower K_i) and low values indicate a weak affinity. To obtain the affinities on their original scale, just apply the inverse function (from pK_i to K_i) to convert the pK_i values predicted by the model back to molarity units. The inverse formula to recover K_i from pK_i is

$$K_i = 10^{9-pK_i} \quad (3)$$

4.2. *Models's Implementation*

The regression algorithms implemented in this study corresponded to supervised learning, where the following regression models were applied: Gradient Boosting (GB), AdaBoost (AB), Random Forest (RF) and Support Vector Regressor (SVR) [73]. XGBRegressor from the XGBoost library [74] was also applied, as was LGBMRegressor from the LightGBM library [75]. A brief theoretical description of each is given below:

- XGBoost Regressor

This algorithm represents an advanced decision tree-based boosting method designed to continuously improve its predictions. This model adjusts multiple trees in sequence, with each tree attempting to correct the errors of its predecessors. XGBRegressor's ability to perform fine-tuning, through gradient optimization, allows for excellent accuracy in complex regression problems. Its flexibility and robustness to overfitting make it a preferred choice for high-dimensional regression scenarios and heterogeneous data.

- LightGBM Regressor

This is a boosting algorithm that is based on decision trees and distinguished by its unique focus on growing leaves instead of levels. This approach allows for greater accuracy in less time, optimizing memory usage and reducing training times. Its ability to handle large data volumes and high dimensionality makes it especially valuable in regression problems where a balance between speed and accuracy is sought. In addition, its dynamic fitting and sparse data handling capabilities position it as an effective tool in advanced applications.

- Gradient Boosting Regressor

This is a sequential boosting algorithm that optimizes performance by combining multiple simple models. Each subsequent model corrects the errors of the previous one through gradient descent, allowing for a steady improvement in prediction accuracy. This model is especially effective in regression problems that require a high level of accuracy and is able to capture complex relationships in the data without extensive pre-processing.

- AdaBoost Regressor

This is an iterative boosting algorithm that consecutively fits simple models, paying more attention to mispredicted observations at each iteration. This method allows for continuous model fitting, resulting in improved accuracy without the need for complex configurations. AdaBoost is a reliable choice in regression applications where an adaptive and fast-fitting model is needed, providing effective solutions to problems of moderate complexity.

- Random Forest Regressor

This is an ensemble algorithm that builds multiple decision trees and averages their results to obtain a stable and accurate prediction. This model is highly effective in reducing overfitting and is especially valuable for handling high-dimensional data and noise. Its ability to maintain a balance between accuracy and efficiency makes it a reliable and versatile tool in a wide range of regression applications.

- Support Vector Regressor

This is a support vector machine-based algorithm designed to find the optimal hyperplane that minimizes prediction error. This approach allows SVR to efficiently handle non-linear relationships and produce accurate predictions even in datasets with high variability. Thanks to its ability to control the sensitivity of the model to extreme data, SVR is a powerful tool in regression scenarios where accuracy is required under conditions of complexity and non-linearity.

Hyperparameter Optimization and Cross-Validation

To maximize the accuracy of the models and avoid overfitting, hyperparameter optimization was performed with 100 evals for each model along with thorough cross-validation at 80% of the data. Hyperparameter optimization was implemented using Hyperopt [76], a Bayesian optimization method that searches for the best set of parameters for each model by iteratively evaluating different combinations of hyperparameters. The search spaces for each parameter were defined in terms of appropriate ranges and values for each specific model, maximizing the accuracy and stability of the predictions.

To evaluate the performance of each set of hyperparameters, Scikit-Learn's cross-validate feature was used with cross-validation of 10 partitions (cv = 10). This technique divides the training data into 5 subsets, where each partition is used once as a test set, while the remaining partitions are used for model training. The results obtained from each partition were averaged to obtain a global performance metric, using the RMSE, R^2 and MAE [77] as the main metrics. This combination of optimization and cross-validation provides a robust model and prevents the fit from being influenced by a single test dataset, which improves the generalizability and accuracy of the final model. The optimal hyperparameters that were used in this experiment to develop the prediction models are shown in Table 2.

Table 2. Hyperparameters explored and optimal values for models implemented.

Model	Parameters	Hyperparameter Search	Optimal Value
XGBoostRegressor	n_estimators	[700, 1200]	800
	learning_rate	[0.009, 0.03]	0.0188610
	max_depth	[10, 15]	10
	min_child_weight	[5, 10]	9
	gamma	[0.00, 0.005]	0.002323
	colsample_bytree	[0.3, 0.6]	0.392232
	subsample	[0.6, 0.9]	0.564263
	reg_alpha	[0.5, 1.0]	0.683823
	reg_lambda	[1.5, 2.0]	1.711287
LightGBMRegressor	n_estimators	[700, 1200]	900
	learning_rate	[0.009, 0.03]	0.0222877
	max_depth	[10, 20]	13
	num_leaves	[20, 150]	148
	min_child_weight	[5, 10]	7
	subsample	[0.5, 1.0]	0.661123
	colsample_bytree	[0.3, 0.8]	0.305414
	reg_alpha	[0, 2]	0.316713
	reg_lambda	[0, 3]	1.579887
GradientBoostingRegressor	n_estimators	[100, 500]	400
	learning_rate	[0.009, 0.03]	0.0265516
	max_depth	[5, 15]	5
	Subsample	[0.5, 1.0]	0.687177
	min_samples_split	[2, 10]	10
	min_samples_leaf	[1, 10]	1
	max_features	[0.1, 0.5]	0.260379
AdaBoostRegressor	n_estimators	[100, 500]	100
	learning_rate	[0.009, 0.03]	0.00996526
RandomForestRegressor	n_estimators	[700, 1200]	1000
	max_depth	[3, 20]	16
	min_samples_split	[2, 20]	5
	min_samples_leaf	[1, 20]	1
SupportVectorRegressor	C	[1000, 5000]	1000
	epsilon	[0.009, 0.03]	0.029727
	degree	[1, 15]	13

4.3. Models' Performance Evaluation

4.3.1. Root-Mean-Square Error (RMSE)

The Root-Mean-Square Error (RMSE) is a robust metric that measures the accuracy of prediction models by assessing the average deviation of predictions from actual values. This value, expressed in the same units as the target variable, facilitates direct and comparative interpretation of the error. A low RMSE [77] indicates that the model fits the data well, thus representing an accurate and reliable prediction. This metric is notable for its sensitivity to large errors, helping to identify areas for improvement in model accuracy.

$$\mathrm{RMSE} = \sqrt{\frac{1}{n}\sum_{i=1}^{n}(y_i - \hat{y}_i)^2} \quad (4)$$

- n: total number of observations.
- y_i: real value of the observation i.
- $\hat{y}_i$: predicted value for the observation i.

4.3.2. Coefficient of Determination

The Coefficient of Determination (R^2) is a metric that assesses the proportion of variability in the target variable that the model is able to explain. This indicator, expressed as a value between 0 and 1, provides a clear understanding of the effectiveness of the model. A high R^2 [77] suggests a satisfactory fit, demonstrating that the model adequately captures the trends in the data.

$$R^2 = 1 - \frac{\sum_{i=1}^{n}(y_i - \hat{y}_i)^2}{\sum_{i=1}^{n}(y_i - \overline{y})^2} \quad (5)$$

- y_i: real value of the observation i.
- $\hat{y}_i$: predicted value for the observation i.
- $\overline{y}$: mean of all real values y_i.
- n: total number of observations (256).

4.3.3. Mean Absolute Error

The Mean Absolute Error (MAE) measures the accuracy of the model by calculating the average of the absolute differences between predictions and actual values. This metric provides a clear and direct view of model performance as it reflects the average error in absolute terms and is less sensitive to outliers. A low MAE [77] reflects the model's ability to consistently make accurate predictions, inspiring confidence in the robustness and applicability of the model in real-world scenarios.

$$\text{MAE} = \frac{1}{n}\sum_{i=1}^{n}|y_i - \hat{y}_i| \quad (6)$$

- n: total number of observations.
- y_i: real value of the observation i.
- $\hat{y}_i$: predicted value for the observation i.

4.3.4. Confidence Interval

The Confidence Interval (CI) assesses the accuracy of the model predictions and provides a range of uncertainty in the estimates. The Confidence Interval [78,79] was calculated based on the standard deviation of the residuals. This interval allows us to estimate the range within which most predictions are expected to lie, given a specific confidence level.

This approach assumes that the residuals are approximately normally distributed and therefore allows for the construction of a symmetric confidence interval around each prediction. By plotting these intervals alongside the predictions, it is possible to visualize the expected variability in the model and to assess whether its predictions are concentrated close to the true values or whether there is high dispersion. Figure 2 compares the predictions of each model against the actual values, with a confidence interval of 90%.

4.4. SHAP Values

SHAP (Shapley Additive exPlanation) analysis was used to interpret the importance and impact of key features, specifically aiming to identify which chemical or protein descriptors have the greatest impact on the binding affinity predicted by the model. In this study, SHAP values were calculated for the XGBoostRegressor model, considering its performance and compatibility with the library.

5. Conclusions

The characterization of binding proteins such as PBPs and GOBPs, and their interaction with VOCs in lepidopterans, show their relevance in odorant detection as well as their potential application within an integrated pest management strategy. In this study, the performed analysis integrated both chemical and protein descriptors along with functional data K_i, evidencing that approaches based on ML are effective tools for decoding complex protein–ligand relationships.

Notably, this study represents the first effort to focus exclusively on predicting K_i binding affinity for odor-binding proteins specifically associated with pheromones and general odors in Lepidoptera. This novel approach highlights an underexplored area within protein–drug ligand research, providing a foundation for future investigations into the unique molecular interactions of these specialized proteins.

Notably, these findings highlight the value of computational methodologies to overcome some limitations in traditional experimental approaches, such as high dependency on live insects, related economical costs and challenges associated with the throughput identification of bioactive chemicals (e.g., semiochemicals). The capacity of prediction that has been demonstrated could allow the identification of novel bioactive compounds, which could be used as attractants or repellents for pest control. Finally, this study reinforces the utility of predictive models to integrate both molecular and functional data, helping to understand the depth with which insects can detect chemicals from the environment. It is believed that this approach will form the basis for future research around chemical ecology and applied biotechnology in the context of insect pest control.

Supplementary Materials: This supplementary material can be accessed through the link https://github.com/Glarah453/ML_obps_vocs (accessed on 18 January 2025). All files are accompanied by detailed documentation and practical examples to facilitate the reproducibility of the results. The source code, processed datasets and scripts used to reproduce the experiments presented in this study are available in the public GitHub repository "ML_obps_vocs".

Author Contributions: Conceptualization, X.A.L.-C. and H.V.; methodology, X.A.L.-C., J.M.M.-T., G.L. and N.F.; software, G.L., N.F. and J.M.M.-T.; validation, X.A.L.-C., G.L. and N.F.; formal analysis, G.L. and N.F.; investigation, X.A.L.-C., G.L. and N.F.; resources, X.A.L.-C.; data curation, G.L. and N.F.; writing—original draft preparation, H.V. and X.A.L.-C.; writing—review and editing, H.V. and X.A.L.-C.; visualization, X.A.L.-C., J.M.M.-T., G.L. and N.F.; supervision, X.A.L.-C.; project administration, H.V.; funding acquisition, H.V. All authors have read and agreed to the published version of the manuscript.

Funding: This research was funded by Agencia Nacional de Investigación y Desarrollo (ANID), Fondecyt Regular, grant number 1240459.

Institutional Review Board Statement: Not applicable.

Informed Consent Statement: Not applicable.

Data Availability Statement: Available on request from the corresponding author.

Acknowledgments: We would like to thank the Research Projects ANID Fondecyt de iniciación en Investigación 2022 (No. 11220897) and Fondecyt Regular (No. 1240459).

Conflicts of Interest: The authors declare no conflicts of interest.

References

1. Hill, M.P.; Clusella-Trullas, S.; Terblanche, J.S.; Richardson, D.M. Drivers, impacts, mechanisms and adaptation in insect invasions. *Biol. Invasions* **2016**, *18*, 883–891. [CrossRef]
2. Bertelsmeier, C. Globalization and the anthropogenic spread of invasive social insects. *Curr. Opin. Insect Sci.* **2021**, *46*, 16–23. [CrossRef] [PubMed]

3. Yang, L.H.; Gratton, C. Insects as drivers of ecosystem processes. *Curr. Opin. Insect Sci.* **2014**, *2*, 26–32. [CrossRef]
4. Balaško, M.K.; Bažok, R.; Mikac, K.M.; Lemic, D.; Živković, I.P. Pest Management Challenges and Control Practices in Codling Moth: A Review. *Insects* **2020**, *11*, 38. [CrossRef]
5. Zhang, D.-D.; Löfstedt, C. Moth pheromone receptors: Gene sequences, function, and evolution. *Front. Ecol. Evol.* **2015**, *2*, 00105. [CrossRef]
6. Sullivan, B.T.; Clarke, S.R. Semiochemicals for management of the southern pine beetle (Coleoptera: Curculionidae: Scolytinae): Successes, failures, and obstacles to progress. *Can. Entomol.* **2020**, *153*, 36–61. [CrossRef]
7. Vassiliou, V.A. Effectiveness of Insecticides in Controlling the First and Second Generations of the Lobesia botrana (Lepidoptera: Tortricidae) in Table Grapes. *J. Econ. Entomol.* **2011**, *104*, 580–585. [CrossRef] [PubMed]
8. Mori, N.; Noge, K. Recent advances in chemical ecology: Complex interactions mediated by molecules. *Biosci. Biotechnol. Biochem.* **2021**, *85*, 33–41. [CrossRef]
9. Zhou, J. Odorant-binding proteins in insects. *Vitam. Horm.* **2010**, *83*, 241–272. [CrossRef] [PubMed]
10. Leal, W.S. Odorant Reception in Insects: Roles of Receptors, Binding Proteins, and Degrading Enzymes. *Annu. Rev. Entomol.* **2013**, *58*, 373–391. [CrossRef]
11. Pelosi, P.; Iovinella, I.; Zhu, J.; Wang, G.; Dani, F.R. Beyond chemoreception: Diverse tasks of soluble olfactory proteins in insects. *Biol. Rev.* **2018**, *93*, 184–200. [CrossRef] [PubMed]
12. Rihani, K.; Ferveur, J.-F.; Briand, L. The 40-Year Mystery of Insect Odorant-Binding Proteins. *Biomolecules* **2021**, *11*, 509. [CrossRef] [PubMed] [PubMed Central]
13. Sweeney, J.D.; McLean, J.A.; Friskie, L.M. Roles of minor components in pheromone-mediated behavior of western spruce budworm male moths. *J. Chem. Ecol.* **1990**, *16*, 1517–1530. [CrossRef]
14. Vogt, R.G.; Riddiford, L.M. Pheromone binding and inactivation by moth antennae. *Nature* **1981**, *293*, 161–163. [CrossRef] [PubMed]
15. Pelosi, P.; Maida, R. Odorant-binding proteins in insects. *Comp. Biochem. Physiol. Part B Biochem. Mol. Biol.* **1995**, *111*, 503–514. [CrossRef] [PubMed]
16. Venthur, H.; Mutis, A.; Zhou, J.; Quiroz, A. Ligand binding and homology modelling of insect odorant-binding proteins. *Physiol. Entomol.* **2014**, *39*, 183–198. [CrossRef]
17. Venthur, H.; Zhou, J.-J. Odorant Receptors and Odorant-Binding Proteins as Insect Pest Control Targets: A Comparative Analysis. *Front. Physiol.* **2018**, *9*, 1163. [CrossRef]
18. Ha, T.S.; Smith, D.P. Recent Insights into Insect Olfactory Receptors and Odorant-Binding Proteins. *Insects* **2022**, *13*, 926. [CrossRef]
19. Venthur, H.; Machuca, J.; Godoy, R.; Palma-Millanao, R.; Zhou, J.; Larama, G.; Bardehle, L.; Quiroz, A.; Ceballos, R.; Mutis, A. Structural investigation of selective binding dynamics for the pheromone-binding protein 1 of the grapevine moth, *Lobesia botrana*. *Arch. Insect Biochem. Physiol.* **2019**, *101*, e21557. [CrossRef]
20. Luo, Y.; Chen, X.; Xu, S.; Li, B.; Luo, K.; Li, G. Functional Role of Odorant-Binding Proteins in Response to Sex Pheromone Component Z8-14:Ac in *Grapholita molesta* (Busck). *Insects* **2024**, *15*, 918. [CrossRef]
21. Yang, H.-H.; Li, S.-P.; Yin, M.-Z.; Zhu, X.-Y.; Li, J.-B.; Zhang, Y.-N.; Li, X.-M. Functional differentiation of two general odorant-binding proteins to sex pheromones in Spodoptera frugiperda. *Pestic. Biochem. Physiol.* **2023**, *191*, 105348. [CrossRef] [PubMed]
22. Lizana, P.; Mutis, A.; Quiroz, A.; Venthur, H. Insights Into Chemosensory Proteins From Non-Model Insects: Advances and Perspectives in the Context of Pest Management. *Front. Physiol.* **2022**, *13*, 924750. [CrossRef]
23. El-Sayed, A.M.; Ganji, S.; Gross, J.; Giesen, N.; Rid, M.; Lo, P.L.; Kokeny, A.; Unelius, C.R. Climate change risk to pheromone application in pest management. *Sci. Nat.* **2021**, *108*, 47. [CrossRef]
24. Campanacci, V.; Krieger, J.; Bette, S.; Sturgis, J.N.; Lartigue, A.; Cambillau, C.; Breer, H.; Tegoni, M. Revisiting the Specificity of Mamestra brassicaeand Antheraea polyphemus Pheromone-binding Proteins with a Fluorescence Binding Assay. *J. Biol. Chem.* **2001**, *276*, 20078–20084. [CrossRef] [PubMed]
25. Ban, L.; Scaloni, A.; D'Ambrosio, C.; Zhang, L.; Yan, Y.; Pelosi, P. Biochemical characterization and bacterial expression of an odorant-binding protein from Locusta migratoria. *Cell. Mol. Life Sci.* **2003**, *60*, 390–400. [CrossRef] [PubMed] [PubMed Central]
26. Shukla, S.; Nakano-Baker, O.; Godin, D.; MacKenzie, D.; Sarikaya, M. iOBPdb A Database for Experimentally Determined Functional Characterization of Insect Odorant Binding Proteins. *Sci. Data* **2023**, *10*, 295. [CrossRef]
27. Gong, D.-P.; Zhang, H.-J.; Zhao, P.; Xia, Q.-Y.; Xiang, Z.-H. The Odorant Binding Protein Gene Family from the Genome of Silkworm, Bombyx mori. *BMC Genom.* **2009**, *10*, 332. [CrossRef]
28. Vogt, R.G.; Große-Wilde, E.; Zhou, J.-J. The Lepidoptera Odorant Binding Protein gene family: Gene gain and loss within the GOBP/PBP complex of moths and butterflies. *Insect Biochem. Mol. Biol.* **2015**, *62*, 142–153. [CrossRef] [PubMed]
29. Sandler, B.H.; Nikonova, L.; Leal, W.S.; Clardy, J. Sexual attraction in the silkworm moth: Structure of the pheromone-binding-protein–bombykol complex. *Chem. Biol.* **2000**, *7*, 143–151. [CrossRef] [PubMed]
30. Lautenschlager, C.; Leal, W.S.; Clardy, J. Coil-to-helix transition and ligand release of Bombyx mori pheromone-binding protein. *Biochem. Biophys. Res. Commun.* **2005**, *335*, 1044–1050. [CrossRef] [PubMed]

31. Zhou, J.-J.; Robertson, G.; He, X.; Dufour, S.; Hooper, A.M.; Pickett, J.A.; Keep, N.H.; Field, L.M. Characterisation of Bombyx mori Odorant-binding Proteins Reveals that a General Odorant-binding Protein Discriminates Between Sex Pheromone Components. *J. Mol. Biol.* **2009**, *389*, 529–545. [CrossRef] [PubMed]
32. Maida, R.; Mameli, M.; Müller, B.; Krieger, J.; Steinbrecht, R.A. The expression pattern of four odorant-binding proteins in male and female silk moths, Bombyx mori. *J. Neurocytol.* **2005**, *34*, 149–163. [CrossRef] [PubMed]
33. Torresen, J. A Review of Future and Ethical Perspectives of Robotics and AI. *Front. Robot. AI* **2018**, *4*, 75. [CrossRef]
34. López-Cortés, X.A.; Matamala, F.; Venegas, B.; Rivera, C. Machine-Learning Applications in Oral Cancer: A Systematic Review. *Appl. Sci.* **2022**, *12*, 5715. [CrossRef]
35. López-Cortés, X.A.; Nachtigall, F.M.; Olate, V.R.; Araya, M.; Oyanedel, S.; Diaz, V.; Jakob, E.; Ríos-Momberg, M.; Santos, L.S. Fast detection of pathogens in salmon farming industry. *Aquaculture* **2017**, *470*, 17–24. [CrossRef]
36. Tapia-Castillo, A.; Carvajal, C.A.; López-Cortés, X.; Vecchiola, A.; Fardella, C.E. Novel metabolomic profile of subjects with non-classic apparent mineralocorticoid excess. *Sci. Rep.* **2021**, *11*, 17156. [CrossRef]
37. Olate-Olave, V.R.; Guzmán, L.; López-Cortés, X.A.; Cornejo, R.; Nachtigall, F.M.; Doorn, M.; Santos, L.S.; Bejarano, A. Comparison of Chilean honeys through MALDI-TOF-MS profiling and evaluation of their antioxidant and antibacterial potential. *Ann. Agric. Sci.* **2021**, *66*, 152–161. [CrossRef]
38. López-Cortés, X.A.; Manríquez-Troncoso, J.M.; Kandalaft-Letelier, J.; Cuadros-Orellana, S. Machine learning and matrix-assisted laser desorption/ionization time-of-flight mass spectra for antimicrobial resistance prediction: A systematic review of recent advancements and future development. *J. Chromatogr. A* **2024**, *1734*, 465262. [CrossRef] [PubMed]
39. López-Cortés, X.A.; Manríquez-Troncoso, J.M.; Hernández-García, R.; Peralta, D. MSDeepAMR: Antimicrobial resistance prediction based on deep neural networks and transfer learning. *Front. Microbiol.* **2024**, *15*, 1361795. [CrossRef]
40. Astudillo, C.A.; López-Cortés, X.A.; Ocque, E.; Manríquez-Troncoso, J.M. Multi-label classification to predict antibiotic resistance from raw clinical MALDI-TOF mass spectrometry data. *Sci. Rep.* **2024**, *14*, 31283. [CrossRef]
41. López-Cortés, X.A.; Manríquez-Troncoso, J.M.; Sepúlveda, A.Y.; Soto, P.S. Integrating Machine Learning with MALDI-TOF Mass Spectrometry for Rapid and Accurate Antimicrobial Resistance Detection in Clinical Pathogens. *Int. J. Mol. Sci.* **2025**, *26*, 1140. [CrossRef] [PubMed]
42. Alzubi, J.; Nayyar, A.; Kumar, A. Machine Learning from Theory to Algorithms: An Overview. *J. Phys. Conf. Ser.* **2018**, *1142*, 012012. [CrossRef]
43. Lima, M.C.F.; Leandro, M.E.D.d.A.; Valero, C.; Coronel, L.C.P.; Bazzo, C.O.G. Automatic Detection and Monitoring of Insect Pests—A Review. *Agriculture* **2020**, *10*, 161. [CrossRef]
44. Balduque-Gil, J.; Lacueva-Pérez, F.J.; Labata-Lezaun, G.; Del-Hoyo-Alonso, R.; Ilarri, S.; Sánchez-Hernández, E.; Martín-Ramos, P.; Barriuso-Vargas, J.J. Big Data and Machine Learning to Improve European Grapevine Moth (*Lobesia botrana*) Predictions. *Plants* **2023**, *12*, 633. [CrossRef]
45. Almeyda, E.; Paiva, J.; Ipanaque, W. Pest Incidence Prediction in Organic Banana Crops with Machine Learning Techniques. In Proceedings of the 2020 IEEE Engineering International Research Conference (EIRCON), Lima, Peru, 21–23 October 2020; Publishing House: Lima, Peru, 2020; pp. 1–4.
46. Caballero-Vidal, G.; Bouysset, C.; Grunig, H.; Fiorucci, S.; Montagné, N.; Golebiowski, J.; Jacquin-Joly, E. Machine learning decodes chemical features to identify novel agonists of a moth odorant receptor. *Sci. Rep.* **2020**, *10*, 1655. [CrossRef]
47. Yuvaraj, J.K.; Roberts, R.E.; Sonntag, Y.; Hou, X.-Q.; Grosse-Wilde, E.; Machara, A.; Zhang, D.-D.; Hansson, B.S.; Johanson, U.; Löfstedt, C.; et al. Putative ligand binding sites of two functionally characterized bark beetle odorant receptors. *BMC Biol.* **2021**, *19*, 16. [CrossRef]
48. Sims, C.; Withall, D.M.; Oldham, N.; Stockman, R.; Birkett, M. Computational investigation of aphid odorant receptor structure and binding function. *J. Biomol. Struct. Dyn.* **2022**, *41*, 3647–3658. [CrossRef]
49. Yi, S.-C.; Wu, Y.-H.; Yang, R.-N.; Li, D.-Z.; Abdelnabby, H.; Wang, M.-Q. A Highly Expressed Antennae Odorant-Binding Protein Involved in Recognition of Herbivore-Induced Plant Volatiles in *Dastarcus helophoroides*. *Int. J. Mol. Sci.* **2023**, *24*, 3464. [CrossRef] [PubMed] [PubMed Central]
50. ACS Publications. Bombykol—Molecule of the Week Archive. American Chemical Society. Available online: https://www.acs.org/molecule-of-the-week (accessed on 27 November 2024).
51. Wikipedia Contributors. Bombykol. Wikipedia. Available online: https://en.wikipedia.org/wiki/Bombykol (accessed on 27 November 2024).
52. The Pherobase: Database of Pheromones and Semiochemicals. The World Largest Database of Behavioural Modifying Chemicals. Available online: https://pherobase.com/ (accessed on 12 February 2025).
53. Larsson, M.C. Pheromones and Other Semiochemicals for Monitoring Rare and Endangered Species. *J. Chem. Ecol.* **2016**, *42*, 853–868. [CrossRef]

54. Caballero-Vidal, G.; Bouysset, C.; Gévar, J.; Mbouzid, H.; Nara, C.; Delaroche, J.; Golebiowski, J.; Montagné, N.; Fiorucci, S.; Jacquin-Joly, E. Reverse chemical ecology in a moth: Machine learning on odorant receptors identifies new behaviorally active agonists. *Cell. Mol. Life Sci.* **2021**, *78*, 6593–6603. [CrossRef]
55. Kepchia, D.; Xu, P.; Terryn, R.; Castro, A.; Schürer, S.C.; Leal, W.S.; Luetje, C.W. Use of machine learning to identify novel, behaviorally active antagonists of the insect odorant receptor co-receptor (Orco) subunit. *Sci. Rep.* **2019**, *9*, 4055. [CrossRef] [PubMed]
56. Sharma, A.; Kumar, R.; Ranjta, S.; Varadwaj, P.K. SMILES to Smell: Decoding the Structure–Odor Relationship of Chemical Compounds Using the Deep Neural Network Approach. *J. Chem. Inf. Model.* **2021**, *61*, 676–688. [CrossRef] [PubMed]
57. Bo, W.; Yu, Y.; He, R.; Qin, D.; Zheng, X.; Wang, Y.; Ding, B.; Liang, G. Insight into the Structure–Odor Relationship of Molecules: A Computational Study Based on Deep Learning. *Foods* **2022**, *11*, 2033. [CrossRef]
58. Pugalenthi, G.; Tang, K.; Suganthan, P.; Archunan, G.; Sowdhamini, R. A machine learning approach for the identification of odorant binding proteins from sequence-derived properties. *BMC Bioinform.* **2007**, *8*, 351. [CrossRef] [PubMed]
59. Tan, J.; Zaremska, V.; Lim, S.; Knoll, W.; Pelosi, P. Probe-dependence of competitive fluorescent ligand binding assays to odorant-binding proteins. *Anal. Bioanal. Chem.* **2020**, *412*, 547–554. [CrossRef]
60. The iOBPdb Database of Insect Odorant Binding Proteins. Available online: https://www.iobpdb.com/Home (accessed on 12 February 2025).
61. Herrera-Acevedo, C.; Perdomo-Madrigal, C.; Herrera-Acevedo, K.; Coy-Barrera, E.; Scotti, L.; Scotti, M.T. Machine learning models to select potential inhibitors of acetylcholinesterase activity from SistematX: A natural products database. *Mol. Divers.* **2021**, *25*, 1553–1568. [CrossRef] [PubMed]
62. Goldwaser, E.; Laurent, C.; Lagarde, N.; Fabrega, S.; Nay, L.; Villoutreix, B.O.; Jelsch, C.; Nicot, A.B.; Loriot, M.-A.; Miteva, M.A. Machine learning-driven identification of drugs inhibiting cytochrome P450 2C9. *PLOS Comput. Biol.* **2022**, *18*, e1009820. [CrossRef] [PubMed]
63. Wang, X.; Zhang, Y.; Cheng, J.; Zheng, Y.; Zhang, Y. A point cloud-based deep learning strategy for protein–ligand binding affinity prediction. *Biomolecules* **2021**, *11*, 1078. [CrossRef]
64. Öztürk, H.; Özgür, A.; Ozkirimli, E. DeepDTA: Deep drug–target binding affinity prediction. *Bioinformatics* **2018**, *34*, i821–i829. [CrossRef]
65. Thafar, M.A.; Alshahrani, M.; Albaradei, S.; Gojobori, T.; Essack, M.; Gao, X. Affinity2Vec: Drug-target binding affinity prediction through representation learning, graph mining, and machine learning. *Sci. Rep.* **2022**, *12*, 4751. [CrossRef]
66. Zhang, Q.; Wu, J.; Lin, L.; Wang, Y. DeepRLI: A Multi-Objective Framework for Universal Protein–Ligand Interaction Prediction. *Biomolecules* **2024**, *40*, 89. [CrossRef]
67. ALee, S.; Kim, J.; Kim, S. PIGNet2: A versatile deep learning-based protein–ligand interaction prediction model for binding affinity scoring and virtual screening. *Molecules* **2023**, *28*, 3456. [CrossRef]
68. The Pandas Development Team. Pandas: Python Data Analysis Library. Available online: https://pandas.pydata.org/ (accessed on 21 November 2024).
69. He, T.; Heidemeyer, M.; Ban, F.; Cherkasov, A.; Ester, M. SimBoost: A read-across approach for predicting drug–target binding affinities using gradient boosting machines. *J. Cheminform.* **2017**, *9*, 24. [CrossRef]
70. Yap, C.W. PaDEL-descriptor: An open source software to calculate molecular descriptors and fingerprints. *J. Comput. Chem.* **2011**, *32*, 1466–1474. [CrossRef] [PubMed]
71. Python Software Foundation. Python Programming Language. Available online: https://www.python.org/ (accessed on 21 November 2024).
72. Cao, D.-S.; Xu, Q.-S.; Liang, Y.-Z. propy: A tool to generate various modes of Chou's PseAAC. *Bioinformatics* **2013**, *29*, 960–962. [CrossRef]
73. Scikit-learn Developers. Scikit-learn: Machine Learning in Python. Available online: https://scikit-learn.org/stable/ (accessed on 21 November 2024).
74. Friedman, J.H. Greedy function approximation: A gradient boosting machine. *Ann. Stat.* **2001**, *29*, 1189–1232. [CrossRef]
75. Ke, G.; Meng, Q.; Finley, T.; Wang, T.; Chen, W.; Ma, W.; Ye, Q.; Liu, T.-Y. LightGBM: A highly efficient gradient boosting decision tree. In Proceedings of the 31st International Conference on Neural Information Processing Systems, Long Beach, CA, USA, 4–9 December 2017; Alicioglu, G., Sun, B., Eds.; Publishing House: Long Beach, CA, USA, 2017; pp. 3149–3157.
76. Bergstra, J.; Komer, B.; Eliasmith, C.; Yamins, D.; Cox, D.D. Hyperopt: A Python library for model selection and hyperparameter optimization. *Comput. Sci. Discov.* **2015**, *8*, 014008. [CrossRef]
77. Chicco, D.; Warrens, M.J.; Jurman, G. The coefficient of determination R-squared is more informative than SMAPE, MAE, MAPE, MSE and RMSE in regression analysis evaluation. *PeerJ Comput. Sci.* **2021**, *7*, e623. [CrossRef]

78. James, G.; Witten, D.; Hastie, T.; Tibshirani, R. *An Introduction to Statistical Learning with Applications in R*; Springer: New York, NY, USA, 2013.
79. Draper, N.R.; Smith, H. *Applied Regression Analysis*, 3rd ed.; Wiley: Hoboken, NJ, USA, 1998.

International Journal of *Molecular Sciences*

MDPI

Article

Evaluation of Three Mutations in Codon 385 of Glucose-6-Phosphate Dehydrogenase via Biochemical and In Silico Analysis

Adriana Gálvez-Ramírez [1,†], Abigail González-Valdez [2,†], Beatriz Hernández-Ochoa [3], Luis Miguel Canseco-Ávila [4], Alexander López-Roblero [4], Roberto Arreguin-Espinosa [5], Verónica Pérez de la Cruz [6], Elizabeth Hernández-Urzua [7], Noemi Cárdenas-Rodríguez [8], Sergio Enríquez-Flores [9], Ignacio De la Mora-De la Mora [9], Abraham Vidal-Limon [10,*] and Saúl Gómez-Manzo [1,*]

[1] Laboratorio de Bioquímica Genética, Instituto Nacional de Pediatría, Secretaría de Salud, Mexico City 04530, Mexico; galvezramirezadriana@gmail.com
[2] Departamento de Biología Molecular y Biotecnología, Instituto de Investigaciones Biomédicas, Universidad Nacional Autónoma de México, Mexico City 04510, Mexico; abigaila@biomedicas.unam.mx
[3] Laboratorio de Inmunoquímica, Hospital Infantil de México Federico Gómez, Secretaría de Salud, Mexico City 06720, Mexico; beatrizhb_16@ciencias.unam.mx
[4] Facultad de Ciencias Químicas, Campus IV, Universidad Autónoma de Chiapas, Tapachula City 30580, Mexico; cansecoavila@gmail.com (L.M.C.-Á.); usdalexander@hotmail.com (A.L.-R.)
[5] Departamento de Química de Biomacromoléculas, Instituto de Química, Universidad Nacional Autónoma de México, Mexico City 04510, Mexico; arrespin@unam.mx
[6] Neurobiochemistry and Behavior Laboratory, National Institute of Neurology and Neurosurgery "Manuel Velasco Suárez", Mexico City 14269, Mexico; veped@yahoo.com.mx
[7] Laboratorio de Toxicología Genética, Instituto Nacional de Pediatría, Secretaría de Salud, Mexico City 04530, Mexico; elyzabet91@yahoo.com.mx
[8] Laboratorio de Neurociencias, Instituto Nacional de Pediatría, Secretaría de Salud, Mexico City 04530, Mexico; noemicr2001@yahoo.com.mx
[9] Laboratorio de Biomoléculas y Salud Infantil, Instituto Nacional de Pediatría, Secretaría de Salud, Mexico City 04530, Mexico; sergioenriquez@ciencias.unam.mx (S.E.-F.); ignaciodelamora@ciencias.unam.mx (I.D.l.M.-D.l.M.)
[10] Red de Estudios Moleculares Avanzados, Clúster Científico y Tecnológico BioMimic, Instituto de Ecología A.C. (INECOL), Carretera Antigua a Coatepec 351, El Haya, Xalapa 91073, Mexico
[*] Correspondence: abraham.vidal@inecol.mx (A.V.-L.); saulmanzo@ciencias.unam.mx (S.G.-M.); Tel.: +52-55-1084-0900 (ext. 1442) (S.G.-M.)
[†] These authors contributed equally to this work.

Abstract: Glucose-6-phosphate dehydrogenase (G6PD) deficiency is an enzymopathy that affects approximately 500 million people worldwide. A great number of mutations in the *G6PD* gene have been described. However, three class A G6PD variants known as G6PD Tomah (C385R), G6PD Kangnam (C385G), and G6PD Madrid (C385W) have been reported to be clinically important due to their associations with severe clinical manifestations such as hemolytic anemia. Therefore, this work aimed to perform, for the first time, biochemical and functional characterizations of these variants. The G6PD variants were cloned and purified for this purpose, followed by analyses of their kinetic parameters and thermal stability, as well as in silico studies. The results showed that the mutations induced changes in the proteins. Regarding the kinetic parameters, it was observed that the three variants showed lower affinities for G6P and $NADP^+$, as well as lower thermal stability compared to WT-G6PD. Molecular dynamics simulations showed that C385 mutations induced changes around neighboring amino acids. Metadynamics simulations showed that most remarkable changes account for the binding pocket volumes, particularly in the structural $NADP^+$ binding site, with a concomitant loss of affinity for catalytic processes.

Keywords: mutations; G6PD; molecular dynamics simulations

1. Introduction

The cytosolic enzyme glucose-6-phosphate dehydrogenase (G6PD) is found in both prokaryotic and eukaryotic cells [1–4]. In humans, the *G6PD* gene is positioned neighboring the telomeric region of the X chromosome. The *G6PD* gene encodes a specific protein of 515 amino acids and a molecular mass of 59 kDa. G6PD protein plays a regulatory role in the pentose–phosphate pathway (PPP), a highly conserved metabolic pathway that produces a variety of critical molecules, for example, nucleotide precursors and nicotinamide adenine dinucleotide phosphate reduced (NADPH). G6PD catalyzes the first reaction of the oxidative phase of the PPP, allowing for the reduction of $NADP^+$ at the expense of the dehydrogenation of glucose-6-phosphate (G6P) to 6-phosphogluconolactone (6PGL) and generating NADPH [5].

In erythrocytes, the PPP is the only route to generate reduced forms of NADPH as they do not contain mitochondria. Therefore, G6PD activity has a crucial role in protecting cells from oxidative injury due to increased reactive oxygen species [6,7]. In addition, NADPH plays an important role in redox homeostasis, which is involved in the reduction of glutathione, and glutathione is the primary protection against reactive oxygen species and helps avoid red cell hemolysis [8–10].

A whole of 230 mutations distributed throughout the *G6PD* gene have been reported, including single-point mutations, causing single amino acid substitutions and multiple mutations (two or more substitutions) [11–13]. These mutations lead to decreased activity and stability of the G6PD enzyme, triggering oxidative injury in RBCs and a variety of clinical symptoms known as G6PD deficiency [4,8]. G6PD deficiency is an enzymopathy that affects approximately 500 million people worldwide (7.5% of the world's population). This enzymopathy can be symptomless or can generate severe clinical symptoms such as chronic hemolytic anemia (G6PD variant class A), neonatal jaundice and acute hemolytic anemia generated by some drugs, fava beans, or infection (G6PD variant class B), and no hemolysis (G6PD variant class C) [13].

In the mutant G6PD Tomah, a transition of T $\rightarrow$ C in nucleotide 1153 was identified, which resulted in a Cys $\rightarrow$ Arg substitution at position 385 [14]. This mutation was identified in two unrelated Spanish G6PD-deficient males with chronic nonspherocytic hemolytic anemia (CNSHA) located in exon 10. On the other hand, in a genetic description of Korean patients diagnosed with glucose-6-phosphate dehydrogenase deficiency, the mutant G6PD Kangnam was identified. The patient in which the Kangnam mutant was identified was a 5-year-old boy who had fever and paleness. The patient visited the hospital after recuperating from a critical clinical condition. The patient presented a G6PD enzyme level of 2.1 U/g Hb, and both the Coombs and the osmotic fragility tests were negative. The G6PD deficiency was categorized as class I, corresponding to the World Health Organization (WHO) classification, due to the drastic decreased G6PD enzyme activity and development of CNSHA. The child had no family genetic history of hematological disorders and had a new T > G mutation at position 1153, which gave rise to a Cys $\rightarrow$ Gly amino acid replacement at position 385 (Figure 1). According to the WHO, this variant belongs to G6PD class I (now variant class A) [15]. Finally, the mutation G6PD Madrid was identified in two Spanish males with G6PD deficiency and hemolytic anemia. One of the patients was diagnosed at birth since he showed signs of neonatal jaundice. At 6 years old, the child still showed moderate to severe CNSHA. G6PD Madrid variant showed undetectable residual activity in RBC. A transition of C to G was identified at nucleotide 1155, causing a Cys $\rightarrow$ Trp replacement in the amino acid residue at position 385. According to the WHO classification, this mutation corresponds to the G6PD variant class A [16]. These G6PD mutations in exon 10 result in variants related to neonatal jaundice and life-long CNSHA [17]. These reports highlight the significance of exon 10 of the *G6PD* gene in the stabilization of the G6PD enzyme.

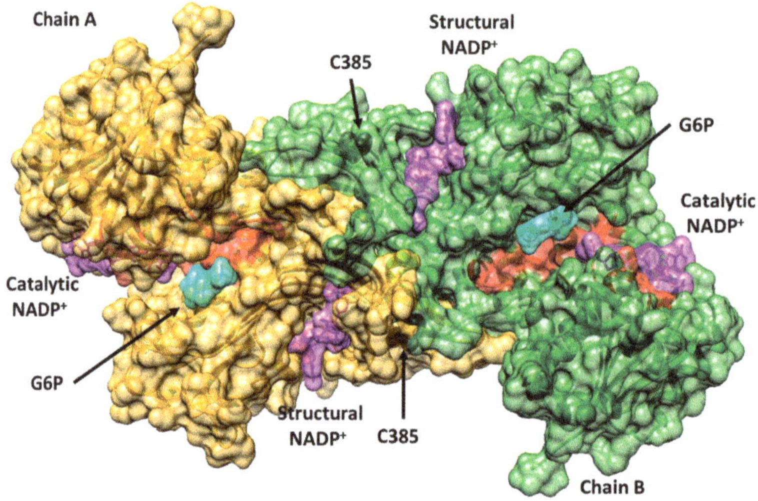

Figure 1. The structure of the human glucose-6-phosphate dehydrogenase dimer (G6PD; PDB entries 2BH9 and 2BHL). The locations of the class A mutations Tomah (C385R), Kangnam (C385G), and Madrid (C385W) are indicated in chain B (green chain and black spheres), while in chain A (yellow chain), the Cys residue is marked as a black sphere. Glucose-6-phosphate (G6P) is drawn in cyan, while catalytic nicotinamide adenine dinucleotide phosphate ($NADP^+$) and structural $NADP^+$ are drawn in purple.

It is important to emphasize that in G6PD deficiency, various mutations in the same codon have been reported where the native amino acid residue was changed by different amino acid residues, producing different G6PD variants that have been reported in different countries around the world. In this sense, three mutations in codon 385, located in exon 10, have been reported. They give rise to the class A G6PD variants (previously class I variants), G6PD Tomah (C385R), G6PD Kangnam (C385G), and G6PD Madrid (C385W) (Figure 1). All these variants are related to clinical manifestations such as chronic hemolytic anemia. In addition, it is essential to mention that these mutations are distant from the active site and the catalytic $NADP^+$ region; however, they are near the "structural" $NADP^+$ site.

Based on the above, in this work, we report the cloning, purification, thermostability assays, and detailed steady-state kinetics for the three clinical variants G6PD Tomah, Kangnam, and Madrid for the first time, which were registered in the same codon (385) and identified in various parts of the worldwide. In addition, based on the three-dimensional (3D) structure of the G6PD protein and using in silico approaches to mine structural information on the effects of the G6PD variants, we identified modifications in the protein structure to understand the clinical symptoms of these mutations in order to explain the molecular mechanisms observed in patients with G6PD deficiency.

2. Results and Discussion

2.1. Construction and Expression of Recombinant G6PD Variants

Three clinical single mutants, G6PD Tomah (C385R), Kangnam (C385G), and Madrid (C385W), were obtained via site-directed mutagenesis. The mutation into the *G6PD* gene was confirmed via DNA sequence in the generated plasmids. As shown in Supplementary Figure S1A, a single nucleotide change of thymine to cytosine ($T \rightarrow C$) at nt 1153 was observed (Figure S1A). In electropherograms for the single mutant G6PD Kangnam, a change of thymine for guanine ($T \rightarrow G$) at nt 1153 was observed (Figure S1B). In addition, for G6PD Madrid, the nucleotide guanine was replaced by thymine ($G \rightarrow T$) at nucleotide (nt) 1154 (Figure S1C). Therefore, we confirmed the desired mutations in each of the variants that were obtained.

2.2. Purification of Recombinant G6PD Variants

To analyze the effects of the mutations on the activity and native structure of the wild-type (WT-G6PD) enzyme, recombinant WT-G6PD, and the three G6PD variants were purified via the anion-exchange chromatography (Q-Sepharose) method and affinity (2′5′ADP Sepharose 4B) methods [4,5,12,18–21]. It was observed that lower protein concentrations were obtained from 2 L of culture for each variant compared to WT-G6PD (4.8 mg/mL). The variant with the highest protein concentration was G6PD Tomah (0.8 mg/mL), followed by the Madrid variant (0.57 mg/mL) and the Kangnam variant (0.23 mg/mL). It is important to highlight that although the three variants purified in this study occurred at the same structural position (amino acid residue 385), different total protein concentrations were obtained for each of them. One of the factors involved in this low protein expression was probably the physicochemical properties of the residues involved in each of the variants (Arg, Gly, and Trp), which have different characteristics than the Cys in the WT-G6PD. These results are in agreement with the report by our working group, in which the concentrations of recombinant proteins were lower than the WT-G6PD enzyme [4,5,12,18–21], likely suggesting poor folding in the native structures of these variants or a loss of stability.

2.3. Kinetic Analysis

Kinetic analysis was performed to determine the effects of the three mutations in the WT-G6PD on the Michaelis–Menten affinity constant (K_m) for the physiological substrates (G6PD and $NADP^+$) and the catalytic constant (k_{cat}). As seen in Supplementary Figure S2, for the WT-G6PD, a K_m of 38.4 and 6.1 μM were obtained for the G6P and $NADP^+$ substrates, respectively, with a k_{cat} of 223 s^{-1} (Table 1). Subsequently, as shown in Figure S2B, the variant G6PD Tomah showed a K_m of 64.9 μM for the G6P substrate and a K_m of 13.8 μM for the $NADP^+$ substrate (Figure S2E), with a k_{cat} of 15 s^{-1}. Figure S2C shows kinetic plots for the variant G6PD Kangnam, where a K_m of 55.2 μM for the G6P substrate was obtained. Similarly, in Figure S2G, a K_m of 10.3 μM for the $NADP^+$ substrate was calculated, with a k_{cat} of 0.4 s^{-1}. Finally, the variant G6PD Madrid showed a K_m of 78.2 μM and a K_m of 12.2 μM for the G6P and $NADP^+$ substrates, respectively (Figure S2D,H), with a k_{cat} of 40.5 s^{-1} (Table 1). For the three class A G6PD variants analyzed in this study, their affinities for the physiological substrates ($NADP^+$ and G6P) decreased regarding native G6PD, and the catalytic constant (k_{cat}) was also affected by the mutations. The three mutants analyzed in this study presented reductions in catalysis (k_{cat}), suggesting that the catalytic site was affected. The variant G6PD Kangnam was the most affected, with a catalysis loss of 99% regarding the WT-G6PD. The variant G6PD Tomah had 91% less activity, followed by the variant G6PD Madrid, which showed a loss of activity of 76% regarding the WT-G6PD. These results suggest that the G6PD variants had reduced affinities for physiological substrates and showed reductions in catalysis, suggesting that this loss of activity was closely related to the clinical manifestations of CNSHA found in each of the patients where these mutations were identified.

Similar results were obtained for the class A G6PD variants Zacatecas, Durham, Veracruz, Nashville, Guadalajara, and Yucatán, which showed greater affinity losses for their substrates regarding class II and III variants. Specifically, it is interesting to mention that the positions of the mutations of the three variants G6PD Tomah (C385R), Kangnam (C385G), and Madrid (C385W) are located near the position of the G6PD Guadalajara variant (R387C, class A), where the mutation occurs near the structural $NADP^+$ binding site and dimer interface. The results of this work are similar to the report by Martínez-Rosas et al. [20] for the variant G6PD Guadalajara. In addition, it should be noted that the three mutants of interest in this study are not located near the catalytic binding sites of $NADP^+$ and G6P; however, the affinities for the substrate and the catalytic binding site were affected.

Table 1. The kinetic parameters of WT-G6PD and Tomah, Kangnam, and Madrid variants.

Kinetic Constants	WT-G6PD	Variants		
		Tomah	Kangnam	Madrid
K_m G6P (μM)	37.5 ± 1.8	64.9 ± 3.2	66.2 ± 3.1	78.2 ± 3.4
K_m $NADP^+$ (μM)	7.2 ± 0.3	13.8 ± 0.6	10.3 ± 0.5	12.2 ± 0.5
k_{cat} (s^{-1})	223 ± 8.9	15 ± 0.6	0.4 ± 0.01	40.5 ± 1.2
k_{cat}/K_m G6P ($s^{-1}{\cdot}M^{-1}$)	5.9×10^6	0.2×10^6	0.006×10^6	0.52×10^6
k_{cat}/K_m $NADP^+$ ($s^{-1}{\cdot}M^{-1}$)	32.3×10^6	1.1×10^6	0.04×10^6	3.3×10^6

The kinetic parameters were obtained from three independent experiments.

2.4. Thermal Inactivation Analysis

To evaluate the effects of the mutations on the stability of the native protein, thermal stability trials were carried out for WT-G6PD and the variants. This assay has been widely used as an indicator of alterations in protein stability [4,5,10,12,18–23]. The residual activities of each of the variants were plotted at different temperatures, and it was observed that the catalytic activity decreased as the temperature increased (Figure 2). For the variant G6PD Tomah, a T_{50} value of 40.6 °C was observed, while for the variants G6PD Kangnam and G6PD Madrid, T_{50} values of 43.5 °C and 42.4 °C were obtained. These T_{50} values were 4 to 7 °C lower regarding the WT G6PD, which showed a T_{50} value of 47.5 °C. These results suggest that the mutations in codon 385 lead to a more unstable G6PD protein. In addition, it was identified that the Tomah variant presented the greatest instability, with a 7 °C decrease in the T_{50} value, followed by the variants Kangnam and Madrid. These results are in agreement with the obtained for the G6PD variant Guadalajara (R387C), where a T_{50} value of 43.3 °C was obtained. The mutations analyzed in this study were located near the structural $NADP^+$ binding site and dimer interface in the 3D structure of the native G6PD protein, which is involved in dimer formation and enzyme stability, as observed in the class A G6PD variants Nashville, Fukaya, Campinan, Wisconsin, Durham, Guadalajara, and Mount Sinai [18,20,23,24]. Therefore, it was proposed that G6PD instability could be a frequent effect of these deleterious mutations. It is noteworthy that all results showed that the variants analyzed in this study presented reductions in catalytic activity and alterations in their stability, suggesting that these two alterations are responsible for the clinical manifestations in patients with G6PD deficiency.

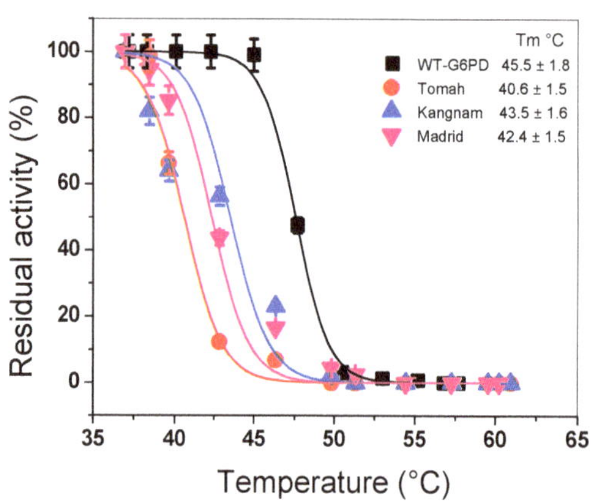

Figure 2. Thermal inactivation trials of three clinical mutants and WT-G6PD. The recombinant proteins were suspended at 0.2 mg/mL and incubated at 37 to 60 °C for 20 min. The residual activity was calculated and expressed as a percentage concerning their controls. The trials were performed in triplicate, with standard errors (SEs) ± <5%.

2.5. *In Silico Analysis*

To evaluate the effects of the mutations on the WT-G6PD protein, we performed in silico mutagenesis. Figure 3 shows the mutations in the protein structure. It can be observed that the amino acid C385 is located near the structural $NADP^+$ binding site, which is important for dimer formation and promotes the stability of the protein [25–27]. As it is a crucial site, any amino acid change at this position could lead to a reduction in the structural binding of the $NADP^+$ molecule and could prevent the generation of the active form of the protein (i.e., dimerization), leading to the CNSHA pathogenicity associated with G6PD deficiency. With the in silico mutations, we set out to explain and understand the influences of the mutations on the structures of the Tomah, Kangnam, and Madrid variants, as well as the causes of their severity, using bioinformatics methods and biochemical trials of these variants was not entirely possible due to difficulties in obtaining enough pure proteins.

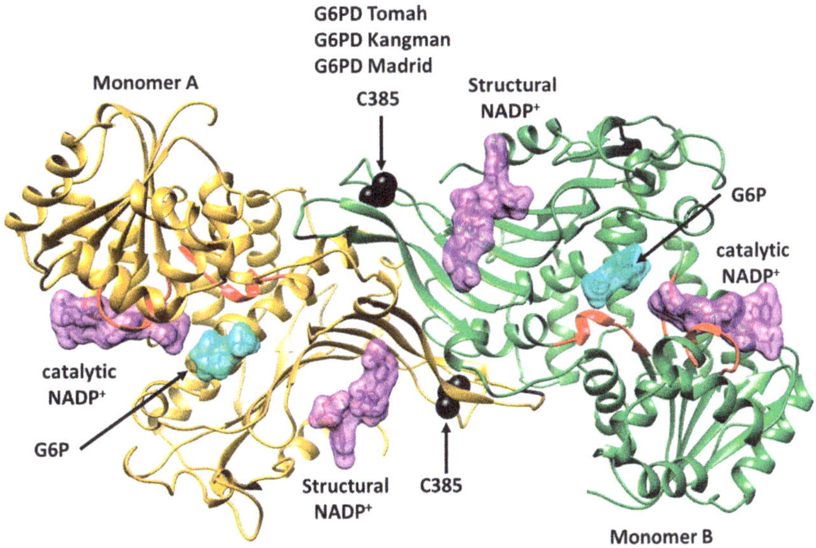

Figure 3. Structure of the WT-G6PD dimer (PDB ID: 2BH9 and 2BHL). The position of the class A G6PD variants Tomah, Kangnam, and Madrid is identified at codon 385 by substituting the amino acid cysteine (black spheres). Catalytic G6P, $NADP^+$, and structural $NADP^+$ are shown in light blue and purple. Monomers A and B are shown in bright yellow and green ribbons.

The Tomah variant of G6PD (C385R) is caused by a change in a nitrogenous base from thymine to cytosine at nucleotide 1153, which leads to the amino acid cysteine being substituted with arginine. Cys is a polar, uncharged amino acid with a molecular weight (MW) of 121 Da (Figure 4A). When it is mutated into Arg, it becomes a polar, positively charged amino acid with a MW of 174 Da. A bulky guanidino-group sidechain occupies the space, forming close interactions with V376 and D379 in chain A (Figure 4B). G6PD Kangnam (C385G) is a single mutation provoked by the substitution of guanine for thymine (1153 T $\rightarrow$ G), which leads to a change from cysteine to glycine, a small nonpolar amino acid with a MW of 75 Da, whose sidechain is a H atom, which probably causes the area to become more flexible and consequently causes the protein to lose stability (Figure 4C). On the other hand, G6PD Madrid (C385W) includes the replacement of cytosine with guanine at nucleotide 1155, which leads to a change from cysteine to tryptophan, whose residue is apolar and hydrophobic, with an indole ring in its sidechain and a molecular weight of 204 Da. The bulky indole sidechain closely interacts with A217 in chain B (Figure 4D).

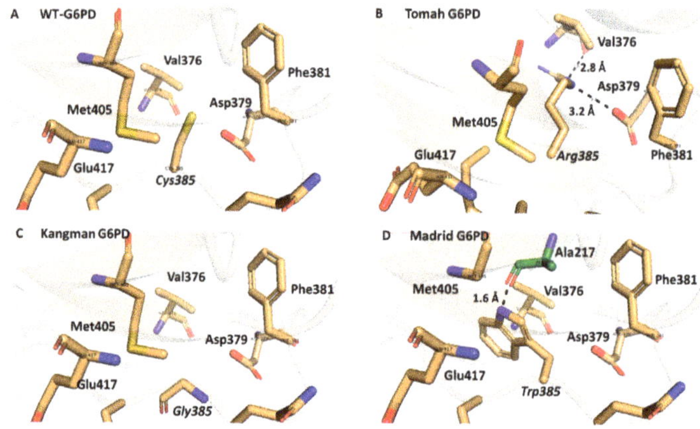

Figure 4. Structural comparison of WT-G6PD (PDB ID: 2BH9 and 2BHL) and minimized G6PD variants class A Tomah, Kangman, and Madrid models. (**A**) WT-G6PD. (**B**) In silico mutation to generate the Tomah variant. (**C**) In silico mutation to generate the Kangman variant. (**D**) In silico mutation to generate the Madrid variant. Interactions are shown as black dotted lines.

Generating in silico mutations allowed us to analyze the interactions between amino acid 385 and neighboring amino acids. The amino acids Arg, Gly, and Trp present in each variant have side chains with different sizes, shapes, polarities, charges, and hydrophobicities about Cys, an amino acid in WT-G6PD. This causes changes in the environment around position 385 and throughout the protein, which probably leads to losses of stability and catalytic efficiency and, consequently, the clinical manifestations of these variants. As interactions between amino acids help the protein to fold and fluctuate to adopt the active structure, any amino acid modification can cause structural alterations that impact the intermolecular interactions in the protein structure.

2.6. Use of In Silico Approaches to Mine Structural Information on the Effects of the G6PD Variants

Bioinformatics approaches have revolutionized the field of molecular biology, enabling researchers to conduct detailed structural analyses of enzymes such as G6PD at the molecular level. Determining the impacts of mutations (insertions, deletions, or missense variations) in protein structures on stability and activity is one of the most exciting areas in protein sciences. In the case of G6PD, extensive studies have differentiated pathogenic variants and determined how these mutations impact overall stability [11,28]. Outstanding efforts were made by several groups working on in silico approximations to predict protein stability from sequences [6,29–31]. For example, different stability predictors have been developed, including some outstanding tools that distinguish between pathogenic variants when combining artificial intelligence (machine learning) or evolutionary information [32]. However, as there is an enormous sequence–annotation gap between the available sequences, which continues to grow, there are gaps, such as the sequence–structure gap, that need to be mined to access more detailed diagnoses, public policies, or even personalized medicine approaches. The most accurate stability predictors for pathogenic mutations are structure-based methods, which require available protein structures [33,34].

With the advent of sequence–structure methodologies based on artificial intelligence such as AlphaFold (AF2) [35] (https://alphafold.ebi.ac.uk, accessed on 1 February 2024), a wide range of problems are being tackled in structural biochemistry. Due to the immediate impacts of these artificial intelligence systems developed by DeepMind, several groups sought to understand the protein stability problem. For example, Pak et al. [36] explored the capacity for AF2 predictions to evaluate the impacts of mutations on protein stability changes ($\Delta\Delta G$) despite the software having a disclaimer stating that the routine "*has not been validated for predicting the effect of mutations*" (https://alphafold.ebi.ac.uk/faq, accessed on 1 February 2024). Using data on experimentally measured effects of mutations on protein stability, their analysis indicated that overall, AF2 metrics can estimate $\Delta\Delta G$ values; however, the effects of mutations on *pLDDT* scores (a superposition-free metric that indicates the extent to which a protein model reproduces a reference structure) were weakly correlated with $\Delta\Delta G$ and protein function. The formulation of a metric to serve as a predictor of the impacts of mutations on protein stability and function is part of the ongoing effort to map the effect of every G6PD mutation.

On the other hand, Keskin et al. [37] reported other examples of the effects of mutations on protein stability using AI methodologies and structural analysis. The overall performances of five protein stability predictors were evaluated, namely, mCSM, MAESTRO, CUPSAT, SAAF2EC-SEQ, and MUpro. This research group also used AF2-computed structures of 26 hereditary cancer-associated proteins to analyze a breast cancer cohort of 355 patients. The results showed that the pathogenicity labels of the missense mutations in the cohorts had an unbalanced distribution. Despite the stability predictors showing moderate performances when discriminating pathogenic variants, an improvement was obtained from an AF2 structure set with high pathogenicity prediction power. These findings suggest that using protein stability predictors in combination with structure-based metrics can aid in predicting the pathogenicity of missense mutations [37]. Further research is needed to improve the performance of stability predictors, as well as to assess changes in the internal dynamics of mutant proteins, such as changes in hydrogen bond patterns, changes in solvation energies, and thermodynamics changes in entropic terms associated with structurally modified protein scaffolds, among others. This type of prospective approach can be applied to the G6PD mutant library as different metrics can be enriched with structural effects of mutations close to the binding site.

Several approximations of G6PD variants' effects on protein structure and stability have been reported. A fine-tuned structural characterization of the presence of structural $NADP^+$ in G6PD dimers was performed by Horikoshi et al. [38]. Despite the large amount of clinical information about these mutations, the atomic and molecular mechanisms underlying the losses of G6PD enzymatic activity and stability due to the absence of structural $NADP^+$ and the role of dimerization in catalysis remained unclear. To gain insight into these mechanisms, the researchers determined the structures of several class A G6PD variants: F381L, R393H, V394L, and P396L. They aim to propose a molecular mechanism for the loss of catalytic activity and stability using full structural characterization (X-ray crystallography, cryogenic electron microscopy (cryo-EM), small-angle X-ray scattering (SAXS), and biophysical analyses). Comparisons of the mutants to the WT-G6PD enzyme through molecular dynamics simulations (MDSs) also exposed a mechanism for severe G6PD deficiencies due to the class A G6PD variants. The mechanism suggests that mutations around the $NADP^+$ binding site alter the b-strand dynamics of dimerization, resulting in inactive scaffolds.

The variants p.(C385R) or Tomah, p.(C385G) or Kangman, and p.(C385W) or Madrid were classified as pathogenic alleles (evidence categories PS3, PM1, PM2, PM5, PP2, PP3, and PP4) according to proposed by the American College of Medical Genetics and Genomics and the Association for Molecular Pathology (ACMG/AMP) [39] (https://www.medschool.umaryland.edu/Genetic_Variant_Interpretation_Tool1.html/, accessed on 1 February 2024). In silico analyses of predictions were performed using PolyPhen (http://genetics.bwh.harvard.edu/pph2/, accessed on 1 February 2024) [40], MutPRed (http://mutpred.mutdb.org/, accessed on 1 February 2024) [31], and Align GVGD (http://agvgd.hci.utah.edu/agvgd_input.php, accessed on 1 February 2024) [41,42] programs. Polyphen classified the Tomah and Kangman variants as benign. In contrast, the Madrid variant was classified as likely damaging. MutPred indicated that the variants had a loss of the allosteric site at N388, and Align GVGD indicated that the variants had altered function (Table S1). The results obtained in the present work through in vivo functional studies and in silico predictions support the damaging effect on the protein.

2.6.1. Molecular Dynamics Calculations of Dimeric G6PD Variants

Different molecular simulation techniques on dimeric G6PD can help reveal numerous limitations posed by docking trials, such as rotamer selection, cofactor charge description, and orientation inside the binding site. In this context, explicit solvent molecular dynamics simulations (MDSs) can sample sidechain orientations and free rotations of cofactors. Moreover, MDSs can improve the sampling of the binding process since solvent waters are explicitly described, resulting in a refining structural outcome with all-atom positions being calculated. These evaluations help to visualize the possible conformational changes associated with C385 point mutations. We aimed to perform conventional MDS simulations to evaluate potential changes in ligand-binding affinity for structural and catalytic $NADP^+$ and the glucose-6-phosphate (G6P) substrate. During the MDS calculations, the G6PD complexes were first relaxed and subsequently clustered with a hierarchical analysis, which assessed highly existent conformations of G6PD dimers and ligands complexes (Figure 5A). The root mean square deviation (RMSD) for stereogenic centers or alpha carbons and root mean square fluctuations for side chains or beta carbons (RMSFs) were monitored. The dominant conformations of the C385 point mutations in complexes with $NADP^+$ and G6P are depicted in Figure 5B.

Introducing an explicit solvent and long chemical simulation times to the complexes slightly refined them, with an average RMSD of cc. 3.0 Å. Severe structural changes could be discarded for overall quaternary dimer structures, as conformational changes were not detected in any simulations despite the point mutations. Moreover, the RMS fluctuations showed that for the Tomah (C385R) variant, the sidechain flexibility increased by cα. 2 Å, in contrast to the Madrid (C385W) and Kangnam (C385G) variants, which showed reduced flexibility in the N-terminal and β-strand domains close to the structural $NADP^+$ (Figure 5C). However, the most remarkable changes were recorded in the volumes of the binding pockets during all chemical simulations, as depicted in Figure 6A–D, where single-point mutations could create relevant changes in the $NADP^+$ environment.

Table 2 shows the $NADP^+$ binding pocket volume changes due to the C385 single-point mutations. The most remarkable changes were found in the Madrid (C385W) variant, which had aromatic and bulky sidechains that could reduce the available volume around the structural $NADP^+$ binding site. This variant showed an increase in the volume of the structural and catalytic $NADP^+$ binding sites of ca. 10 Å^3, implying a fine restructuring of the binding sites that was not perceptible through RMSD calculations.

Table 2. $NADP^+$ binding pocket volume changes of G6PD variants.

G6PD Variant	Structural $NADP^+$ Binding Pocket Volume (Å^3)	Catalytic $NADP^+$ Binding Pocket Volume (Å^3)	Structural $NADP^+$ SASA (Å^2)	Catalytic $NADP^+$ SASA (Å^2)
WT	494.3	490.1	261.1	270.9
Tomah	497.4	500.3	289.9	278.2
Kangnam	490.2	491.7	262.2	261.3
Madrid	505.6	507.2	288.3	285.7

All $NADP^+$ binding pocket volume and solvent-accessible surface area (SASA) were analyzed with Pyvol v1.0 under PyMOL (Schrödinger, New York, NY, USA) and expressed as the sum of the binding sites from both monomers.

Overall, the changes in the volumes of the $NADP^+$ binding sites were remarkable; however, the quaternary structure was not affected in any dimeric complex, suggesting a combinatorial effect resulting in differences in the apparent affinities for ligands and the formation of the dimeric complex (Figure S3) [43].

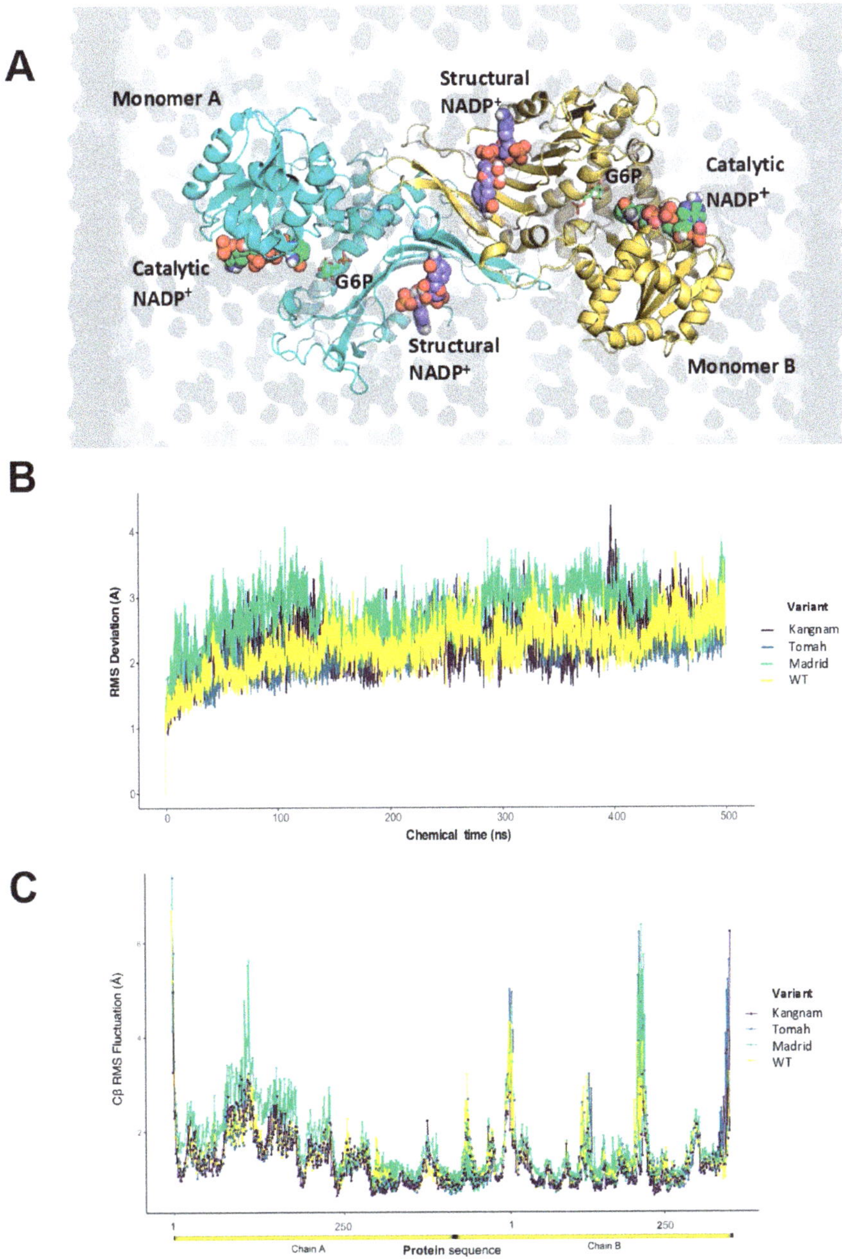

Figure 5. Explicit all-atom molecular dynamics simulations of G6PD variants complexes. (**A**) Evaluated dimeric G6PD model system in explicit solvent. (**B**) Alpha carbon (Cα) RMSD calculations of the G6PD variants. Calculated average RMSD values were 2.3 Å for WT, 2.5 Å for Tomah, 2.7 Å for Kangnam, and 3.0 Å for the Madrid variant. The values were averaged for ±5 aa from N- and C-terminal groups. (**C**) Beta carbon (Cβ) RMSF calculations (sidechains fluctuations) of the G6PD variants. The shown fluctuations were averaged for 500 ns of MDSs.

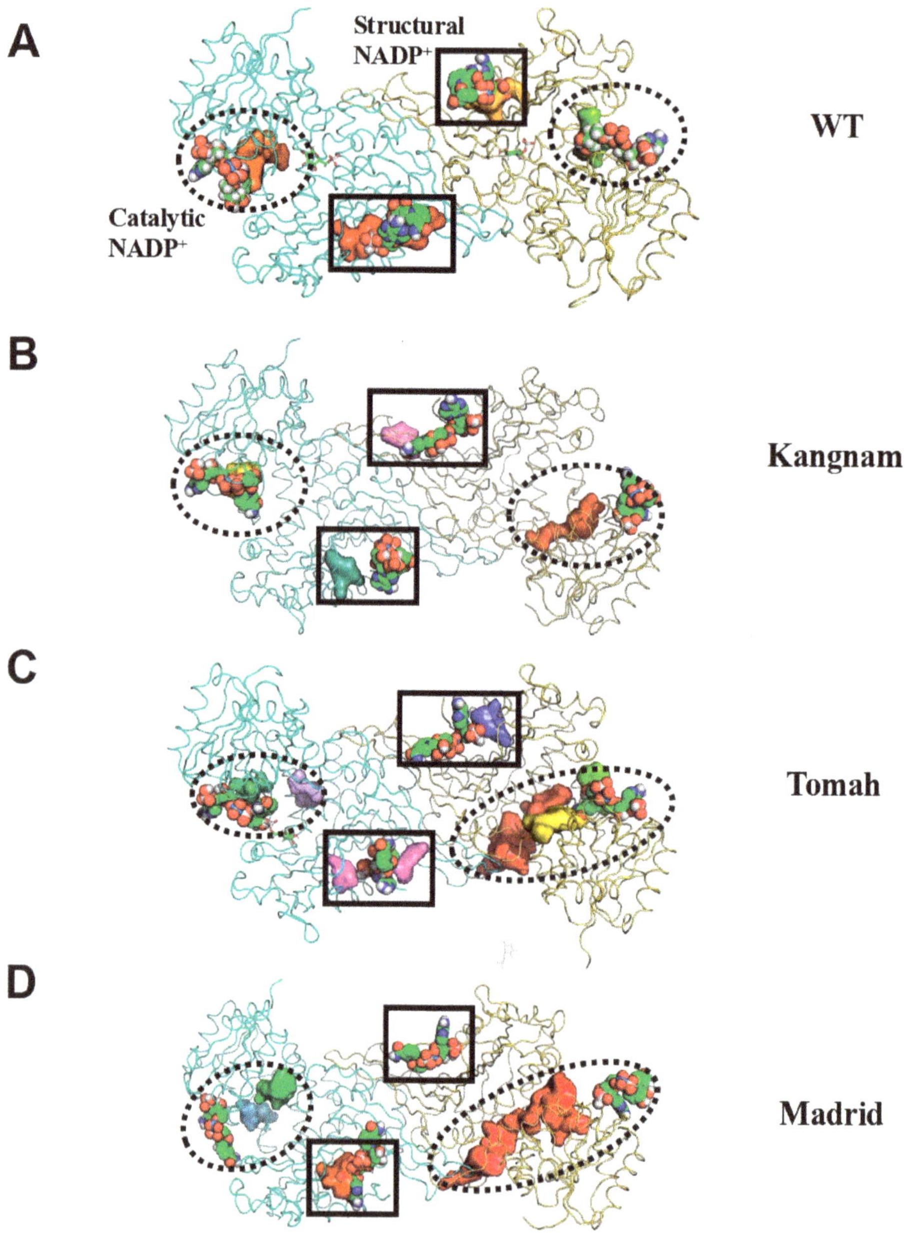

Figure 6. The average changes in the $NADP^+$ binding pockets of the G6PD variants: (**A**) G6PD WT, (**B**) Kangnam, (**C**) Tomah, and (**D**) Madrid. The identified binding pockets are shown as solid surfaces. Protein structures are shown as green ribbons, and $NAPD^+$ is shown as licorice sticks. Binding pockets for $NADP^+$ are identified by circles for catalytic $NADP^+$ and squares for structural $NADP^+$.

2.6.2. Evaluation of Binding and Unbinding of G6PD Dimer Assembly with Metadynamics Simulations

A well-tempered metadynamics methodology was enhanced with a sampling technique for the binding and unbinding process of two monomers for each G6PD variant to evaluate the possible role of the point mutations in forming the active dimeric complex. The potential mean force (PMF) change of the dimeric assembly was averaged from three independent simulation trajectories, where all previously simulated conditions were reproduced, resulting in PMF as a descriptor of quaternary structure stability. In the metadynamics simulations, the maximal value of external force over the center of mass was described as a rupture force (unbinding), and the pulling work (binding) averaged from all trajectories was evaluated over 34 Å of intermonomer distance or a collective variable (CV) (Figure 7A).

Like a ligand-binding process, pulling work can be considered more valuable than the rupture force because it can be associated with free energy via Jarzynski's isobaric–isothermal average, such as in steered molecular dynamics methods [44]. The average pulling forces of the G6PD monomers were time-dependent for an additional 100 ns (Figure 7B). The results indicated that the mean rupture forces for WT-G6PD ranged from −2800 Kcal·mol^{-1} at intermonomer distances of 30–40 Å and during the binding process to −1800 Kcal·mol^{-1} when the dimer was fully dissociated in the simulation box. The data suggested that dimer formation was a very stable process with the most negative values when the center-of-mass (COM) distance reached cα. 35 Å; however, the monomers tended to dissociate in the presence of a significant amount of energy, as depicted for higher CV distances.

The Tomah (C385R) and Kangnam (C385G) variants displayed similar PMF profiles compared to WT-G6PD (Table 3) but with reduced energetic barriers between processes (−490.94 and −108.08, respectively). These processes were not as stable as the dimerization process of WT-G6PD but could lead to a displacement in the equilibrium that could stabilize the dimeric active form. However, both variants displayed changes in their catalytic $NADP^+$ binding sites, indicating that both enzymes lost their apparent affinities for both catalytic processes. The difference between the bound and unbound states displayed changes in the work profiles and was identified as a critical factor for the binding of monomers. According to metadynamics calculations, the average pulling work values of the three tested variants fell within 35.74 Å and ca. −1300 kcal·mol^{-1} (Table 3). Furthermore, the Madrid variant (C385W) was a complex case due partly to a stable dimerization process ($\Delta\Delta G$ = −703.77 kcal·mol^{-1}) but had the most significant changes in the volumes of both $NAPD^+$ binding sites. Fine-tuning the molecular scaffolds can affect the transformation rates of G6P, as it has been reported that small changes in the architectures of redox enzymes can have exponential impacts on electron transfer rates [45].

Table 3. The free energy profiles of the G6DP enzyme variants for bound/unbound dimerization.

G6PD Variant	ΔG_{bound}	$\Delta G_{unbound}$	$\Delta\Delta G$
WT	−2867.57 ± 173.23	−1796.98 ± 122.95	−1070.59
Tomah (C385R)	−1574.32 ± 79.56	−1083.38 ± 101.33	−490.94
Kangnam (C385G)	−1420.67 ± 123.43	−1312.59 ± 102.94	−108.08
Madrid (C385W)	−1556.81 ± 92.30	−853.04 ± 67.90	−703.77

All units were averaged from triplicate metadynamics simulations lasting 100 ns and expressed in kcal mol^{-1}.

Finally, within the therapeutic implications, this work contributes that the three G6PD variants analyzed were identified in male patients that, at the moment of the identification, showed CNSHA. Since no functional data were available for Tomah, Kangman, and Madrid G6PD variants, we determined experimentally the principal biochemical properties of these enzymes. Moreover, it is essential to note that, through understanding the implications of genetic variants for enzyme structure, different groups have shown that structural analysis based on molecular dynamics simulations [44] and artificial intelligence techniques [45] can represent a crucial step toward identifying a potential therapeutic approach to correcting G6PD deficiency.

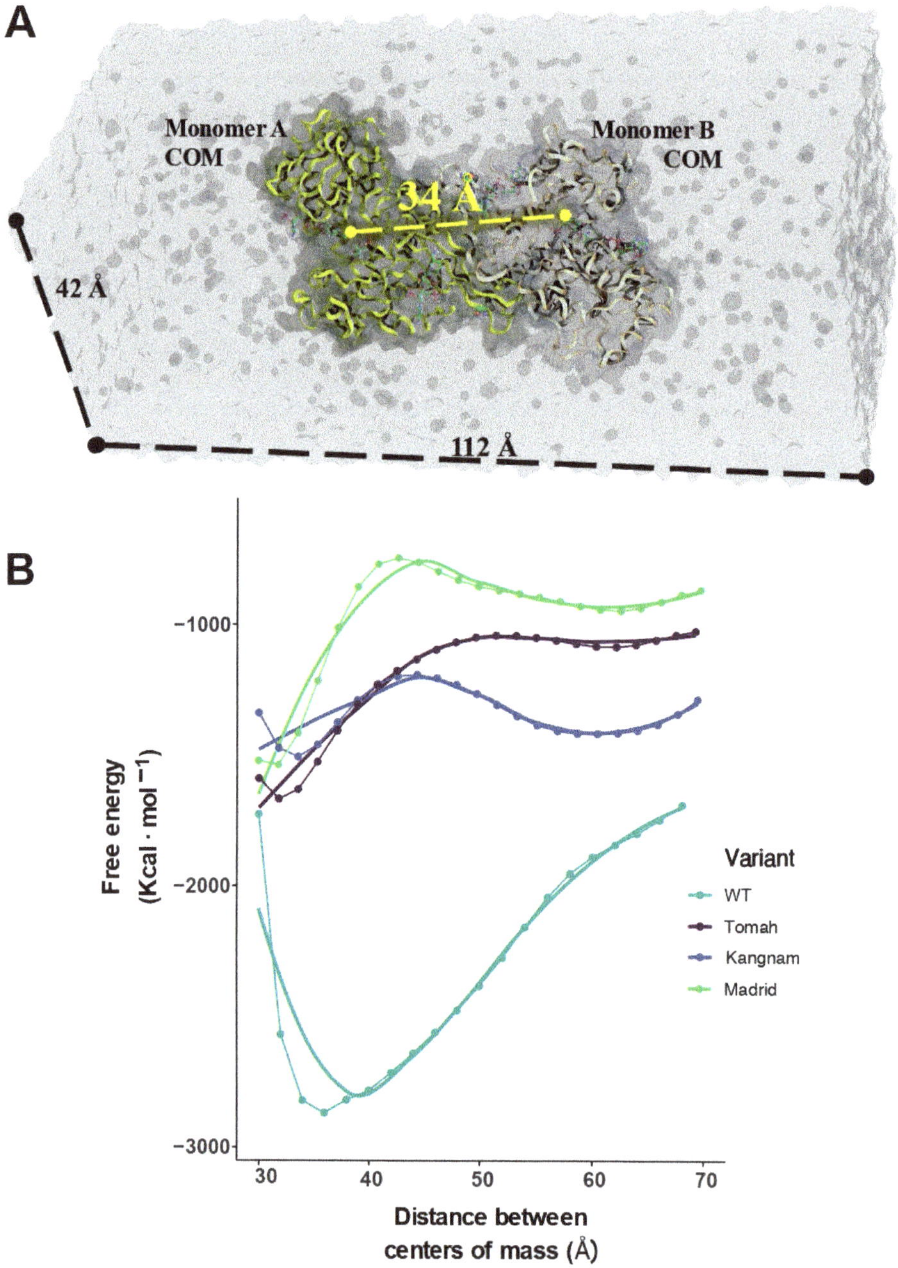

Figure 7. A metadynamics-solvated model system and a collective variable description of the evolution of the bound/unbound states of the G6PD variants. (**A**) The configuration of the metadynamics-enlarged simulation box. (**B**) The potential mean forces of the dimerizations of the G6DP variants. All depicted values were averaged from triplicate calculations for each variant.

Our biochemical results in this study showed that the alterations caused by these mutations at the functional and structural levels in the native structure of the G6PD protein resulted in a loss of catalytic activity and stability of the variants regarding WT-G6PD. Moreover, it is remarkable because the G6PD deficiency is high where malaria is prevalent [8], and it has been proposed that it inhibits the growth of specific types of malaria, giving rise to protective adaptation. However, tafenoquine was approved in 2018 by the Food and Drug Administration for treating malaria. Tafenoquine eliminates the hypnozoites of *Plasmodium vivax* and has a long half-life, where a single dose of this drug is as effective as an entire 14-day course of primaquine [13]. Moreover, it has been observed that in patients with G6PD deficiency, tafenoquine and primaquine can produce acute hemolytic anemia [46]. Moreover, the set of clinical, molecular, and biochemical data and in silico predictions will allow us to classify a variant as "pathogenic", "likely pathogenic", "uncertain significance", "likely benign", and "benign", as recommended by ACMG/AMP. So, when a variant is reported to be pathogenic, healthcare providers are very likely to consider it "applicable" and modify a patient's treatment or monitoring.

At present, blood transfusions are the only treatment option for this disease, particularly for the most severe class A G6PD variants. This study provides insights that could help identify potential therapeutics to correct G6PD deficiency and improve the search for chemical space to develop suitable small-molecule drugs to control this prevalent disease. This breakthrough could pave the way for a more effective and accessible treatment option for G6PD deficiency, improving the lives of millions of people worldwide.

A limitation of our study is that it was carried out at the molecular and biochemical levels, where, through site-directed mutagenesis, we created the three clinical G6PD variants and carried out the biochemical studies involved in this study. In addition, we related the assays obtained in this study with those observed in patients where the mutation was identified, who presented chronic nonspherocytic hemolytic anemia (CNSHA). However, to have a better explanation of these mutations at a physiological level, in the future, studies in animal models will have to be carried out, as was previously performed by Rovira et al. [47] that demonstrated the integration of the *hG6PD* gene in totipotent stem cells using a retroviral vector. The authors determined that the expression of hG6PD in vivo was stable and provided a rescue of G6PD deficiency in stem cells. Moreover, a similar vector containing the human *G6PD* gene (VSV-G vector) was adequate for the expression of hG6PD in Macaque monkeys [48].

3. Materials and Methods

3.1. Construction, Expression, and Purification of Recombinant G6PD Variants

The clinical G6PD mutants Tomah, Kangnam, and Madrid were obtained via site-directed mutagenesis, according to reports by Gómez-Manzo et al. [5]. Mutagenic oligonucleotides were designed for each mutant from the human *G6PD* gene sequence deposited in the Gen Bank database (accession: NM 001042351.2) (Table 4). The polymerase chain reaction (PCR) was reproduced using a previously reported method [4,5,12,18–21]. The PCR products were analyzed via electrophoresis in 1% agarose gel and visualized using GelRed (Nucleic Acid Gel, Biotium, Fremon, CA, USA). Then, the PCR products were digested with a *Dpn*I enzyme for 2 h at 37 °C. Afterward, the final digestion was used to transform competent *E. coli* TOP-10 cells, which were selected on a solid LB agar plate containing 200 μg/μL carbenicillin (CB-200). Then, the plasmid DNA was purified and verified via sequencing to confirm the desired G6PD mutants (pETgC385R, pETgC385G, and pETgC385W). Finally, it was used to transform into competent *E. coli* BL21(DE3)Δ*zwf*::*kan*^{r} cells.

To obtain crude extracts of the G6PD variants Tomah, Kangnam, and Madrid, liquid LB cultures were carried out with CB-200 (200 μg/μL) plus kanamycin (200 μg/μL). The cultures were grown, induced, harvested, and lysed via sonication according to a previously reported method [4,5,12,18–21]. The sonication product was centrifuged, and the crude extract was obtained. Finally, the enzymatic activity of each of the crude extracts was determined with a standard reaction mixture (100 mM Tris-HCl, 1 mM $NADP^+$, 1 mM G6P, and 3 mM $MgCl_2$ at a pH of 8.0).

Table 4. The oligonucleotides and strains used to obtain the G6PD mutants Tomah, Kangnam, and Madrid.

Mutant	Mutagenic Oligonucleotide Sequence	Reference
Tomah	Fw: 5′-CACCAGCAGCGCAAGCGCAA-3′ Rv: 5′-TTGCGCTTGCGCTGCTGGTG-3′	This study
Kangnam	Fw: 5′-CACCAGCAGGGCAAGCGCAA-3′ Rv: 5′-TTGCGCTTGCGCTGCTGGTG-3′	This study
Madrid	Fw: 5′-CACCAGCAGTGGAAGCGCAA-3′ Rv: 5′-TTGCGCTTCCACTGCTGGTG-3′	This study
Strains	**Genetic Characteristics**	**Reference**
E. coli TOP-10	F^-, DE(araD-araB)567, lacZ4787(del)::rrnB-3, LAM^-, rph-1, DE(rhaD-rhaB)568, hsdR514	[49]
E. coli BL21(DE3)Δzwf::kan^r	F^- ompT gal dcm lon hsdSB(r^- m^-) λ(DE3 [lacI lacUV5-T7 gene 1 ind1 sam7 nin5]) *Δzwf-777::kan*	[5]

The recombinant G6PD variants were purified using anion-exchange columns (Q-Toyopearl and Q-Sepharose) and affinity columns (2′,5′-ADP Sepharose 4B). First, the crude extract was loaded in an anion-exchange Q-Toyopearl column. Using a linear gradient from 0 to 0.35 M, the proteins were eluted. The portions with specific activity for G6PD were collected and loaded onto the 2′5′ADP Sepharose 4B affinity column. The column was washed until the absorbance at 280 nm was zero. Proteins were eluted from the column with 100 μM $NADP^+$ in the equilibrium buffer. The G6PD's activity was measured, and the fractions with enzymatic activity were concentrated using Amicon YM-30 tubes (Millipore, Bedford, MA, USA). Subsequently, the concentrate was loaded onto the Q-Sepharose 4B column (5 mL), which was previously equilibrated. The column was washed, and the recombinant protein was eluted with a linear gradient of NaCl from 0 to 0.35 M. The portions that presented enzymatic activity were concentrated, and the purity of the G6PD variants was assessed by 12% SDS-PAGE gels stained with Coomassie brilliant blue.

3.2. Determination of Steady-State Kinetic Parameters

The kinetic parameters of the variants G6PD Tomah, Kangnam, and Madrid were calculated by evaluating the production of NADPH at 340 nm. To obtain the kinetic parameters, one of the substrates was maintained at a saturating concentration ($NADP^+$ or G6P), while the concentration of the second substrate was changed from 2.5 to 200 μM ($NADP^+$ or G6P). The reaction started with 800 ng of total protein. The kinetic parameters (K_m, k_{cat}, and V_{max}) were obtained from the initial velocity and fit of the Michaelis–Menten equation. k_{cat} value was obtained from V_{max}, taking into the molecular mass (59.25 kDa for monomer).

3.3. Thermal Inactivation Analysis

Thermal stability trials were performed to determine the effects of the mutations in the WT-G6PD. The G6PD proteins were suspended at 0.2 mg/mL and incubated at temperatures ranging from 37 to 60 °C by 20 min. The residual G6PD activity was determined using a standard reaction mixture. At the ends of the assays, the residual activities were expressed as percentages concerning their controls. The 100% activity was fixed from the enzymes incubated at 37 °C. The experiments were carried out in triplicate.

3.4. Analysis of In Silico and Site-Directed Mutagenesis

In order to present the main structural elements of the G6PD enzyme, the three-dimensional structure of the G6PD dimer was constructed with the PyMOL software package (v. 2.5.7) (New York, NY, USA) using files deposited in the Protein Data Bank (1QK1, 2BHL, and 2BH9). Subsequently, the mutation was constructed on the G6PD structure using the mutagenesis wizard to generate the G6PD Tomah (C385R), G6PD Kangnam (C385G), and G6PD Madrid (C385W) variants.

3.5. Enhanced Molecular Dynamics Calculations

The enhanced molecular dynamics simulations (MDSs) technique was performed for the three G6PD variants in complex with glucose-6-phosphate and $NADP^+$ under *pmemd.cuda* [29] within the Amber20 package (https://ambermd.org/, accessed on 1 April 2023) [29,30]. The GAFF and Amberff14SB parameter sets [31,32] were applied to model protein and substrate organic molecules. The TIP3P explicit water model was applied to simulate ca. 112,780 solvent. The final dimensions of the orthorhombic water box were $155 \times 155 \times 155$ Å for all systems at a pH of 6.8 with 0.15 M NaCl.

The MDS protocol included a minimization routine of 5000 steps under the steepest descendent minimization algorithm, with subsequent minimization under the conjugated gradient algorithm for 50,000 steps. A second simulation stage under simulated annealing protocol and NvT ensemble was performed. All systems were heated with a linear interpolation scheme for 0.3 ns at 10 K, followed by 1 ns at 100 K, 1 ns at 300 K, and 0.3 ns at 400 K. Finally, the systems were cooled down at 303 K for 2 ns. For pre-equilibration, all systems were simulated under the NpT ensemble during 50 ns and equilibrated for an extra 100 ns of conventional molecular dynamics simulations at 1 atm of constant pressure under the control of Monte Carlo Barostat = 1. The temperature of the systems was controlled with an external bath coupling at 303 K. For calculation of the long-range electrostatic interactions, the Particle Mesh Ewald (PME) method was applied (1×10^{-10} of tolerance) with an integration step of 2 fs for all periodic systems. Gaussian Accelerated Molecular Dynamics simulations encompass different sampling techniques that smooth potential energy surfaces between biomolecular processes and reduce energy barriers between two or more processes [33,34], such as conformational changes, the binding and unbinding of substrates, and peptide dissociations, among others. A dual boost method was applied to add Gaussian harmonic boost potentials to dihedral angles and potential energy for all calculated systems. However, more in-depth descriptions of this methodology have been published elsewhere [35,36]. The GaMD simulations were performed in triplicate for 500 ns for each G6PD system with sigma0P = 6.0 and sigma0D = 6.0 at 303 K at the same integration time. Each GaMD production replica was combined and analyzed via hierarchical clustering methods with the *cpptraj* v5.0 [37] and VMD v1.9.3 [38] packages to identify conformational changes associated with each mutation. Additionally, the RMSDs of the alpha carbons and the RMSFs of the sidechains were averaged.

Furthermore, as the C385 mutations are allocated around the structural $NADP^+$ binding pocket, we suspected that single-point mutations could exert structural changes that modify the substrate affinity. The Python package *Pyvol* [43] was interfaced with PyMOL v. 2.5.7 (New York, NY, USA). It was used to calculate, visualize, and compare the binding pockets volume and SASA in a systematic and reproducible fashion for each G6PD system. *Pyvol* evaluated all the cavities within the dimer models that were accessible to small-radius probes (~2 Å) and stored the information on the identified pockets in the form of sets of coordinates and radii of tangent spheres within a three-dimensional voxel grid. Finally, the volumes of the $NADP^+$ pockets were calculated in $Å^3$ and SASA in $Å^2$.

3.6. Metadynamics Simulations

To evaluate the free energy profile of the dimer dissociation, we performed enhanced sampling methods under the well-tempered metadynamics technique [44] since the spontaneous dimerization process is inaccessible to unbiased MDSs due to timescale limitations. This method enhances sampling using collective variables (CVs), such as dihedral rotation, angle or RMSD perturbation, and distance-based stretching; moreover, these variables are steered, akin to an adaptive umbrella sampling theoretical framework [44]. All potential mean forces ($Kcal \cdot mol^{-1}$) were solved for each CV using a histogram of the differentiable function of the Gaussian potential at a specific width and height (ω and σ, respectively). Without a systematic way to choose CVs due in part to the specific conditions of each biomolecular system, the selection of a CV must rely on physically reasonable variables. Our simulations were carried out using Desmond (v.2022-1), and collective variables were based on the distance between the centers of mass for each G6PD monomer, approx. 34 Å, and the stretching along the x-axis. The evaluation provided insight into the effect of each mutation on dimer stabilization and was of significant importance to our study. The Gaussian function parameters for all simulations were set to 0.03 $Kcal \cdot mol^{-1}$ (ω) and 0.5 Å/ps (σ). A rate of 1 ps for Gaussian functions was applied, and a bias factor of 3 was set for the well-tempered algorithm (the rate at which the Gaussian height decreased until reaching a maximum of 70 Å long or full dimer dissociation). Three replicas were generated for each mutation for an additional 100 ns at 1 fs of integration time, and the potential mean force was derived using the Maestro Metadynamics Analysis plugin inside the Maestro 2022-1 suite from Schrödinger ® (New York, NY, USA).

4. Conclusions

In this work, we report the cloning, purification, biochemical characterization, and molecular dynamics of three clinically relevant class A G6PD variants—named G6PD Tomah, G6PD Kangnam, and G6PD Madrid—for the first time, which are located in the same codon. The kinetic parameters (K_m, V_{max}, and k_{cat}) of the variants were obtained, and their values were decreased compared to native G6PD, showing impairment in their catalytic activity. The thermal stability of the variant proteins was decreased compared to wild-type G6PD, suggesting that the mutations lead to more unstable G6PD protein. In silico and molecular dynamics analyses of the variants revealed significant structural changes, mainly in the structural $NADP^+$ binding site, which could explain the clinical manifestations of G6PD deficiency and open new lines of research to develop new drugs that target these G6PD variants, improving the quality of life of patients.

Supplementary Materials: The following supporting information can be downloaded at https://www.mdpi.com/article/10.3390/ijms252312556/s1.

Author Contributions: Conceptualization, S.G.-M. and A.V.-L.; software, methodology, validation, investigation, formal analysis, writing—original draft preparation, and writing—review and editing, A.G.-R., A.G.-V., B.H.-O., L.M.C.-Á., A.L.-R., R.A.-E., V.P.d.l.C., E.H.-U., N.C.-R., S.E.-F., I.D.l.M.-D.l.M., A.V.-L. and S.G.-M.; supervision, A.G.-V., A.V.-L. and S.G.-M.; project administration, S.G.-M. and A.V.-L.; funding acquisition, B.H.-O., A.V.-L. and S.G.-M. All authors have read and agreed to the published version of the manuscript.

Funding: This research was funded by Recursos Fiscales para la Investigación, E022 Program, from the National Institute of Pediatrics 034/2022 (S.G.-M). B.H.-O. was supported by HIM/2019/036 SSA.

1595, and A.V.-L. acknowledges the National Supercomputing Center—IPICYT for the computational resources provided in support of this research through grant TKII-AMVL001.

Institutional Review Board Statement: Not applicable.

Informed Consent Statement: Not applicable.

Data Availability Statement: Data is contained within the article and Supplementary Materials.

Acknowledgments: We thank Ximena Gomez-Gonzalez and Maria Jose Gomez-Gonzalez for their technical assistance. We thank Javier Gallegos Infante (Instituto de Fisiología Celular, UNAM) for the help with the bibliographic materials. Emanuel Villafán de la Torre for technical support of the Huitzilin supercomputing cluster at INECOL A.C.

Conflicts of Interest: The authors declare no conflicts of interest.

References

1. Wakao, S.; Andre, C.; Benning, C. Functional Analyses of Cytosolic Glucose-6-Phosphate Dehydrogenases and Their Contribution to Seed Oil Accumulation in Arabidopsis. *Plant Physiol.* **2008**, *146*, 277–288. [CrossRef] [PubMed]
2. Hauschild, R.; von Schaewen, A. Differential Regulation of Glucose-6-Phosphate Dehydrogenase Isoenzyme Activities in Potato. *Plant Physiol.* **2003**, *133*, 47–62. [CrossRef] [PubMed]
3. Landi, S.; Nurcato, R.; De Lillo, A.; Lentini, M.; Grillo, S.; Esposito, S. Glucose-6-phosphate dehydrogenase plays a central role in the response of tomato (*Solanum lycopersicum*) plants to short and long-term drought. *Plant Physiol. Biochem.* **2016**, *105*, 79–89. [CrossRef] [PubMed]
4. Cortés-Morales, Y.Y.; Vanoye-Carlo, A.; Castillo-Rodríguez, R.A.; Serrano-Posada, H.; González-Valdez, A.; Ortega-Cuellar, D.; Hernández-Ochoa, B.; Moreno-Vargas, L.M.; Prada-Gracia, D.; Sierra-Palacios, E.; et al. Cloning and biochemical characterization of three glucose 6 phosphate dehydrogenase mutants presents in the Mexican population. *Int. J. Biol. Macromol.* **2018**, *119*, 926–936. [CrossRef]
5. Gómez-Manzo, S.; Terrón-Hernández, J.; de la Mora-de la Mora, I.; González-Valdez, A.; Marcial-Quino, J.; García-Torres, I.; Vanoye-Carlo, A.; López-Velázquez, G.; Hernández-Alcántara, G.; Oria-Hernández, J.; et al. The Stability of G6PD Is Affected by Mutations with Different Clinical Phenotypes. *Int. J. Mol. Sci.* **2014**, *15*, 21179–21201. [CrossRef]
6. Gómez-Manzo, S.; Marcial-Quino, J.; Vanoye-Carlo, A.; Serrano-Posada, H.; González-Valdez, A.; Martínez-Rosas, V.; Hernández-Ochoa, B.; Sierra-Palacios, E.; Castillo-Rodríguez, R.A.; Cuevas-Cruz, M.; et al. Functional and Biochemical Characterization of Three Recombinant Human Glucose-6-Phosphate Dehydrogenase Mutants: Zacatecas, Vanua-Lava and Viangchan. *Int. J. Mol. Sci.* **2016**, *17*, 787. [CrossRef]
7. Alatorre-Salas, M.A.; González-Bizarro, J.I.; López-Venegas, I.F.; Rojo-Contreras, W. Deficiencia de glucosa-6-fosfato-deshidrogenasa. *Sal. Jal.* **2017**, *4*, 178–184.
8. Howes, R.E.; Battle, K.E.; Satyagraha, A.W.; Baird, J.K.; Hay, S.I. Chapter Four—G6PD Deficiency: Global Distribution, Genetic Variants and Primaquine Therapy. *Adv. Parasitol.* **2013**, *81*, 133–201. [CrossRef]
9. García-Magallanes, N.; Luque-Ortega, F.; Aguilar-Medina, E.M.; Ramos-Payán, R.; Galaviz-Hernández, C.; Romero-Quintana, J.G.; del Pozo-Yauner, L.; Rangel-Villalobos, H.; Arámbula-Meraz, E. Glucose-6-phosphate dehydrogenase deficiency in northern Mexico and description of a novel mutation. *J. Genet.* **2014**, *93*, 325–330. [CrossRef]
10. Boonyuen, U.; Chamchoy, K.; Swangsri, T.; Junkree, T.; Day, N.P.; White, N.J.; Imwong, M. A trade off between catalytic activity and protein stability determines the clinical manifestations of glucose-6-phosphate dehydrogenase (G6PD) deficiency. *Int. J. Biol. Macromol.* **2017**, *104*, 145–156. [CrossRef]
11. Minucci, A.; Moradkhani, K.; Hwang, M.J.; Zuppi, C.; Giardina, B.; Capoluongo, E. Glucose-6-phosphate dehydrogenase (G6PD) mutations database: Review of the "old" and update of the new mutations. *Blood Cells, Mol. Dis.* **2012**, *48*, 154–165. [CrossRef] [PubMed]
12. Gómez-Manzo, S.; Marcial-Quino, J.; Vanoye-Carlo, A.; Serrano-Posada, H.; Ortega-Cuellar, D.; González-Valdez, A.; Castillo-Rodríguez, R.A.; Hernández-Ochoa, B.; Sierra-Palacios, E.; Rodríguez-Bustamante, E.; et al. Glucose-6-Phosphate Dehydrogenase: Update and Analysis of New Mutations around the World. *Int. J. Mol. Sci.* **2016**, *17*, 2069. [CrossRef] [PubMed]
13. Luzzatto, L.; Ally, M.; Notaro, R. Glucose-6-phosphate dehydrogenase deficiency. *Blood* **2020**, *136*, 1225–1240. [CrossRef] [PubMed]
14. Rovira, A.; Vulliamy, T.; Pujades, M.A.; Luzzatto, L.; Vives, J.L. Molecular genetics of glucose-6-phosphate dehydrogenase (G6PD) deficiency in Spain: Identification of two new point mutations in the G6PD gene. *Br. J. Haematol.* **1995**, *91*, 66–71. [CrossRef] [PubMed]

15. Lee, J.; Park, J.; Choi, H.; Kim, J.; Kwon, A.; Jang, W.; Chae, H.; Kim, M.; Kim, Y.; Lee, J.W.; et al. Genetic Profiles of Korean Patients with Glucose-6-Phosphate Dehydrogenase Deficiency. *Ann. Lab. Med.* **2017**, *37*, 108–116. [CrossRef]
16. Luzzatto, L.; Bancone, G.; Dugué, P.-A.; Jiang, W.; Minucci, A.; Nannelli, C.; Pfeffer, D.; Prchal, J.; Sirdah, M.; Sodeinde, O.; et al. New WHO classification of genetic variants causing G6PD deficiency. *Bull. World Heal. Organ.* **2024**, *102*, 615–617. [CrossRef]
17. Zarza, R.; Pujades, A.; Rovira, A.; Saavedra, R.; Fernandez, J.; Aymerich, M.; Corrons, J.L.V. Two new mutations of the glucose-6-phospate dehydrogenase (G6PD) gene associated with haemolytic anaemia: Clinical, biochemical and molecular relationships. *Br. J. Haematol.* **1997**, *98*, 578–582. [CrossRef]
18. Gómez-Manzo, S.; Marcial-Quino, J.; Vanoye-Carlo, A.; Enríquez-Flores, S.; De la Mora-De la Mora, I.; González-Valdez, A.; García-Torres, I.; Martínez-Rosas, V.; Sierra-Palacios, E.; Lazcano-Pérez, F.; et al. Mutations of Glucose-6-Phosphate Dehydrogenase Durham, Santa-Maria and A+ Variants Are Associated with Loss Functional and Structural Stability of the Protein. *Int. J. Mol. Sci.* **2015**, *16*, 28657–28668. [CrossRef]
19. Ramírez-Nava, E.J.; Ortega-Cuellar, D.; Serrano-Posada, H.; González-Valdez, A.; Vanoye-Carlo, A.; Hernández-Ochoa, B.; Sierra-Palacios, E.; Hernández-Pineda, J.; Rodríguez-Bustamante, E.; Arreguin-Espinosa, R.; et al. Biochemical Analysis of Two Single Mutants that Give Rise to a Polymorphic G6PD A-Double Mutant. *Int. J. Mol. Sci.* **2017**, *18*, 2244. [CrossRef]
20. Martínez-Rosas, V.; Juárez-Cruz, M.V.; Ramírez-Nava, E.J.; Hernández-Ochoa, B.; Morales-Luna, L.; González-Valdez, A.; Serrano-Posada, H.; Cárdenas-Rodríguez, N.; Ortiz-Ramírez, P.; Centeno-Leija, S.; et al. Effects of Single and Double Mutants in Human Glucose-6-Phosphate Dehydrogenase Variants Present in the Mexican Population: Biochemical and Structural Analysis. *Int. J. Mol. Sci.* **2020**, *21*, 2732. [CrossRef]
21. Alcántara-Ortigoza, M.A.; Hernández-Ochoa, B.; Angel, A.G.-D.; Ibarra-González, I.; Belmont-Martínez, L.; Gómez-Manzo, S.; Vela-Amieva, M. Functional characterization of the p. (Gln195His) or Tainan and novel p. (Ser184Cys) or Toluca glucose-6-phosphate dehydrogenase (G6PD) gene natural variants identified through Mexican newborn screening for glucose-6-phosphatedehydrogenase deficiency. *Clin. Biochem.* **2022**, *109–110*, 64–73. [CrossRef] [PubMed]
22. Praoparotai, A.; Junkree, T.; Imwong, M.; Boonyuen, U. Functional and structural analysis of double and triple mutants reveals the contribution of protein instability to clinical manifestations of G6PD variants. *Int. J. Biol. Macromol.* **2020**, *158*, 884–893. [CrossRef] [PubMed]
23. Hernández-Ochoa, B.; Ortega-Cuellar, D.; González-Valdez, A.; Martínez-Rosas, V.; Morales-Luna, L.; Rojas-Alarcón, M.A.; Vázquez-Bautista, M.; Arreguin-Espinosa, R.; Pérez de la Cruz, V.; Castillo-Rodríguez, R.A.; et al. An Overall View of the Functional and Structural Characterization of Glucose-6-Phosphate Dehydrogenase Variants in the Mexican Population. *Int. J. Mol. Sci.* **2023**, *24*, 12691. [CrossRef] [PubMed]
24. Wang, X.-T.; Lam, V.M.; Engel, P.C. Functional properties of two mutants of human glucose 6-phosphate dehydrogenase, R393G and R393H, corresponding to the clinical variants G6PD Wisconsin and Nashville. *Biochim. Biophys. Acta (BBA)—Mol. Basis Dis.* **2006**, *1762*, 767–774. [CrossRef] [PubMed]
25. Wang, X.-T.; Engel, P.C. Clinical mutants of human glucose 6-phosphate dehydrogenase: Impairment of NADP+ binding affects both folding and stability. *Biochim. Biophys. Acta (BBA)—Mol. Basis Dis.* **2009**, *1792*, 804–809. [CrossRef]
26. Alakbaree, M.; Abdulsalam, A.H.; Ahmed, H.H.; Ali, F.H.; Al-Hili, A.; Omar, M.S.S.; Alonazi, M.; Jamalis, J.; Ab Latif, N.; Hamza, M.A.; et al. A computational study of structural analysis of Class I human glucose-6-phosphate dehydrogenase (G6PD) variants: Elaborating the correlation to chronic non-spherocytic hemolytic anemia (CNSHA). *Comput. Biol. Chem.* **2023**, *104*, 107873. [CrossRef]
27. Garcia, A.A.; Mathews, I.I.; Horikoshi, N.; Matsui, T.; Kaur, M.; Wakatsuki, S.; Mochly-Rosen, D. Stabilization of glucose-6-phosphate dehydrogenase oligomers enhances catalytic activity and stability of clinical variants. *J. Biol. Chem.* **2022**, *298*, 101610. [CrossRef]
28. Beutler, E. Glucose-6-phosphate dehydrogenase deficiency: A historical perspective. *Blood* **2008**, *111*, 16–24. [CrossRef]
29. Bernhofer, M.; Dallago, C.; Karl, T.; Satagopam, V.; Heinzinger, M.; Littmann, M.; Olenyi, T.; Qiu, J.; Schütze, K.; Yachdav, G.; et al. PredictProtein—Predicting Protein Structure and Function for 29 Years. *Nucleic Acids Res.* **2021**, *49*, W535–W540. [CrossRef]
30. Alakbaree, M.; Amran, S.; Shamsir, M.; Ahmed, H.; Hamza, M.; Alonazi, M.; Warsy, A.; Ab Latif, N. Construction of a complete human glucose-6-phosphate dehydrogenase dimer structure bound to glucose-6-phosphate and nicotinamide adenine dinucleotide phosphate cofactors using molecular docking approach. In Proceedings of the 1st Samarra International Conference for Pure and Applied Sciences, Samarra, Iraq, 23–24 March 2021; AIP: Melville, NY, USA, 2021; p. 030006.
31. Pejaver, V.; Urresti, J.; Lugo-Martinez, J.; Pagel, K.A.; Lin, G.N.; Nam, H.-J.; Mort, M.; Cooper, D.N.; Sebat, J.; Iakoucheva, L.M.; et al. Inferring the molecular and phenotypic impact of amino acid variants with MutPred2. *Nat. Commun.* **2020**, *11*, 5918. [CrossRef]
32. Mahlich, Y.; Steinegger, M.; Rost, B.; Bromberg, Y. HFSP: High speed homology-driven function annotation of proteins. *Bioinformatics* **2018**, *34*, i304–i312. [CrossRef] [PubMed]
33. Gerasimavicius, L.; Liu, X.; Marsh, J.A. Identification of pathogenic missense mutations using protein stability predictors. *Sci. Rep.* **2020**, *10*, 15387. [CrossRef]

34. Olenyi, T.; Marquet, C.; Heinzinger, M.; Kröger, B.; Nikolova, T.; Bernhofer, M.; Sändig, P.; Schütze, K.; Littmann, M.; Mirdita, M.; et al. LambdaPP: Fast and accessible protein-specific phenotype predictions. *Protein Sci.* **2023**, *32*, e4524. [CrossRef] [PubMed]
35. Jumper, J.; Evans, R.; Pritzel, A.; Green, T.; Figurnov, M.; Ronneberger, O.; Tunyasuvunakool, K.; Bates, R.; Žídek, A.; Potapenko, A.; et al. Highly accurate protein structure prediction with AlphaFold. *Nature* **2021**, *596*, 583–589. [CrossRef] [PubMed]
36. Pak, M.A.; Markhieva, K.A.; Novikova, M.S.; Petrov, D.S.; Vorobyev, I.S.; Maksimova, E.S.; Kondrashov, F.A.; Ivankov, D.N. Using AlphaFold to predict the impact of single mutations on protein stability and function. *PLoS ONE* **2023**, *18*, e0282689. [CrossRef] [PubMed]
37. Karakoyun, H.K.; Yüksel, K.; Amanoglu, I.; Naserikhojasteh, L.; Yeşilyurt, A.; Yakıcıer, C.; Timuçin, E.; Akyerli, C.B. Evaluation of AlphaFold structure-based protein stability prediction on missense variations in cancer. *Front. Genet.* **2023**, *14*, 1052383. [CrossRef]
38. Horikoshi, N.; Hwang, S.; Gati, C.; Matsui, T.; Castillo-Orellana, C.; Raub, A.G.; Garcia, A.A.; Jabbarpour, F.; Batyuk, A.; Broweleit, J.; et al. Long-range structural defects by pathogenic mutations in most severe glucose-6-phosphate dehydrogenase deficiency. *Proc. Natl. Acad. Sci. USA* **2021**, *118*, e2022790118. [CrossRef]
39. Richards, S.; Aziz, N.; Bale, S.; Bick, D.; Das, S.; Gastier-Foster, J.; Grody, W.W.; Hegde, M.; Lyon, E.; Spector, E.; et al. Standards and guidelines for the interpretation of sequence variants: A joint consensus recommendation of the American College of Medical Genetics and Genomics and the Association for Molecular Pathology. *Genet. Med.* **2015**, *17*, 405–424. [CrossRef]
40. Adzhubei, I.A.; Schmidt, S.; Peshkin, L.; Ramensky, V.E.; Gerasimova, A.; Bork, P.; Kondrashov, A.S.; Sunyaev, S.R. A method and server for predicting damaging missense mutations. *Nat. Methods* **2010**, *7*, 248–249. [CrossRef]
41. Tavtigian, S.V.; Deffenbaugh, A.M.; Yin, L.; Judkins, T.; Scholl, T.; Samollow, P.B.; de Silva, D.; Zharkikh, A.; Thomas, A. Comprehensive statistical study of 452 BRCA1 missense substitutions with classification of eight recurrent substitutions as neutral. *J. Med. Genet.* **2006**, *43*, 295–305. [CrossRef]
42. Mathe, E.; Olivier, M.; Kato, S.; Ishioka, C.; Hainaut, P.; Tavtigian, S.V. Computational approaches for predicting the biological effect of p53 missense mutations: A comparison of three sequence analysis based methods. *Nucleic Acids Res.* **2006**, *34*, 1317–1325. [CrossRef] [PubMed]
43. Zhou, Y.; Pan, Q.; Pires, D.E.V.; Rodrigues, C.H.M.; Ascher, D.B. DDMut: Predicting effects of mutations on protein stability using deep learning. *Nucleic Acids Res.* **2023**, *51*, W122–W128. [CrossRef] [PubMed]
44. Hamza, M.A. Determination of structural stability and catalytic activity of G6PD variants using molecular dynamic simulation (MDS) approach. *FASEB J.* **2020**, *34*, 1. [CrossRef]
45. Saddala, M.S.; Lennikov, A.; Huang, H. Discovery of Small-Molecule Activators for Glucose-6-Phosphate Dehydrogenase (G6PD) Using Machine Learning Approaches. *Int. J. Mol. Sci.* **2020**, *21*, 1523. [CrossRef] [PubMed]
46. Rueangweerayut, R.; Bancone, G.; Harrell, E.J.; Beelen, A.P.; Kongpatanakul, S.; Möhrle, J.J.; Rousell, V.; Mohamed, K.; Qureshi, A.; Narayan, S.; et al. Hemolytic Potential of Tafenoquine in Female Volunteers Heterozygous for Glucose-6-Phosphate Dehydrogenase (G6PD) Deficiency (G6PD Mahidol Variant) versus G6PD-Normal Volunteers. *Am. J. Trop. Med. Hyg.* **2017**, *97*, 702–711. [CrossRef]
47. Rovira, A.; De Angioletti, M.; Camacho-Vanegas, O.; Liu, D.; Rosti, V.; Gallardo, H.F.; Notaro, R.; Sadelain, M.; Luzzatto, L. Stable in vivo expression of glucose-6-phosphate dehydrogenase (G6PD) and rescue of G6PD deficiency in stem cells by gene transfer. *Blood* **2000**, *96*, 4111–4117. [CrossRef]
48. Shi, P.A.; De Angioletti, M.; Donahue, R.E.; Notaro, R.; Luzzatto, L.; Dunbar, C.E. In vivo gene marking of rhesus macaque long-term repopulating hematopoietic cells using a VSV-G pseudotyped versus amphotropic oncoretroviral vector. *J. Gene Med.* **2004**, *6*, 367–373. [CrossRef]
49. Datsenko, K.A.; Wanner, B.L. One-step inactivation of chromosomal genes in *Escherichia coli* K-12 using PCR products. *Proc. Natl. Acad. Sci. USA* **2000**, *97*, 6640–6645. [CrossRef]

International Journal of
Molecular Sciences

MDPI

Article

Peptide Activator Stabilizes DJ-1 Structure and Enhances Its Activity

Jing-Yuan Shih and Yuan-Hao Howard Hsu *

Department of Chemistry, Tunghai University, No. 1727, Sec. 4, Taiwan Boulevard, Xitun District, Taichung 40704, Taiwan; rrichaard1998@gmail.com
* Correspondence: howardhsu@thu.edu.tw

Abstract: DJ-1 is a vital enzyme involved in the maintenance of mitochondrial health, and its mutation has been associated with an increased risk of Parkinson's disease (PD). Effective regulation of DJ-1 activity is essential for the well-being of mitochondria, and DJ-1 is thus a potential target for PD drug development. In this study, two peptides (^{15}EEMETIIPVDVMRRA^{29} and ^{47}SRDVVICPDA^{56}) were utilized with the aim of enhancing the activity of DJ-1. The mechanisms underlying the activity enhancement by these two peptides were investigated using hydrogen/deuterium exchange mass spectrometry (HDXMS). The HDXMS results revealed distinct mechanisms. Peptide 1 obstructs the access of solvent to the dimer interface and stabilizes the α/β hydrolase structure, facilitating substrate binding to a stabilized active site. Conversely, peptide 2 induces a destabilization of the α/β hydrolase core, enhancing substrate accessibility and subsequently increasing DJ-1 activity. The binding of these two peptides optimizes the activity site within the dimeric structure. These findings offer valuable insights into the mechanisms underlying the activity enhancement of DJ-1 by the two peptides, potentially aiding the development of new drugs that can enhance the activity of DJ-1 and, consequently, advance PD treatment.

Keywords: Parkinson's disease; DJ-1; hydrogen/deuterium mass spectrometry; *PARK7*

1. Introduction

Parkinson's disease is a neurodegenerative disease, mainly affecting the central nervous system. In patients with Parkinson's disease (PD), α-synuclein protein aggregates and dopamine-generating cells die in the substantia nigra [1]. Although the exact cause of PD remains unclear, environmental toxins and genetic disorders are recognized as two risk factors for the disease. Both factors can lead to mitochondrial dysfunction, initiating neuronal cell death and triggering PD [2,3]. Dysregulated mitochondria contribute to the overproduction of reactive oxygen species (ROS), significantly impairing mitochondrial function and reducing the efficiency of the electron transport chain and ATP production, particularly in electron transport chains I and III [4]. Mutations in α-synuclein [5], Parkin (*PARK2*) [6], PINK1 (*PARK6*) [7], DJ-1 (*PARK7*) [8], and iPLA$_2\beta$ (*PARK14*) [9] have been identified as being associated with mitochondrial impairment and the underlying causes of PD.

DJ-1 is a Potential Drug Target for PD

DJ-1, a deglycase encoded by *PARK7* [10], has been associated with PD, with the association due to the loss of its enzymatic activity and its ability to interact with other proteins [8]. DJ-1 forms complexes with Parkin and PINK1 to repair mitochondrial function in response to oxidative stress [11]. Both Parkin and PINK1 exhibit neuroprotective activity against oxidative stress [12,13]. PINK1, also known as PTEN-induced putative kinase, localizes in mitochondria to activate Parkin [14]. Together, these three proteins collaborate to regulate ROS in neuronal cells and maintain mitochondrial functionality. Furthermore, DJ-1 protein was demonstrated to directly coaggregate with α-synuclein through hydrophobic interactions, suggesting a potential connection to the initiation of α-synuclein aggregation [15]. Moreover, DJ-1 was shown to play a role in the transcriptional regulation of dopamine synthesis [16] and to interact with a large protein complex involved in the regulation of catecholamine homeostasis [17]. Dysfunction of DJ-1 may lead to dopamine oxidation, contributing to the aggregation of α-synuclein [18–22]. This implies that the oxidation of dopamine by ROS causes the precipitation of α-synuclein in the substantia nigra.

The endogenous antioxidant DJ-1 exhibits redox-activated chaperone activity and functions as a cytoprotective agent [23–25]. DJ-1 can interact with NADPH oxidase, thereby regulating ROS production in macrophages [26]. Additionally, DJ-1 may upregulate glutathione synthesis to counteract α-synuclein toxicity [27]. DJ-1 was demonstrated to have specific protective effects on astrocytes against rotenone-induced neurotoxicity in a rat model of PD [28]. DJ-1 has become a potential drug target for PD. The small molecules, compound **23** [29], compound **B** [30] and the ND-13 peptide [31] have been shown to enhance the cellular functions through DJ-1 activity.

Structure and Cellular Function of DJ-1

DJ-1, a cytosolic protein containing 189 amino acids and with a molecular weight of 20 kDa, can localize to mitochondria in response to oxidative stress [32]. Through cysteine palmitoylation, DJ-1 can also localize to lipid rafts, thereby regulating endocytosis in astrocytes [33]. DJ-1 exhibits multiple enzymatic activities, including deglycase activity [34], chaperone activity [35], proteolytic activity [36], and glyoxalase activity [37]. Of the three cysteine residues in DJ-1 (Cys46, Cys53, and Cys106), Cys106 serves as the active site [38]. The antioxidative defense mechanism has been associated with the oxidation of Cys106 into sulfinic acid [38]. Both the crystal structure of DJ-1 and various cellular results support the understanding that DJ-1 exists in dimeric form [39,40]. Mutations such as L166P and M26I on the interaction interface hinder DJ-1 activity, underscoring the crucial role of the dimer structure in its functionality [41].

Because of the observed decrease in DJ-1 activity resulting from Cys106 oxidation and mutation variants, our intention was to design peptides that can interact with the dimer interface, thereby enhancing the enzymatic activities of DJ-1. The effects of peptide binding and the mechanisms underlying activity enhancement were comprehensively analyzed using hydrogen/deuterium exchange mass spectrometry (HDXMS).

2. Results

2.1. Peptide Design

In pursuit of therapeutic outcomes for PD, Daniel Offen and colleagues designed multiple peptides derived from the primary sequence of DJ-1. These peptides could protect cells against oxidative and neurotoxic damage, thereby reducing intracellular ROS accumulation [31]. One such peptide, ND-13 (^{12}KGAEEMETVIPVD^{24}), was also demonstrated to exhibit cardioprotective efficacy [42]. However, the mechanisms underlying these cellular effects remain unclear. Because the active dimeric structure of DJ-1 influences its activity, DJ-1's dimeric nature can be effectively disrupted by binding peptides to it to alter its functionality. The dimerization configuration of DJ-1 was shown by the crystal structure (Figure 1). There are three cysteine residues on DJ-1, including the active site Cys106 and two critical residues, Met26 and Lys166, for the dimerization of DJ-1. An analysis of the crystal structure of DJ-1 (PDB: 2OR3) revealed that the dimerization interface comprises an α-helix segment (Glu15–Ala29) and a β-sheet segment (Ser47–Ala56). In the α-helix contact region (Glu15–Ala29), the interaction between Glu15 and Arg28 takes the form of intermolecular hydrogen bonds and salt bridges. Hydrophobic amino acids—such as Ile, Pro, Ala, Met, and Val—contribute to hydrophobic forces at the binding site, facilitating dimerization. To preserve the α-helix structure, the designed short peptide, named peptide 1 (P1; Glu15–Ala29), includes the entire α-helix.

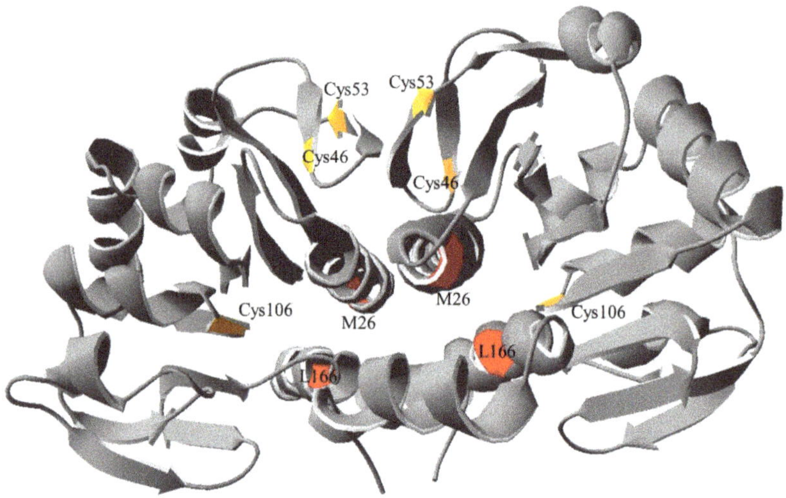

Figure 1. The dimerization configuration of DJ-1. The crystal structure of DJ-1 (PDB: 2OR3) showing the three cysteine residues in yellow and two critical residues in red related to the dimerization of DJ-1.

At the contact interface of the β-sheet (Ser47–Ala56), hydrogen bonds can form between Val51 and Cys53 as well as between Arg28 and Arg48 due to their close proximity. These hydrophobic interactions and hydrogen bonds contribute to the stability of the dimeric structure. To preserve the β-sheet secondary structure, the designed peptide, named peptide 2 (P2; Ser47–Ala56), includes the entire β-sheet. In experiments involving hydrogen–deuterium exchange, the sequence of P1 was observed to overlap with the pepsin-digested sequence of DJ-1. To comprehensively analyze hydrogen–deuterium (H/D) exchange, Val20 was substituted with the structurally similar amino acid Ile. Two purchased peptide segments, namely P1 (V20I) and P2, were employed, and their interactions with DJ-1 were investigated.

2.2. DJ-1 Purification and Activity Measurement

The DJ-1 protein with the 6X-His tag was overexpressed using pET3a-His-DJ1-transformed *Escherichia coli* and subsequently purified using Ni–nitrilotriacetic acid agarose beads. The sodium dodecyl sulfate–polyacrylamide gel electrophoresis (SDS-PAGE) patterns of eluted DJ-1 had either one or two bands across different fractions (Figure 2A). Western blot analysis confirmed both bands as corresponding to DJ-1 (Figure 2B). Following a 1 M DTT reduction of the sample with double bands, the SDS-PAGE pattern contained a single band. This confirmed the lower band as the oxidized form of DJ-1. The deglycation activity of the purified DJ-1 was then assessed. A substrate was prepared through the reaction of 10 mM *N*-acetyl-L-cysteine with 10 mM methylglyoxal solution. After 10 min of reaction, the reaction plateaued. DJ-1 demonstrated the deglycase activity to glyoxal-glycated cysteine, as indicated by the reduction in UV absorbance at 288 nm.

The active form of DJ-1 has been demonstrated to form a homodimer conformation in its crystal structure. The purified DJ-1 was confirmed to exhibit the dimeric conformation through DSS cross linking (Figure 2C). Two peptides, designated P1 and P2, were specifically designed to disrupt the dimerization interfaces of DJ-1 based on its crystal structure. The designed peptides, P1 and P2, were found to enhance the activity of DJ-1, as indicated by the MGO activity assay. Upon peptide binding, further analysis of DJ-1 dimerization was conducted using DSS cross linking and SDS-PAGE. The results of both indicated that peptide binding did not disrupt the dimer structure of DJ-1 (Figure 2C).

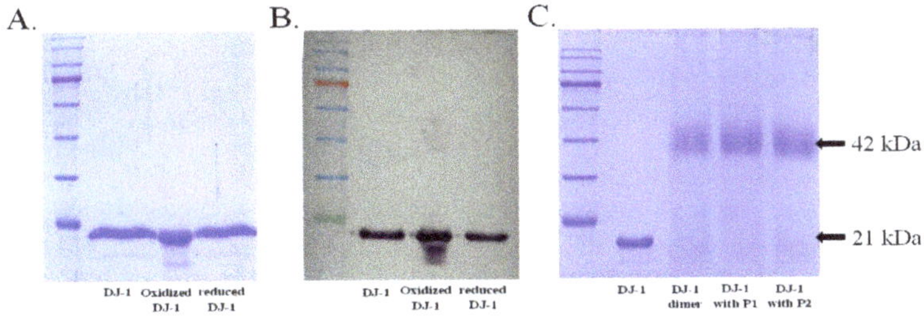

Figure 2. Purification and verification of DJ-1. (**A**) SDS-PAGE analysis of DJ-1, stained with Coomassie blue, revealing distinct bands at approximately 21 kDa. From left to right, the samples are DJ-1, the oxidized form of DJ-1 with two bands, and DTT-reduced DJ-1. (**B**) Western blot analysis by using anti-DJ-1 as the primary antibody revealed clearly defined bands at approximately 21 kDa. The samples are DJ-1, the oxidized form of DJ-1, and reduced DJ-1. (**C**) DJ-1 was treated with the DSS cross-linker for SDS-PAGE. Following DSS cross-linking, prominent dimer bands were observed at 42 kDa. The samples, from left to right, are DJ-1, DSS-crosslinked DJ-1, and DSS-crosslinked DJ-1 following the addition of P1 and P2.

Under anaerobic conditions, DJ-1 broke down glyoxal-glycated cysteine, resulting in a reduction in UV absorbance, thus confirming the deglycase activity of DJ-1 (Figure 3A). The activity was calculated to be 2.25 mmol/μg DJ-1/min. However, under aerobic conditions, the activity of DJ-1 was much lower at 1.00 mmol/μg/min; this prompted the assay to be conducted in a deoxygenated solution (Figure 3B). The oxidized form of DJ-1, characterized by two bands on SDS-PAGE patterns, exhibited activity of 1.06 mmol/μg/min. Following DTT reduction of DJ-1, its activity was restored to 1.60 mmol/μg/min. These findings indicate that oxidation reduces the activity of DJ-1 and that DTT reduction of oxidized DJ-1 can partially restore its functionality. The addition of P1 resulted in a notable 36% increase in the activity of DJ-1, which reached 3.28 mmol/μg/min. Similarly, the addition of P2 led to a 30% improvement in the activity of DJ-1, which reached 3.14 mmol/μg/min (Figure 3C,D).

2.3. *Identification of Proteolyzed DJ-1 Fragments*

Purified DJ-1 protein was obtained and validated using SDS-PAGE and Western blotting. The activity of DJ-1 was subsequently confirmed through the deglycase activity assay. The protein was then injected into a high-performance liquid chromatography system with a pepsin digestion column and a C18 reverse-phase column, and ion-trap mass spectrometry was then performed. The DJ-1 was digested by pepsin, and the digested peptides of DJ-1 were separated using a C18 column. The eluted peptides were then subject to tandem mass spectroscopy (MS/MS) by using ion-trap mass spectrometry, and the sequences of the fragments were identified using the X! Tandem-parser-1.7.7 program. The identified fragments were mapped onto the primary sequence of DJ-1 (Figure 4). In total, 49 peptides were identified, and these 49 covered 99.5% of the protein sequence (Supplemental Table S1). Notably, the initiator methionine was not identified in the map, which may be attributable to posttranslational excision. The His tag at the C-terminus was also identified.

2.4. *H/D Analysis of DJ-1 and the Effects of Peptide Binding*

The HDXMS analysis of DJ-1 highlighted its solvent accessibility and the flexibility of its backbone. H/D exchange was performed on the DJ-1 protein for 10–10,000 s. To ensure the homogeneity of DJ-1, the protein samples were treated with 1 M DTT, and residual DTT was removed using a PALL centrifugal filtration device. The deuteration levels of each pepsin-digested peptide were quantified at seven time points (Supplemental Figure S1). Because deuteration can be back-exchanged during mass spectrometry analysis, the data were normalized based on the 24 h full exchanged data (Supplemental Figure S2). Although 48 h exchange experiments were also conducted, unfortunately, the protein was precipitated. The deuteration percentage levels were then calculated based on the maximal deuteration of each peptide, and representing peptides were selected and color-coded to illustrate the exchange levels (Figure 5). Peptides 17–25, 18–24, 18–26, and 19–26 exhibited the lowest H/D exchanging region, with approximately 40% deuteration after 10,000 s of exchange. By contrast, peptides 59–69, 73–82, 76–83, 138–154, 147–158, and 154–163 exhibited relatively rapid exchange, reaching approximately 65% deuteration after 10,000 s of exchange.

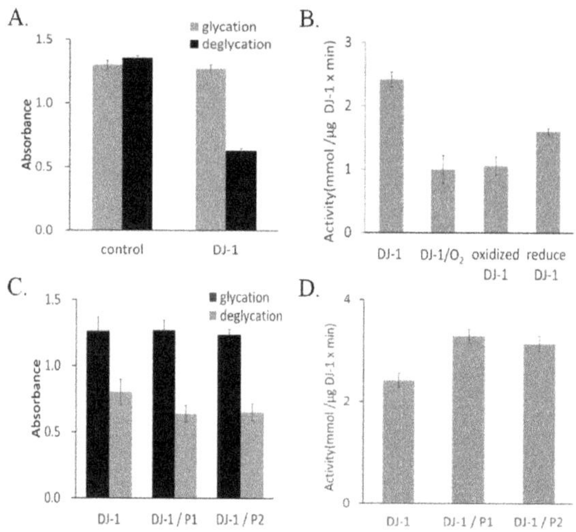

Figure 3. DJ-1 activity measurement. (**A**) The deglycation assay of DJ-1 involved the breakdown of hemithioacetal and resulted in a significant decrease in absorbance. The control corresponds to no DJ-1 protein. (**B**) The activity of DJ-1 was measured using a deglycation assay. The samples included were degassed assay DJ-1, DJ-1 without degassed assay, oxidized DJ-1, and reduced DJ-1; variation was discovered in deglycation activity under different conditions. (**C**) DJ-1 with or without P1 and P2 was added to the prepared substrate, and the resultant mixture was incubated at 37 °C for 30 min. (**D**) Calculated activities from (**C**).

The exposed active site on the α/β hydrolase core exhibited a higher level of deuteration when the period of H/D exchange was longer. The active site triad comprises Glu18, His126, and Cys106, with the oxidation status of Cys106 being linked to the activity of DJ-1. His126 was located in a domain exhibiting a low rate of H/D exchange, whereas Glu18 resided in a highly conserved acidic region, including Glu15, 16, and 18, and experienced a relatively high rate of H/D exchange in the N-terminus. Peptide 104–112 included the active site Cys106 in its sequence, making it ideal for representing the corresponding changes in the active site. The mass spectrum displayed in Figure 6A reveals a clear shift of the mass envelope, indicating an increase in deuteration over time. After 10,000 s of exchange, peptide 104–112 exhibited a mass increase of 4.4 Da. In the P1 and P2 addition experiments (Figure 6B,C), the mass shift further increased to 4.8 Da after P1 and P2 binding.

P1 or P2 was combined with DJ-1 in a molar ratio of 10:1 and allowed to react at 37 °C for 10 min. H/D exchange reactions were then conducted over a time range of 10–10,000 s. The changes in H/D exchange levels after P1 or P2 binding were assessed. Following the addition of P1 or P2, the exchange levels of most sequences were slightly lower. In the P1 experimental group, a significant reduction in H/D exchange was observed at 10,000 s for peptides 19–26 and 27–38. Critical regions for enzymatic activity—the active catalytic triad Glu18, His126, and Cys106—exhibited noteworthy decreases in H/D exchange.

Peptides 17–25 and 9–16, containing Glu18, exhibited noticeable decreases, whereas peptide 123–133, containing His126 in its sequence, exhibited a decrease at 10,000 s. Notably, although the overall results revealed a reduction in H/D exchange, regions 78–91 and 92–96 demonstrated an increase. In the P2 experimental group, the trends in H/D exchange for most sequences were similar to those in the P1 experimental group. Peptides 19–26, 27–38, and 31–38 as well as peptides containing the active catalytic triad (peptides 17–26 and 123–133) exhibited varying degrees of decrease in H/D exchange. Furthermore, regions 78–91 and 92–96 demonstrated varying degrees of increase in H/D exchange. Notably, unlike the P1 experimental group, an increase in H/D exchange was noted in peptide 2–10 at 10,000 s.

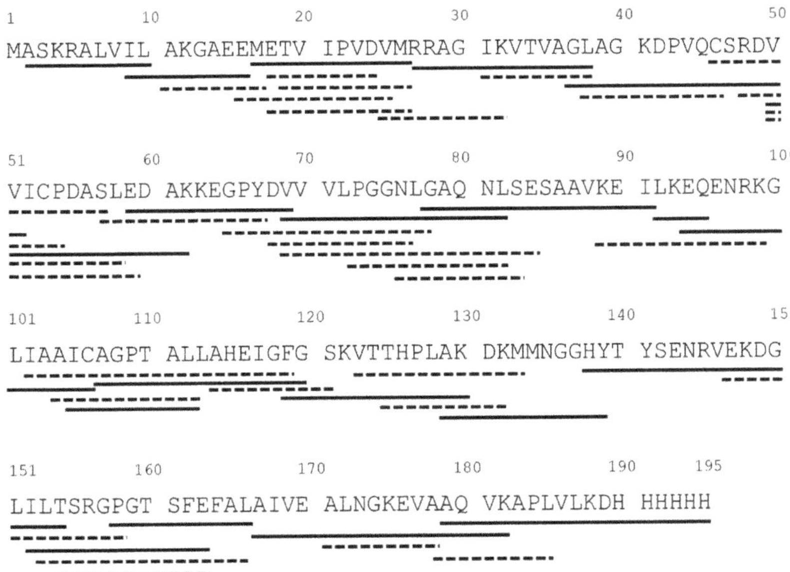

Figure 4. Peptide map of DJ-1 protein. DJ-1 protein was digested using a self-packed pepsin column on ice, and the resulting peptic fragments were retained on a peptide trap. Subsequently, the peptides were eluted through a C18 column by using a gradient for mass spectrometry analysis. The MS/MS data were exported to X! Tandem for sequence identification. The solid and the dashed lines are the identified peptides and the solid lines are selected for further structural presentation.

To further examine the interaction of and structural changes in DJ-1 induced by P1 and P2, the differences in H/D exchange percentages between P1 and P2 adsorption and the control group were mapped onto the crystal structure (Figure 7). Following the addition of P1 or P2, at 10 s of H/D exchange, slight reductions in exchange were observed in the αG and αH-helices, whereas increases in exchange were noted in the αD and αE-helices. However, the overall change in H/D exchange at 10 s was limited. At 10,000 s of H/D exchange, both P1 and P2 exhibited evident reductions in exchange at the interface of the dimer in the αA-helix, β2, and β4 sheets, with a more pronounced decrease in the β2, β3, and β4 sheets in the P2 experimental group. Additionally, on the backside of the active site in the αD-helix, a significant increase in H/D exchange was observed in both the P1 and P2 experimental groups. By contrast, in the P2 experimental group, the β1-sheet exhibited an increase in H/D exchange, and the β5 sheet exhibited a decrease, whereas in the P1 experimental group, no notable change was observed in the β1 and β5 sheets.

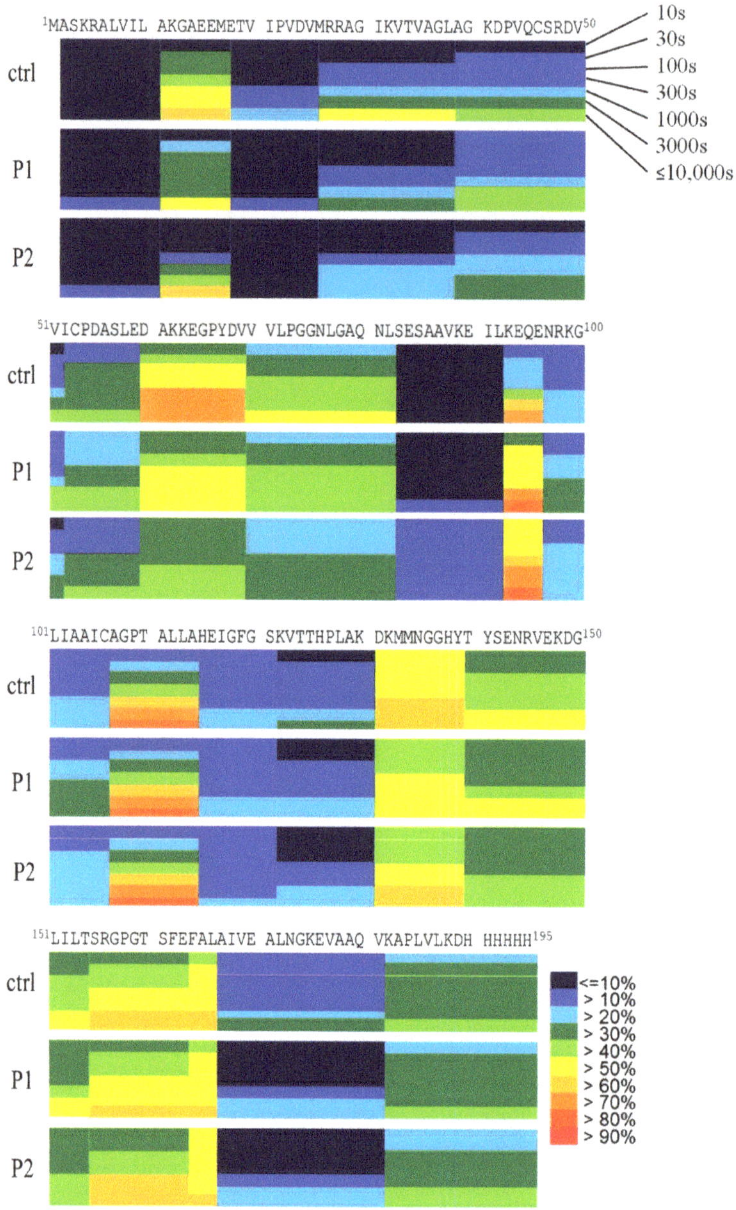

Figure 5. Deuteration levels of DJ-1. DJ-1 was incubated with or without P1 and P2 at 37 °C for 10 min, followed by deuteration in an H/D exchange reaction from 10 to 10,000 s. The reactions were quenched using ice-cold quench solution containing formic acid and guanidine hydrochloride. The deuterated protein was then subjected to mass spectrometry analysis. The number of deuterons incorporated in each peptide was measured, and the deuteration levels were subsequently calculated.

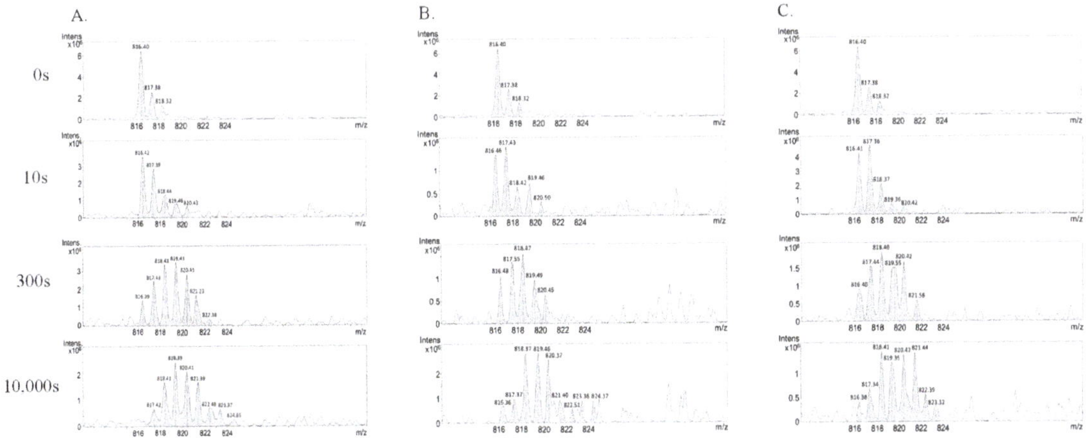

Figure 6. H/D exchange mass spectrum in the region 104–112 upon P1 and P2 binding. DJ-1 underwent deuteration from 0 to 10,000 s, and the results at 0, 10, 300, and 10,000 s are presented. (**A**) Mass spectrum of the sequence 104–112 after HDXMS. (**B**) HDXMS results of DJ-1 interacting with P1. (**C**) HDXMS results of DJ-1 interacting with P2.

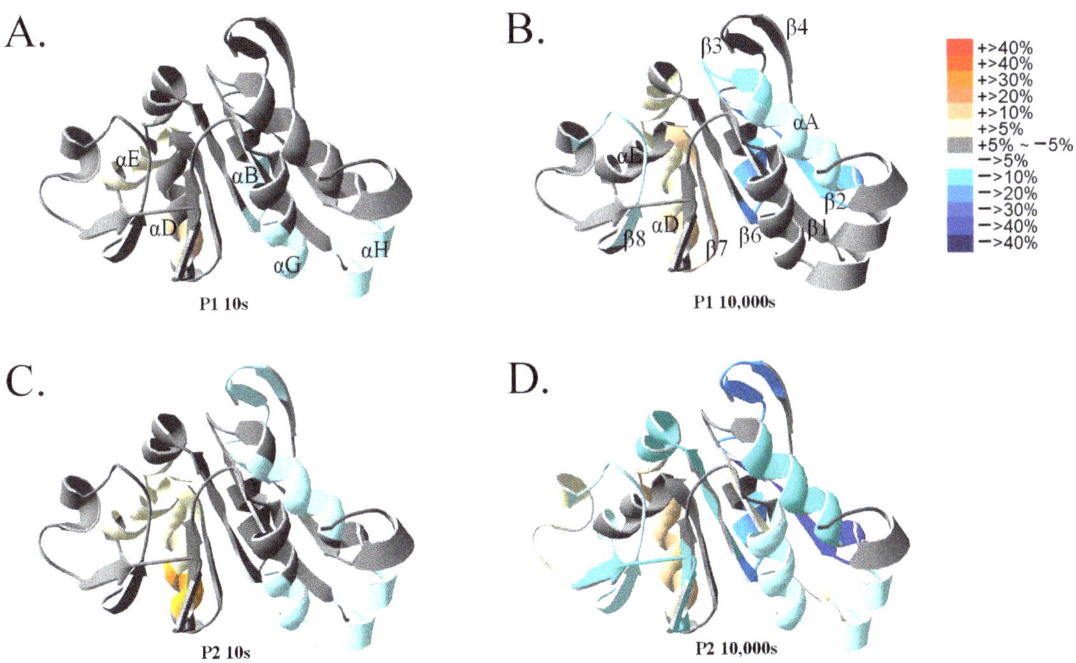

Figure 7. Effects of P1 and P2 peptide binding on DJ-1, as analyzed using HDXMS. DJ-1 protein was incubated with P1 and P2 peptides for 10 min. The changes in HDXMS levels at 10 s and 10,000 s were mapped onto the DJ-1 crystal structure (2OR3.PDB). The representations of P1 treatment at 10 s (**A**), P1 treatment at 10,000 s (**B**), P2 treatment at 10 s (**C**), and P2 treatment at 10,000 s (**D**) are shown to illustrate the changes in HDXMS levels.

2.5. H/D Exchange of Peptide 1 Activated DJ-1

The addition of P1 resulted in 36% activation of the deglycation activity of DJ-1, indirectly confirming interactions between DJ-1 and peptide P1. However, the tertiary structure of peptide-bound DJ-1 remained the dimer conformation. Figure 8 presents the HDXMS results of selected peptides after the reaction of DJ-1 with P1, along with the 10,000 s-HDXMS mapped structure. The detailed HDXMS results of all peptides before and after back exchange correction are shown in Supplemental Figures S1 and S2. In peptide 94–106, which contains the active site Cys106, an increase in the deuteration level was observed, and near the dimer interface region—particularly in peptides 17–25, 17–26, 18–24, 18–26, and 19–26, assumed to be binding sites for P1—a trend of decreased H/D exchange was evident. The H/D exchange in peptide 19–26 exhibited decreased slightly starting at 30 s and reached the maximum decrease of 21.5% at 10,000 s. At peptide 27–38, located at the center of the dimer interface, the H/D exchange changed slightly starting from 100 s and reached the maximum decrease of 23.3% at 10,000 s. Peptide 31–38 in the same region also exhibited a trend of decreased H/D exchange. At 10,000 s, peptides 9–16 (−11.0%) and 119–130 (−5.2%) in the catalytic region exhibited decreased H/D exchange. A particularly noteworthy observation was that regarding the region opposite the active site, including peptides 78–91, 92–96, 94–106, and 102–119, because H/D exchange was increased at all of these. The H/D exchange at peptide 78–91 started to increase at 3000 s, reaching a 9% increase at 10,000 s. Peptide 92–96 exhibited the greatest increase in H/D exchange, with a 19.1% increase at 10 s and a 20.4% increase at 10,000 s. At peptide 94–106, the H/D exchange increased slightly starting from 30 s, with a 9.8% increase at 10,000 s. At peptide 102–119, the H/D exchange had increased by 15.8% at 10,000 s.

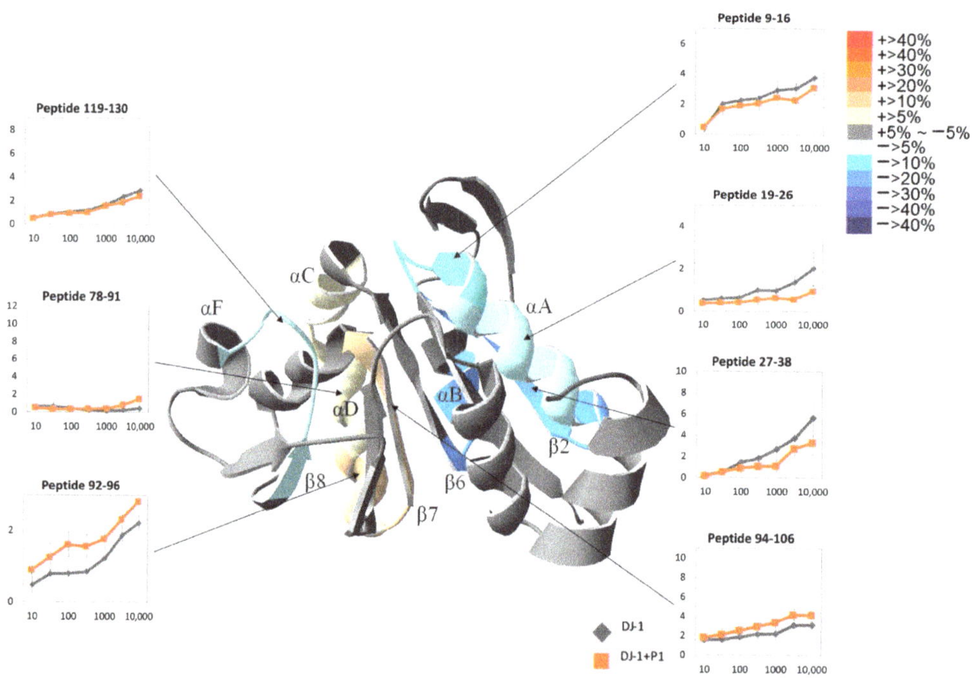

Figure 8. Changes in H/D exchange observed in HDXMS after binding of DJ-1 with peptide 1. DJ-1 bound to P1 was deuterated for 10, 30, 100, 300, 1000, 3000, and 10,000 s. The HDXMS results of the selected peptides are presented in the linear graph. The structural graph depicts the changes in each fragment at 10,000 s. The assorted colors in the heat map represent the various levels of changes. All experiments were conducted in triplicate, and the errors represent the standard deviation.

2.6. H/D Exchange in DJ-1 Activated by Peptide 2

The alterations in HDXMS after the reaction of DJ-1 with P2 are presented on the structure of DJ-1 in Figure 9. The detailed HDXMS results of all peptides before and after back exchange correction are shown in Supplemental Figures S3 and S4. Notably, a slight increase in H/D exchange was observed at peptides 78–91 and 92–96. By contrast, decreases in exchange were noted in peptide 19–26 at the dimer interface and peptide 119–130 at the positions of Glu18 and His126 in the catalytic triad. As the exchange time increased, the decrease in H/D exchange extended to the entire protein. In the DJ-1 dimeric interface region, speculated to be the site of the interaction with P2, a slight 5% decrease in H/D exchange was observed from 300 s in peptide 50–58, increasing to a 10.0% decrease at 10,000 s (Supplemental Figure S3). Similarly, the exchange at peptide 19–26, containing αA at the dimeric binding interface, exhibited a slight decrease from 7.2% at 300 s to 14.7% at 10,000 s. Peptide 37–51, in the middle of the binding interface, exhibited deuterium exchange decreases of 14.3% at 10 s and 13.1% at 10,000 s. Moreover, peptide 59–69 exhibited a 26.2% H/D exchange decrease at 10,000 s. In the catalytic region, peptide 119–130 exhibited a 6.7% decrease in deuterium exchange at 10,000 s. Conversely, peptides 78–91 and 92–96, located in the opposite direction to the active site, exhibited increases of 5.1% and 31.8% at 10 s, respectively, reaching 14.9% and 11.1% at 10,000 s. However, in the central β-sheet of the αβ-sandwich, the deuterium exchange differed from that observed in the P1 experimental group. In peptide 2–10 of the β1 sheet, a 7.8% increase in deuterium exchange was observed at 10,000 s, whereas in peptide 69–82 of the β6 sheet, a 10.9% decrease in deuterium exchange was observed at 10,000 s.

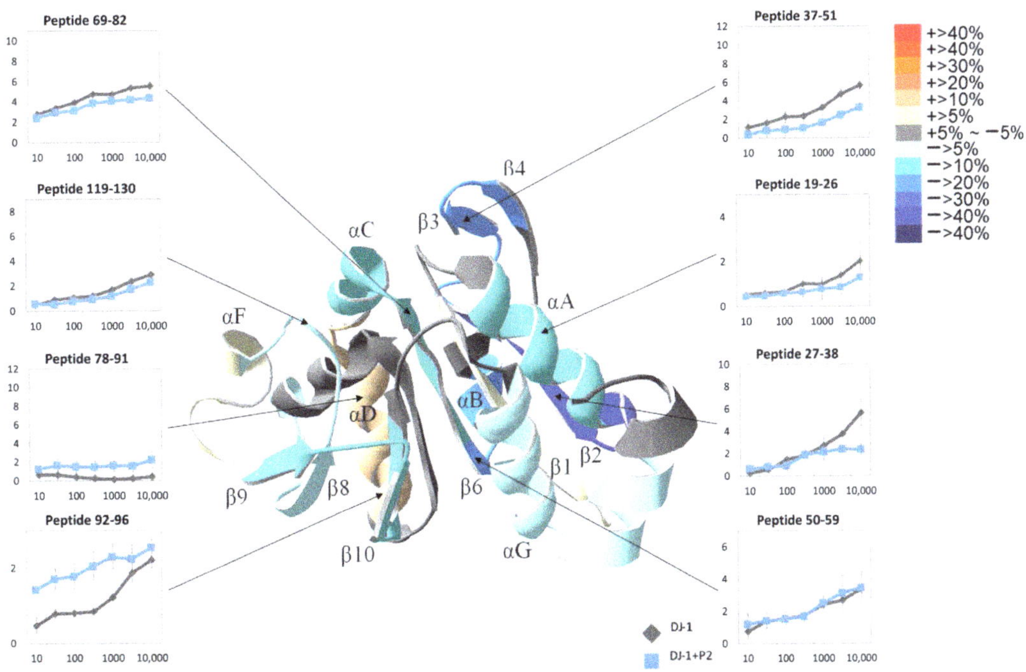

Figure 9. Changes in H/D exchange observed in HDXMS after binding of DJ-1 with peptide 2. DJ-1 bound to P2 was deuterated for 10, 30, 100, 300, 1000, 3000, and 10,000 s. The HDXMS results of the selected peptides are presented in the linear graph. The structural graph depicts the changes in each fragment at 10,000 s. The assorted colors in the heat map depict various levels of changes. All experiments were conducted in triplicate, and the errors represent the standard deviation.

2.7. H/D Exchange in Nonreduced DJ-1

Reduction with 1 M DTT led to partial recovery of the activity of oxidized DJ-1; however, its activity remained lower than that of the originally reduced form. Therefore, we compared the structural differences of these forms through HDXMS. The results revealed that the DTT-reduced DJ-1 protein exhibited a significant increase in deuterium exchange at both the dimeric binding interface and the active catalytic triad, resulting in an overall increase in the deuterium exchange rate. The percentage difference for each fragment was computed by subtracting the H/D exchange percentage of DJ-1 before DTT reduction from that after DTT reduction, and these results are mapped onto the crystal structure in Figure 10A. After DTT reduction, increased deuterium exchange was observed in DJ-1, specifically in peptides 17–25 (17.6%), 17–26 (17.6%), 18–26 (16.9%), and 19–26 (23.3%) in the αA helix. An increase in deuterium exchange was also discovered in the central β2 sheet at the dimeric interface, with sequences 27–38 exhibiting a 27.3% increase and 31–38 exhibiting a 16.1% increase. Peptides related to the catalytic region—such as sequences 18–26 (16.9%) and 119–130 (6.2%)—as well as those near Cys106—including sequences 94–106 (6.1%), 104–112 (20.7%), and 105–112 (14.3%)—exhibited varying degrees of increased deuterium exchange. In the β6 sheet, peptides 69–82 exhibited an 18.2% increase, and peptides 69–84 exhibited a 19.5% increase. Furthermore, peptides 138–154 in sheets β9, β10, and β11 exhibited a 10.1% increase, and peptides 154–163 exhibited a 14.7% increase. Finally, in the hydrophobic core region, peptides 167–181 in the αG and αH helices exhibited a 5.8% increase, and peptides 179–195 exhibited a 6.6% increase. Overall, the H/D exchange of DJ-1 after DTT reduction was significantly higher than that of unreduced DJ-1. To confirm the effectiveness of DTT reduction, we conducted the reduction reaction to the protein samples, which were stored apart in the freezer for four months, at two separate times (Supplemental Figure S5). The HDXMS results at 10,000 s showed no difference.

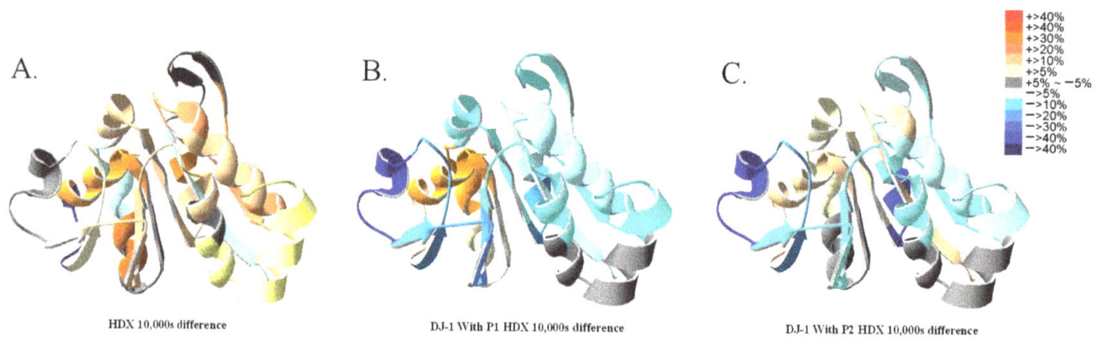

Figure 10. Peptide binding effects of the reduced DJ-1, as observed through HDXMS. (**A**) Differences in HDXMS results between the originally reduced DJ-1 and the DTT-reduced DJ-1. Peptides P1 (**B**) and P2 (**C**) were bound to the original DJ-1 without further reduction at 37 °C. The differences in deuteration levels at 10,000 s after H/D exchange of DJ-1 with or without peptide binding are shown. The deuteration differences are mapped onto the DJ-1 structure.

An analysis of the variation in H/D exchange through the addition of P1 and P2 to the originally reduced DJ-1 protein revealed greater differences compared with the results for the DTT treated protein (Supplemental Figure S6). This enabled the analysis of activation disparities between P1 and P2 (Figure 10B,C). Although both P1 and P2 perturb DJ-1 dimerization, these two peptides yielded distinct levels of perturbation. The overall results are consistent with the results obtained through HDXMS for the DTT-reduced protein, but more prominent differences were found for P1 and P2 treatments. Both P1 and P2 resulted in reduced H/D exchange at the dimerization interface and increased H/D exchange surrounding the active site. By contrast, at the central β sheets, P1 addition led to a decrease in deuteration on the β1 and β6 sheets, whereas P2 addition enhanced deuteration in the same region. We further confirmed that DTT treatment did not have any structural effects on the originally reduced DJ-1 by HDXMS (Supplemental Figure S7).

3. Discussion

3.1. DJ-1 Dimerization and Activity

DJ-1 activity is essential for maintaining mitochondrial health during periods of oxidative stress. Experimental findings and crystal structure analyses revealed the dimeric nature of DJ-1. After the deglycation activity of purified DJ-1 was confirmed, DSS cross-linking experiments were conducted to verify the dimeric conformation of DJ-1 (Figure 2C). The results of H/D exchange indicated relatively low deuterium exchange levels in the outer regions of the protein, specifically in the αA helix and β4 sheet. The X-ray crystal structure of DJ-1 in its dimeric form (PDB: 2OR3) revealed that the αA helix forms the core of the dimeric interface, whereas the β4 sheet is situated on the side of the dimeric binding interface. These regions precisely align with the dimeric binding interface of DJ-1, thereby reducing the solvent accessibility of the αA helix and β4 sheet. The L166P mutation destabilizes the αG-helix and disrupts the homodimer, resulting in activity loss. However, the impact of this point mutation on the overall structure of the DJ-1 monomer remains unclear, and whether the monomeric form of DJ-1 exhibits lower activity is yet to be confirmed. The αG-helix plays a critical role in the activity of DJ-1. Although the monomer form of DJ-1 could not be obtained for a direct HDXMS comparison with the dimeric form, our findings suggest that the fast-exchanging form with high solvent accessibility may have lower activity (Figure 10A). The DJ-1 monomer can be assumed to have higher solvent accessibility than the DJ-1 dimer.

Cys106, Glu18, and His126, situated on distinct loops, collectively form the catalytic triad of DJ-1, a characteristic feature shared with other enzymes with an α/β hydrolase domain. The active site of Cys106 is a crucial residue governing the deglycation activity of DJ-1, and the microenvironment surrounding this active site plays a crucial role in ensuring the optimal activity of DJ-1. The oxidation of Cys106 enhances deglycation activity, but excessive oxidation of Cys106 leads to loss of activity [35]. Further analysis of the dimeric form of DJ-1 revealed hydrogen bonding between Cys106 and Glu18. Any disruption in the dimerization can induce a change in the shape of the active site, thereby altering catalytic activity. Thus, alterations in the conformation around the active site can differentially influence the catalytic activity of DJ-1.

3.2. Dimeric DJ-1 Binding with P1 and P2

The dimeric configuration of DJ-1 was demonstrated to exhibit superior deglycosylation activity and neuroprotective effects compared with the monomeric form [41]. To modulate DJ-1's deglycation activity, we designed two peptides targeting the dimeric structure of DJ-1. Initially, our hypothesis suggested that these peptides, upon binding to DJ-1, would disrupt the dimeric structure, thereby altering DJ-1's activity. However, subsequent cross-linking reactions conducted using DSS and the analysis of SDS-PAGE results revealed that the DJ-1 remained in its dimeric form. Further investigation through HDXMS analysis enabled quantification of the impact of the designed peptides on the enzyme. Remarkably, the HDXMS results not only confirmed the binding of the peptides to the dimeric structure of DJ-1 but also revealed an unexpected outcome. Instead of destabilizing the DJ-1 dimer, these peptides had a stabilizing effect and enhanced the activity of DJ-1.

The HDXMS data of DJ-1 with P1 and P2 revealed substantial changes, confirming the binding of these peptides to DJ-1. If P1 and P2 could disrupt the dimeric structure of DJ-1, their addition should result in an overall increase in deuterium exchange, potentially leading to a decrease in deuteration near the αA helix (16E–28R) at the P1 binding site. The HDXMS results revealed a decrease in deuterium exchange at the αA helix positions (16E–28R), with decreases of 10.1% and 31.5% at peptides 17–25 and 19–26, respectively, at 10,000 s. However, at the binding interface on the opposite side of the dimer, near the β4 sheet (50V–53C), a clear difference was not found at peptides 37–51 and 46–57. Importantly, no increase in deuteration of the entire enzyme was observed, particularly around the active sites. This finding indicated that although P1 can interact with DJ-1, this interaction may not be sufficient to completely disrupt the dimeric structure of DJ-1. Similarly, in the P2 experiment, where P2 is expected to bind to the β4 sheet (V50–C53), the results revealed decreases in deuterium exchange at 10,000 s for sequences 37–51 (−19.8%), 48–54 (−6.4%), and 50–58 (−10.1%). However, at the αA helix (16E–28R) positions, no increase was found; instead, a decrease was observed in the HDXMS analysis. This indicates that both P1 and P2 are unable to disrupt the dimeric structure of DJ-1. The validation of the DSS cross-linking reactions and the results of H/D exchange indicate that P1 and P2 do not disrupt the dimeric structure of DJ-1 and that they induce structural changes in the protein, enhancing its deglycation activity.

3.3. Change in Activity Mechanisms of DJ-1 after P1 and P2 Binding

The activity tests of DJ-1 revealed that both P1 and P2 effectively enhanced the deglycation activity of DJ-1. The addition of P1 resulted in a decrease in deuterium exchange at 10,000 s for peptides 27–38 (−23.3%) and 31–38 (−18.2%). These decreases may be attributable to the binding of P1 to the αA helix, which made the surrounding structure more compact and led to structural hindrance and a decrease in deuterium exchange in these two regions. Similarly, the addition of P2 led to a decrease in deuterium exchange at 10,000 s for peptides 37–51 (−19.8%) and 19–26 (−27.4%). This decrease may be attributable to the binding of P2 to the β4 sheet, which induced structural hindrance and reduced deuterium exchange in these sequences. Additionally, both P1 and P2 resulted in a decrease in deuterium exchange in peptides 11–17, 17–25, 119–130, and 129–139. These regions are associated with the catalytic triad structure containing Cys106, Glu18, and His126. Studies have indicated that the formation of bonds with SO_2^- by E18, H126, and oxidized Cys106 after oxidation results in enhancement of the activity of DJ-1 [35]. The binding of SO_2^- stabilized Cys106 in the active sites and facilitated hydrogen dissociation of the thiol on cysteine. This stabilized structure led to improved solvent accessibility and reactivity, enhancing substrate binding, and increasing the activity of DJ-1.

Although both P1 and P2 were found to effectively enhance the deglycation activity of DJ-1, they employ distinct mechanisms to regulate this activity. The HDXMS results indicated similar effects on the active sites, dimer-binding region, and hydrophobic core, suggesting that both peptides activate DJ-1 by binding to the dimer-binding regions, inducing crowding effects in the dimer and hydrophobic core regions. This, in turn, stabilizes Glu18, His126, and Cys106, increasing substrate affinity and enhancing the deglycation activity of DJ-1. However, on the β sheet core of DJ-1, P1 tightens the structure of the β1 and β6 sheets, whereas P2 exerts a loosening effect on these sheets. This discrepancy may be attributable to the distinct binding regions of P1 and P2. When P1 binds to the dimer-binding interface at the αA helix through hydrophobic interaction, crowding effects are created in the central structure of the dimer-binding interface and exert an inward force that compresses the adjacent β1 and β6 sheet structures. By contrast, P2, located on the side of the dimer-binding interface, exerts binding interactions that function as a pulling force toward the α/β hydrolase core. This simultaneous action stretches the structures of the β1 and β6 sheets outward, resulting in loosening of the structure.

3.4. Oxidized DJ-1 Reduced by DTT

After the protein was purified, its SDS-PAGE pattern contained two distinct bands, with the upper band indicating significantly higher deglycation activity than that of the two-band DJ-1. Notably, the lower band of the protein could be reduced to a single band of DJ-1 through the addition of a high concentration of DTT, resulting in increased activity. However, despite the SDS-PAGE patterns of both the original DJ-1 and the DTT-reduced DJ-1 containing a single band, differences in activity persisted, as evidenced by the H/D exchange data. In the DTT-reduced DJ-1, the overall protein structure became more relaxed, including the αA helix; the β2 and β4 sheets in the dimeric functional region; the catalytic triad (C106, E18, H126) in the active catalytic region; the αC, αD, and αE helices; the β6, β8, β9, β10, and β11 sheets in the protein backbone region; and the αG and αH helices in the hydrophobic core region. Deuterium exchange increased in all these regions. However, a decrease in deuterium exchange was discovered in specific regions, such as the αD helix in the protein backbone (sequences 78–91, −6.2%), the β1 sheet (sequences 2–10, −7.1%), and a loop structure in the αE helix to the middle of the β8 sheet (sequences 107–120, −47.9%). The H/D exchange data revealed that although DTT can reduce the amount of oxygen in DJ-1, oxidation may induce irreversible changes in the structure of DJ-1. In particular, in the catalytic mechanism, DJ-1 is hypothesized to adjust its structure for activation by contracting the dimeric and catalytic functional regions. However, in the DTT-reduced DJ-1, these functional regions were in a more relaxed state, indicating a strong correlation between the activity regulation of DJ-1 and these two functional regions.

In the DTT-reduced DJ-1 protein, the structure near the active site C106 was found to be relaxed. When C106 was oxidized, steric hindrance occurred and was speculated to create an outward pushing force and disrupt the catalytic triad structure. Removal of oxygen atoms from the oxidized Cys106 failed to restore the correct catalytic triad, leading to relaxation of structures near C106, such as the αC and αE helices, β6 and β11 sheets, and the surrounding loop structures. Simultaneously, the dimeric structure's αA helix and β2 sheet also became relaxed, resulting in the overall alteration of the protein backbone structure.

3.5. DJ-1 Peptide Activation

Given the crucial role of dimerization in regulating DJ-1 activity, disrupting dimerization is a natural strategy for modulating DJ-1 activity. Two peptide sequences from the dimerization interface were designed as potential modulators. Notably, both peptides were found to enhance the deglycation activity of DJ-1 in vitro, but they exhibited different effects on the catalytic site.

P1, located in region 15–29 and containing the critical Glu18, competes with the dimerization interface and rearranges the geometry of the catalytic triad. HDXMS analysis revealed a significant reduction in deuteration at the direct binding region, which extends to the entire 11–38 region. Notably, the other dimerization region remains unaffected or, at least, does not loosen, corresponding to the maintained dimer structure. The catalytic triad forms a rigid structure with Glu18 and His126, but Cys106 seems to be in a more relaxed loop.

P2, located in the 47–56 region, forms a β-sheet in the homodimer structure, interacting antiparallelly with the same β-sheet of the dimer partner protein. Although the competition of P2 toward the dimerization interface does not break the dimer, it leads to a decrease in H/D exchange, indicating a more compact structure at the interface. Although P2 insertion may tighten the dimer and optimize the catalytic triad, the overall α/β hydrolase domain remains in a relaxed conformation.

4. Material and Methods

4.1. Materials

pET3a-His-DJ1 was purchased from Addgene, Watertown, MA, USA (plasmid number 51488), originally a gift from the Michael J Fox Foundation, New York City, NY, USA. Imidazole, tween 20, DTT were purchased from Amresco LLC, Solon, OH, USA. A 30% LB broth powder, lysozyme, and IPTG were purchased from Chumeia, Hsinchu City, Taiwan. β-Mercaptoethanol(β-ME) was purchased from Calbiochem, St. Louis, MO, USA. BCIP (5-bromo-4-chloro-3-indolyl phosphate) and guanidine hydrochloride (GuHCl) were purchased from Cyrusbioscience, New Taipei City, Taiwan. Glycerol, TEMED, and DJ-1 antibody were purchased from Invitrogen, Waltham, MA, USA. Triton, SDS, and ammonium persulfate (APS) were purchased from OmniPur, Radnor, PA, USA. Ni-NTA agarose was purchased from QIAGEN, Hilden, Germany. *N*-acetyl-l-cysteine, methylglyoxal solution (MGO), pepsin, NBT (Nitro blue tetrazolium chloride), phenylmethyl sulfonyl fluoride

(PMSF), formic acid (FA), anti-rabbit IgG, and D2O were purchased from Sigma-Aldrich, St. Louis, MO, USA. Protein assay dye was purchased from Bio-Rad, Hercules, CA, USA. Immobilized pepsin and protein stain were purchased from Thermo Fisher Scientific, Waltham, MA, USA. Peptide 1 (^{15}EEMETIIPVDVMRRA^{29}) and peptide 2 (^{47}SRDVVICPDA^{56}) were synthesized by Yao-Hong Biotechnology, New Taipei City, Taiwan.

4.2. Protein Purification

pET3a-His-DJ1 transformed BL21 and was grown at 37 °C to OD = 0.45. Protein expression was induced by 0.1 mM IPTG at 30 °C for 2 h. The cell lysate was collected by 10,000 rpm, centrifuged at 4 °C for 30 min and stored at −80 °C. The cell lysate was defrosted for 10 min on ice before addition of 20 mL of lysis buffer, containing 50 mM Tris pH 8, 150 mM NaCl, 1 mg/mL lysozyme, 1 mM PMSF, 20 mM β-ME, and 0.1% Triton.

After being thoroughly vortexed, the cell lysate was homogenized at 65% 125 W sonication on ice (QSonica Q125, 6.4 mm probe, from QSonica LLC., Newtown, CT, USA). The sonication was in 10 cycles of 30 s sonication and 30 s stops. After 10,000 rpm centrifugation for 30 min, the collected supernatant was flowed through 2 mL Ni-NTA column. The Ni-NTA column was pre-washed with 2 mL of washing buffer (50 mM Tris pH 8, 150 mM NaCl, 20 mM β-ME, 10 mM Imidazole). The column with bound protein was washed with 40 mL of washing buffer. After washing, the protein was eluted by elution buffer (50 mM Tris pH 7.5, 125 mM NaCl, 3 mM DTT, 150 mM Imidazole, 50% Glycerol). The eluate fractions were collected every 0.5 mL. The protein concentrations were determined by Bradford assay (Bio-RAD Laboratories, Hercules, CA, USA) and quantitated by a 96-well plate reader at 595 nm absorbance. Protein was verified by Western blotting and the purity were estimated by SDS-PAGE.

4.3. DJ-1 Activity Assay

To measure the activity of DJ-1, the fresh substrate was prepared immediately before the assay. PBS buffer was bubbled by nitrogen gas for 3 min before experiments. The glycation reactions were initiated by mixing 0.5 mL of PBS buffer, 5 μL of 1M *N*-acetyl-L-cysteine (10 mM final concentration), and 7.82 μL of 0.65 M methylglyoxal solution (8 mM final concentration). After mixing, the sample was purged by nitrogen gas for 15 s and sealed by parafilm. The samples were incubated in 37 °C water baths for 10 min. After adding 23 μL (1.266 μg/μL) of DJ-1, the samples were further incubated for 30 min, and then the absorbance at 288 nm was detected.

4.4. Electrophoresis of DJ-1 Dimer

Twelve milligrams of DJ-1 protein was treated by 25 mM DSS crosslinker (Disuccinimidyl substrate) at 25 °C for 30 min. Fifteen milligrams of 15-mer peptide 1 (P1) and 10 μg of 10-mer peptide 2 (P2) were added to interact with DJ-1. DSS-treated and peptide-added protein samples were loaded onto a 12% polyacrylamide gel for SDS-PAGE. For the native-PAGE, protein samples were loaded onto a 12% polyacrylamide gel without SDS. NativePAGE cathode buffer additive (20×) was added to the NativePAGE running buffer to prepare the buffer for the cathode chamber. The running buffer was used without additive for the anode.

4.5. Sequence Identification of the DJ-1 Proteolytic Fragment

Before analyzing the HDXMS data, the sequences of the detected pepsin-digested fragments were first identified by MS/MS. To prepare the pepsin column, a 2.1 mm × 35 mm empty column was packed by Thermo Scientific Pierce immobilized pepsin agarose, (Thermo Scientific, Waltham, MA, USA). The column was flushed with 0.1% formic acid at a flow rate of 0.1 mL/min. The above two steps were repeated 3 to 4 times until the pepsin column was filled. After DJ-1 was proteolyzed into fragments by pepsin, the fragments were collected by a micro RP peptide trap (OPTI-TRAPTM from optimize technology, Oregon City, OR, USA) and then separated by a reverse phase HPLC in a C18 column

(BioBasic 18 LC Columns, Thermo Scientific, Waltham, MA, USA). The mobile phase was the gradient between Buffer A: 0.1% formic acid and Buffer B: 80% acetonitrile/0.02% formic acid. The flow rate was set to 0.15 mL/min and most peptides were eluted in 30 min. The peptides were fragmented by tandem mass spectrometry (MS/MS), and the MS/MS data were imported into the X! Tandem-parser-1.7.7 software to calculate the sequences of the fragments.

4.6. Sequence Identification by X! Tandem

DJ-1 MS/MS data were exported from Bruker DataAnalysis and the intensity threshold was set at 700. The MS/MS data and the sequence were imported to X! Tandem for sequence identification. The mass error was set at 500 ppm. The sequence of each fragment was further manually verified based on the primary mass and the matched product ions under X! tandem-parser-1.7.7.

4.7. Hydrogen/Deuterium Exchange Experiments

A 20× Tris buffer (1 M Tris-base, 2.5 M NaCl, pH 7.5) was diluted by 99.9% D_2O to prepare a 1× D_2O buffer. Then, 15 µL (50 µg) of DJ-1 protein and D_2O buffer was incubated in a 37 °C water bath for 30 min. H/D exchange was initiated by mixing 15 µL DJ-1 protein with 45 µL D_2O buffer in a 37 °C water bath for 10, 30, 100, 300, 1000, 3000, and 10,000 s. The reactions were quenched by adding 140 µL of quench buffer (0.5% formic acid, 1 M GuHCl).

4.8. Mass Spectrometry Analysis

The pepsin column connected with the peptide trap was precooled on ice and the pepsin was activated by flushing the column with 0.1% formic acid at 0.1 mL/min. The quenched deuterated samples were withdrawn by syringe and injected into the pepsin column. The column was washed with 0.1% formic acid at 0.5 mL/min in HPLC for 1.5 min. The pepsin column was then disconnected, and the peptide trap was reconnected to a C18 column. The peptides were then separated by HPLC and analyzed by mass spectrometer. The HPLC mobile phase was the gradient between Buffer A: 0.1% formic acid and Buffer B: 80% acetonitrile/0.02% formic acid. The flow rate was set to 0.15 mL/min and most peptides were eluted in 30 min. The C18 column, LC running buffer, and the tubing were covered by ice. The peptides of the protein were detected on the positive mode of the mass spectrometry and the mass detection range was 400 to 2000 m/z. The data were processed on Bruker DataAnalysis (ver.4.1).

4.9. DJ-1 HDXMS Calculation

The sequence, retention time, and charge of the identified peptide fragments were uploaded to the H/D exchange software HDExaminer 1.2 as the peptide pool. The non-deuterated MS data of DJ-1 were further imported to identify the peptic peak. The peptides with low intensity were excluded for further calculation, and this step served as a second filter for peptide identification. The masses of the fragment of the non-deuterated DJ-1 were the standards for all other H/D exchange samples. Then, the MS data of the H/D exchange samples at seven time points were imported into the HDExaminer. The average mass of each deuterated peptide that appeared at each time point was calculated. The mass shift at a specific time point after H/D exchange was the amount of deuteration.

5. Conclusions

Two peptides have been identified as enhancers of DJ-1 activity. To understand the mechanisms behind this enhancement, we employed hydrogen/deuterium exchange mass spectrometry. The findings revealed distinct binding mechanisms for each peptide with the DJ-1 protein, ultimately contributing to the observed increase in activity.

Supplementary Materials: The following supporting information can be downloaded at https://www.mdpi.com/article/10.3390/ijms252011075/s1.

Author Contributions: Conceptualization, Y.-H.H.H.; methodology, J.-Y.S.; validation, J.-Y.S.; formal analysis, J.-Y.S.; investigation, J.-Y.S.; writing–original draft, Y.-H.H.H.; supervision, Y.-H.H.H. All authors have read and agreed to the published version of the manuscript.

Funding: This work was supported by Ministry of Science and Technology Council, Taiwan (NSTC 112-2113-M-029-002, NSTC 113-2923-M-029-002-MY3, NSTC 113-2113-M-029-011 and NSTC 113-2113-M-029-003).

Institutional Review Board Statement: Not applicable.

Informed Consent Statement: Not applicable.

Data Availability Statement: Data is contained within the article and Supplementary Materials.

Conflicts of Interest: The authors declare no conflict of interest.

References

1. Davis, G.C.; Williams, A.C.; Markey, S.P.; Ebert, M.H.; Caine, E.D.; Reichert, C.M.; Kopin, I.J. Chronic Parkinsonism secondary to intravenous injection of meperidine analogues. *Psychiatry Res.* **1979**, *1*, 249–254. [CrossRef] [PubMed]
2. Exner, N.; Lutz, A.K.; Haass, C.; Winklhofer, K.F. Mitochondrial dysfunction in Parkinson's disease: Molecular mechanisms and pathophysiological consequences. *EMBO J.* **2012**, *31*, 3038–3062. [CrossRef] [PubMed]
3. Winklhofer, K.F.; Haass, C. Mitochondrial dysfunction in Parkinson's disease. *Biochim. Biophys. Acta* **2010**, *1802*, 29–44. [CrossRef] [PubMed]
4. Lambert, A.J.; Brand, M.D. Reactive oxygen species production by mitochondria. *Methods Mol. Biol.* **2009**, *554*, 165–181. [PubMed]
5. Polymeropoulos, M.H.; Lavedan, C.; Leroy, E.; Ide, S.E.; Dehejia, A.; Dutra, A.; Pike, B.; Root, H.; Rubenstein, J.; Boyer, R.; et al. Mutation in the alpha-synuclein gene identified in families with Parkinson's disease. *Science* **1997**, *276*, 2045–2047. [CrossRef]
6. Lucking, C.B.; Durr, A.; Bonifati, V.; Vaughan, J.; De Michele, G.; Gasser, T.; Harhangi, B.S.; Meco, G.; Denefle, P.; Wood, N.W.; et al. Association between early-onset Parkinson's disease and mutations in the parkin gene. *N. Engl. J. Med.* **2000**, *342*, 1560–1567. [CrossRef]
7. Valente, E.M.; Abou-Sleiman, P.M.; Caputo, V.; Muqit, M.M.K.; Harvey, K.; Gispert, S.; Ali, Z.; Del Turco, D.; Bentivoglio, A.R.; Healy, D.G.; et al. Hereditary early-onset Parkinson's disease caused by mutations in PINK1. *Science* **2004**, *304*, 1158–1160. [CrossRef]
8. Bonifati, V.; Rizzu, P.; van Baren, M.J.; Schaap, O.; Breedveld, G.J.; Krieger, E.; Dekker, M.C.; Squitieri, F.; Ibanez, P.; Joosse, M.; et al. Mutations in the DJ-1 gene associated with autosomal recessive early-onset parkinsonism. *Science* **2003**, *299*, 256–259. [CrossRef]
9. Malik, I.; Turk, J.; Mancuso, D.J.; Montier, L.; Wohltmann, M.; Wozniak, D.F.; Schmidt, R.E.; Gross, R.W.; Kotzbauerg, P.T. Disrupted membrane homeostasis and accumulation of ubiquitinated proteins in a mouse model of infantile neuroaxonal dystrophy caused by PLA2G6 mutations. *Am. J. Pathol.* **2008**, *172*, 406–416. [CrossRef]
10. Taira, T.; Takahashi, K.; Kitagawa, R.; Iguchi-Ariga, S.M.; Ariga, H. Molecular cloning of human and mouse DJ-1 genes and identification of Sp1-dependent activation of the human DJ-1 promoter. *Gene* **2001**, *263*, 285–292. [CrossRef]
11. Cookson, M.R. Parkinsonism Due to Mutations in PINK1, Parkin, and DJ-1 and Oxidative Stress and Mitochondrial Pathways. *Cold Spring Harb. Perspect. Med.* **2012**, *2*, a009415. [CrossRef] [PubMed]
12. Plun-Favreau, H.; Klupsch, K.; Moisoi, N.; Gandhi, S.; Kjaer, S.; Frith, D.; Harvey, K.; Deas, E.; Harvey, R.J.; McDonald, N.; et al. The mitochondrial protease HtrA2 is regulated by Parkinson's disease-associated kinase PINK1. *Nat. Cell Biol.* **2007**, *9*, 1243–1252. [CrossRef] [PubMed]
13. Pridgeon, J.W.; Olzmann, J.A.; Chin, L.S.; Li, L. PINK1 protects against oxidative stress by phosphorylating mitochondrial chaperone TRAP1. *PLoS Biol.* **2007**, *5*, 1494–1503. [CrossRef]
14. Narendra, D.P.; Jin, S.M.; Tanaka, A.; Suen, D.F.; Gautier, C.A.; Shen, J.; Cookson, M.R.; Youle, R.J. PINK1 Is Selectively Stabilized on Impaired Mitochondria to Activate Parkin. *PLoS Biol.* **2010**, *8*, e1000298. [CrossRef] [PubMed]
15. Zhu, M.; Patel, S.H.; Han, S. DJ-1, a Parkinson's disease related protein, aggregates under denaturing conditions and co-aggregates with alpha-synuclein through hydrophobic interaction. *Biochim. Biophys. Acta Gen. Subj.* **2017**, *1861*, 1759–1769. [CrossRef] [PubMed]
16. Zhong, N.; Kim, C.Y.; Rizzu, P.; Geula, C.; Porter, D.R.; Pothos, E.N.; Squitieri, F.; Heutink, P.; Xu, J. DJ-1 transcriptionally up-regulates the human tyrosine hydroxylase by inhibiting the sumoylation of pyrimidine tract-binding protein-associated splicing factor. *J. Biol. Chem.* **2006**, *281*, 20940–20948. [CrossRef] [PubMed]
17. Piston, D.; Alvarez-Erviti, L.; Bansal, V.; Gargano, D.; Yao, Z.; Szabadkai, G.; Odell, M.; Puno, M.R.; Bjorkblom, B.; Maple-Grodem, J.; et al. DJ-1 is a redox sensitive adapter protein for high molecular weight complexes involved in regulation of catecholamine homeostasis. *Hum. Mol. Genet.* **2018**, *27*, 576. [CrossRef] [PubMed]

18. Jain, M.K.; Bhat, R. Modulation of human alpha-synuclein aggregation by a combined effect of calcium and dopamine. *Neurobiol. Dis.* **2014**, *63*, 115–128. [CrossRef]
19. Jethva, P.N.; Kardani, J.R.; Roy, I. Modulation of alpha-synuclein aggregation by dopamine in the presence of MPTP and its metabolite. *FEBS J.* **2011**, *278*, 1688–1698. [CrossRef]
20. Latawiec, D.; Herrera, F.; Bek, A.; Losasso, V.; Candotti, M.; Benetti, F.; Carlino, E.; Kranjc, A.; Lazzarino, M.; Gustincich, S.; et al. Modulation of alpha-synuclein aggregation by dopamine analogs. *PLoS ONE* **2010**, *5*, e9234. [CrossRef]
21. Leong, S.L.; Cappai, R.; Barnham, K.J.; Pham, C.L. Modulation of alpha-synuclein aggregation by dopamine: A review. *Neurochem. Res.* **2009**, *34*, 1838–1846. [CrossRef] [PubMed]
22. Pham, C.L.; Leong, S.L.; Ali, F.E.; Kenche, V.B.; Hill, A.F.; Gras, S.L.; Barnham, K.J.; Cappai, R. Dopamine and the dopamine oxidation product 5,6-dihydroxylindole promote distinct on-pathway and off-pathway aggregation of alpha-synuclein in a pH-dependent manner. *J. Mol. Biol.* **2009**, *387*, 771–785. [CrossRef] [PubMed]
23. Lev, N.; Barhum, Y.; Pilosof, N.S.; Ickowicz, D.; Cohen, H.Y.; Melamed, E.; Offen, D. DJ-1 Protects Against Dopamine Toxicity: Implications for Parkinson's Disease and Aging. *J. Gerontol. A-Biol.* **2013**, *68*, 215–225. [CrossRef] [PubMed]
24. Lev, N.; Ickowicz, D.; Barhum, Y.; Lev, S.; Melamed, E.; Offen, D. DJ-1 protects against dopamine toxicity. *J. Neural Transm.* **2009**, *116*, 151–160. [CrossRef] [PubMed]
25. Taira, T.; Saito, Y.; Niki, T.; Iguchi-Ariga, S.M.M.; Takahashi, K.; Ariga, H. DJ-1 has a role in antioxidative stress to prevent cell death. *Embo Rep.* **2004**, *5*, 213–218. [CrossRef] [PubMed]
26. Liu, W.; Wu, H.; Chen, L.; Wen, Y.; Kong, X.; Gao, W.Q. Park7 interacts with p47(phox) to direct NADPH oxidase-dependent ROS production and protect against sepsis. *Cell Res.* **2015**, *25*, 691–706. [CrossRef]
27. Zhou, W.; Freed, C.R. DJ-1 up-regulates glutathione synthesis during oxidative stress and inhibits A53T alpha-synuclein toxicity. *J. Biol. Chem.* **2005**, *280*, 43150–43158. [CrossRef]
28. De Miranda, B.R.; Rocha, E.M.; Bai, Q.; El Ayadi, A.; Hinkle, D.; Burton, E.A.; Timothy Greenamyre, J. Astrocyte-specific DJ-1 overexpression protects against rotenone-induced neurotoxicity in a rat model of Parkinson's disease. *Neurobiol. Dis.* **2018**, *115*, 101–114. [CrossRef]
29. Takahashi-Niki, K.; Inafune, A.; Michitani, N.; Hatakeyama, Y.; Suzuki, K.; Sasaki, M.; Kitamura, Y.; Niki, T.; Iguchi-Ariga, S.M.; Ariga, H. DJ-1-dependent protective activity of DJ-1-binding compound no. 23 against neuronal cell death in MPTP-treated mouse model of Parkinson's disease. *J. Pharmacol. Sci.* **2015**, *127*, 305–310. [CrossRef]
30. Niki, T.; Endo, J.; Takahashi-Niki, K.; Yasuda, T.; Okamoto, A.; Saito, Y.; Ariga, H.; Iguchi-Ariga, S.M.M. DJ-1-binding compound B enhances Nrf2 activity through the PI3-kinase-Akt pathway by DJ-1-dependent inactivation of PTEN. *Brain Res.* **2020**, *1729*, 146641. [CrossRef]
31. De Miguel, C.; Kraus, A.C.; Saludes, M.A.; Konkalmatt, P.; Ruiz Domínguez, A.; Asico, L.D.; Latham, P.S.; Offen, D.; Jose, P.A.; Cuevas, S. ND-13, a DJ-1-Derived Peptide, Attenuates the Renal Expression of Fibrotic and Inflammatory Markers Associated with Unilateral Ureter Obstruction. *Int. J. Mol. Sci.* **2020**, *21*, 7048. [CrossRef] [PubMed]
32. Junn, E.; Jang, W.H.; Zhao, X.; Jeong, B.S.; Mouradian, M.M. Mitochondrial localization of DJ-1 leads to enhanced neuroprotection. *J. Neurosci. Res.* **2009**, *87*, 123–129. [CrossRef] [PubMed]
33. Kim, K.S.; Kim, J.S.; Park, J.Y.; Suh, Y.H.; Jou, I.; Joe, E.H.; Park, S.M. DJ-1 associates with lipid rafts by palmitoylation and regulates lipid rafts-dependent endocytosis in astrocytes. *Hum. Mol. Genet.* **2013**, *22*, 4805–4817. [CrossRef] [PubMed]
34. Richarme, G.; Mihoub, M.; Dairou, J.; Bui, L.C.; Leger, T.; Lamouri, A. Parkinsonism-associated Protein DJ-1/Park7 Is a Major Protein Deglycase That Repairs Methylglyoxal- and Glyoxal-glycated Cysteine, Arginine, and Lysine Residues. *J. Biol. Chem.* **2015**, *290*, 1885–1897. [CrossRef]
35. Zhou, W.B.; Zhu, M.; Wilson, M.A.; Petsko, G.A.; Fink, A.L. The oxidation state of DJ-1 regulates its chaperone activity toward alpha-synuclein. *J. Mol. Biol.* **2006**, *356*, 1036–1048. [CrossRef]
36. Lee, S.J.; Kim, S.J.; Kim, I.K.; Ko, J.; Jeong, C.S.; Kim, G.H.; Park, C.; Kang, S.O.; Suh, P.G.; Lee, H.S.; et al. Crystal structures of human DJ-1 and Escherichia coli Hsp31, which share an evolutionarily conserved domain. *J. Biol. Chem.* **2003**, *278*, 44552–44559. [CrossRef]
37. Wei, Y.; Ringe, D.; Wilson, M.A.; Ondrechen, M.J. Identification of functional subclasses in the DJ-1 superfamily proteins. *PLoS Comput. Biol.* **2007**, *3*, 120–126. [CrossRef]
38. Blackinton, J.; Lakshminarasimhan, M.; Thomas, K.J.; Ahmad, R.; Greggio, E.; Raza, A.S.; Cookson, M.R.; Wilson, M.A. Formation of a Stabilized Cysteine Sulfinic Acid Is Critical for the Mitochondrial Function of the Parkinsonism Protein DJ-1. *J. Biol. Chem.* **2009**, *284*, 6476–6485. [CrossRef]
39. Tao, X.; Tong, L. Crystal structure of human DJ-1, a protein associated with early onset Parkinson's disease. *J. Biol. Chem.* **2003**, *278*, 31372–31379. [CrossRef]
40. Honbou, K.; Suzuki, N.N.; Horiuchi, M.; Niki, T.; Taira, T.; Ariga, H.; Inagaki, F. The crystal structure of DJ-1, a protein related to male fertility and Parkinson's disease. *J. Biol. Chem.* **2003**, *278*, 31380–31384. [CrossRef]

41. Huai, Q.; Sun, Y.; Wang, H.; Chin, L.S.; Li, L.; Robinson, H.; Ke, H. Crystal structure of DJ-1/RS and implication on familial Parkinson's disease. *FEBS Lett.* **2003**, *549*, 171–175. [CrossRef] [PubMed]
42. Lu, S.J.; Liu, P.; Resendiz, S.H.; Chua, J.T.L.; Offen, D.; Hausenloy, D. Novel nanoparticle improves intracellular uptake and cardioprotective efficacy of DJ-1 mimic peptide. *J. Mol. Cell Cardiol.* **2022**, *173*, S127. [CrossRef]

International Journal of
Molecular Sciences

MDPI

Article

Genome-Wide Identification, Expression, and Interaction Analysis of the Auxin Response Factor and *AUX/IAA* Gene Families in *Vaccinium bracteatum*

Xuan Gao [1,†], Xiaohui Liu [1,†], Hong Zhang [2], Li Cheng [1], Xingliang Wang [2], Cheng Zhen [1], Haijing Du [1], Yufei Chen [2], Hongmei Yu [2], Bo Zhu [1,*] and Jiaxin Xiao [1,*]

1 Anhui Provincial Key Laboratory of Molecular Enzymology and Mechanism of Major Metabolic Diseases, College of Life Sciences, Anhui Normal University, Wuhu 241000, China; gaoxuan@ahnu.edu.cn (X.G.); liuxiaohui@ahnu.edu.cn (X.L.); cl17855368613@163.com (L.C.); c2320288389@163.com (C.Z.); 18355632985@163.com (H.D.)

2 Anhui Provincial Engineering Research Centre for Molecular Detection and Diagnostics, College of Life Sciences, Anhui Normal University, Wuhu 241000, China; 18672312335@163.com (H.Z.); wangxingliang2000@163.com (X.W.); cyf2130035271@163.com (Y.C.); yhm1201@ahnu.edu.cn (H.Y.)

* Correspondence: zhubo@ahnu.edu.cn (B.Z.); xjx0930@ahnu.edu.cn (J.X.)

† These authors contributed equally to this work.

Abstract: Background: Auxin, a plant hormone, plays diverse roles in the modulation of plant growth and development. The transport and signal transduction of auxin are regulated by various factors involved in shaping plant morphology and responding to external environmental conditions. The auxin signal transduction is primarily governed by the following two gene families: the auxin response factor (*ARF*) and auxin/indole-3-acetic acid (*AUX/IAA*). However, a comprehensive genomic analysis involving the expression profiles, structures, and functional features of the *ARF* and *AUX/IAA* gene families in *Vaccinium bracteatum* has not been carried out to date. Results: Through the acquisition of genomic and expression data, coupled with an analysis using online tools, two gene family members were identified. This groundwork provides a distinguishing characterization of the chosen gene families in terms of expression, interaction, and response in the growth and development of plant fruits. In our genome-wide search of the *VaARF* and *VaIAA* genes in *Vaccinium bracteatum*, we identified 26 *VaARF* and 17 *VaIAA* genes. We analyzed the sequence and structural characteristics of these *VaARF* and *VaIAA* genes. We found that 26 *VaARF* and 17 *VaIAA* genes were divided into six subfamilies. Based on protein interaction predictions, *VaIAA1* and *VaIAA20* were designated core members of *VaIAA* gene families. Moreover, an analysis of expression patterns showed that 14 *ARF* genes and 12 *IAA* genes exhibited significantly varied expressions during fruit development. Conclusion: Two key genes, namely, *VaIAA1* and *VaIAA20*, belonging to a gene family, play a potentially crucial role in fruit development through 26 *VaARF-IAAs*. This study provides a valuable reference for investigating the molecular mechanism of fruit development and lays the foundation for further research on *Vaccinium bracteatum*.

Keywords: auxin response factor; cis-elements; domain analysis; expression profiling

1. Introduction

Auxin is a pivotal hormone in plants that is crucial for the regulation of growth and development across various stages of the plant lifecycle [1]. It plays a central role in numerous biological processes including the development of various organs, regulation of shoot apical dominance, and differentiation of conducting tissues like xylem and phloem [2]. In *Arabidopsis*, *YUCCA2 (YUC2)* regulates chloroplast RNA editing by modulating auxin levels, affecting the expression of editing factors through ARF1-dependent pathways [3]. Genome-wide analysis in *Mikania micrantha* identified 18 *YUCCA* genes with diverse expression patterns and significant roles in plant growth and auxin biosynthesis [4]. Auxin also modulates responses to both biotic and abiotic stresses, making its signaling pathways critical targets for agricultural innovation [5].

At the heart of auxin signaling are the following two key families of transcription factors: the auxin response factor (*ARF*) family and the *AUX/IAA* repressor family [6]. These proteins are essential for the transcriptional regulation of auxin-responsive genes. *ARF*s recognize auxin-responsive elements (AuxREs) in the promoter regions of target genes, which typically include a TGTCTC motif or its variations [7]. Structurally, *ARF*s are characterized by a DNA-binding domain (DBD) incorporating a B3-like DNA-binding domain [8], a middle region (MR) that acts either as a repression or an activation domain, and a carboxyl-terminal domain (CTD) that mediates oligomerization with other *ARF*s or AUX/IAA proteins [9]. The modulation of *ARF* activity is intricately linked to their interaction with AUX/IAA proteins. In the absence of auxin, AUX/IAA proteins bind to *ARF*s, inhibiting the expression of auxin-responsive genes. Upon auxin binding, AUX/IAA proteins undergo ubiquitination and subsequent degradation via the 26S proteasome, thereby freeing *ARF*s to activate their target genes [1,10].

In the two decades since the identification of the inaugural member of the auxin response factor (*ARF*) family, 23 *ARF* members have been identified and characterized within the genome of *Arabidopsis thaliana* (L.). With the advancements in whole-genome sequencing, researchers have been able to ascertain the presence of 22 [11], 25 [12], 23 [13], 25 [14], and 36 [15] *ARF* genes within the genomic structures of *Solanum lycopersicum* L., *Oryza sativa*, *Triticum aestivum* L., *Sorghum bicolor* (L.) *Moench*, and *Zea mays* L., respectively. It is noteworthy that reports elucidating certain biological functionalities associated with *ARF*s have also begun to garner attention. An interesting aspect of these *ARF* genes is their predilection to perform functions in different plants, offering a fascinating dimension to their biological capacities.

Previous studies have highlighted the complexity of auxin signaling, noting that the specificity of *ARF* and AUX/IAA interactions can lead to diverse developmental and physiological outcomes [16]. The cell-type specificity of *ARF-AUX/IAA* expression in relation to different local auxin sources is a critical area of study in plant development. Research highlights how distinct cell types interpret auxin signals to mediate various developmental processes. Rademacher [17] provided a cellular expression map of *ARF* genes, showing complex overlapping patterns during embryogenesis and primary root meristem development. The study underscores the developmental specificity generated by the transcriptional regulation of *ARF* genes. Bargmann and Estelle [18] speculated on the tissue specificity of auxin responses through the SCF(TIR1)-*AUX/IAA-ARF* pathway, addressing how different cell types respond to the same auxin gradient. Ding [19] mapped the tissue-specific contributions of auxin signals in root growth, demonstrating significant roles of specific cell types like the endodermis. Specific *ARF*s regulate processes ranging from leaf senescence and floral organ abscission to vascular patterning and reproductive development in model plants such as *Arabidopsis thaliana* [3,20]. This underscores the potential of similar genes in *Vaccinium bracteatum* to affect key traits such as fruit development and stress tolerance [21].

*ARF*s and other TF families such as WRKY and MADS have recently been defined as instrumental in regulating various auxin-mediated responses [22]. Functional characterization of *AUX/IAA*-encoding genes has been testified through mutant and over-expressed studies conducted in *Arabidopsis thaliana* (L.). For instance, loss-of-function mutations in *AtIAA3* and *AtIAA28* resulted in lateral root formation defects [23,24]. It is discerned that *AtIAA12* and *AtIAA18* [25] can modulate embryonic apical configuration and root meristem formation by inhibiting *AtARF5* [26] activity. In *Arabidopsis*, the *AtARF7* gene participates in water absorption and root growth and, similarly [27], maize genes *ZmARF25* and *ZmARF34* [28] play significant roles in the regulation of lateral and crown root development. In the case of tomatoes, *ARF* genes occur as pivotal players in controlling chloroplast development and the accumulation of sugar and carotenoids, as well as fruit ripening. Simultaneously, *SlARF2A* and *SlARF2B* are also implicated in fruit ripening regulation.

Upon silencing of *SlARF2A*, *SlARF2B* [5], or both, fruit ripening is severely inhibited and cannot be restored by exogenous ethylene treatment. Furthermore, *SlARF2* is involved in regulating the senescence of tomato floral organs. In rice, the *OsARF11* and *OsARF15* [29] genes, and in kiwi, the *AcARF4*, *AcARF5*, *AcARF23a*, and *AcARF28a* genes are significantly upregulated after being induced by salt stress.

Importantly, *Vaccinium bracteatum Thunb.*, an Ericaceae plant [30], is known for its berries that turn purple-black when ripe, bearing flowers in June–July and fruits in August–October. These small, deep blue fruits are naturally high in anthocyanin content, making this species a valuable ornamental fruit-greening plant. The plant's leaves contain a substantial amount of plant polyphenols, which bestow high dietary and health preservation value. In *Vaccinium bracteatum* [31], a species renowned for its medicinal and nutritional value [32], understanding the role and regulation of *ARF* and IAA genes could provide significant insights into its growth and stress response mechanisms [30]. This study aims to provide a comprehensive analysis of the *ARF* and IAA gene families in *Vaccinium bracteatum*, exploring their gene structures, expression profiles, and evolutionary relationships within and across species [20]. By elucidating the phylogenetic relationships and functional annotations of the *ARF* and IAA genes in *Vaccinium bracteatum*, this study aims to contribute to the broader understanding of auxin-mediated regulation in plants [33], potentially guiding future agricultural strategies to exploit these pathways for crop improvement [34–36].

2. Results

2.1. Identification of VaARF-IAA Families in Vaccinium bracteatum

Through a meticulous HMM analysis followed by BLASTP comparison against the NCBI database, we successfully identified a total of 26 *VaARF* and 17 *VaIAA* genes in *Vaccinium ashei* (*Vaccinium bracteatum*) (Table S1), representing a comprehensive catalog of genes implicated in the auxin response and signaling pathways. The open reading frames of these *VaARF*-IAA genes range from 173 to 531 amino acids in length, reflecting a significant diversity in protein size and potential function within the family. The predicted isoelectric points (PIs) of the resultant proteins span from 5.45 to 9.08, with molecular weights (MWs) ranging from 22,120.92 to 133,143.06 Da, highlighting the biochemical diversity of these auxin response factors and auxin-induced proteins in *Vaccinium bracteatum*.

2.2. Chromosomal Localization and Collinearity Analysis of VaARF-IAA Gene Families in Vaccinium bracteatum

2.2.1. Chromosomal Distribution and Gene Duplication Events

Chromosome 9 has the highest density of *VaARF* genes, with a total of five identified, whereas chromosomes 1, 3, 4, 5, and 10 each harbor a single *VaARF* gene (Figure 1A). The *VaIAA* genes exhibit a similarly uneven distribution, with chromosome 5 containing the most *VaIAA* genes (four), and chromosomes 2 and 9 showing no *VaIAA* gene presence. Further analysis revealed several instances of gene clustering, suggesting potential gene duplication events. This distribution pattern suggests a non-uniform duplication and divergence history across different chromosomes, possibly influenced by specific evolutionary pressures or chromosomal environments.

2.2.2. Synteny Relationships Highlighting Evolutionary Conservation

The collinearity analysis of *ARF* and *IAA* gene families in *Vaccinium bracteatum* was conducted using TBtools (version 2.019) software, focusing on assessing the chromosomal distribution and potential gene duplication events within the genome. Gene families can be generated by tandem and segmental duplications of chromosomal regions as well as whole-genome duplications (WGDs) [37]. Two homologous repeats within the same chromosome that are less than 200 kb apart are usually referred to as tandem repeats [38]. A significant finding from this study was the observation of tandem duplications and segmental duplications that contribute to the expansion of the *ARF* and *IAA* gene families. For instance, within chromosome 9, two pairs of *VaARFs* (*VaARF19-2* and *VaARF19-4*;

VaARF31-1 and *VaARF31-2*) were identified, spaced 137 kb and 18 kb apart, respectively. The nucleotide homology between *VaARF19-2* and *VaARF19-4* was calculated at 68.0%, and between *VaARF31-1* and *VaARF31-2* at 85.7%, as detailed in Supplementary Table S2. Such closely linked gene pairs underscore the role of tandem duplications in the local expansion of gene families within the genome.

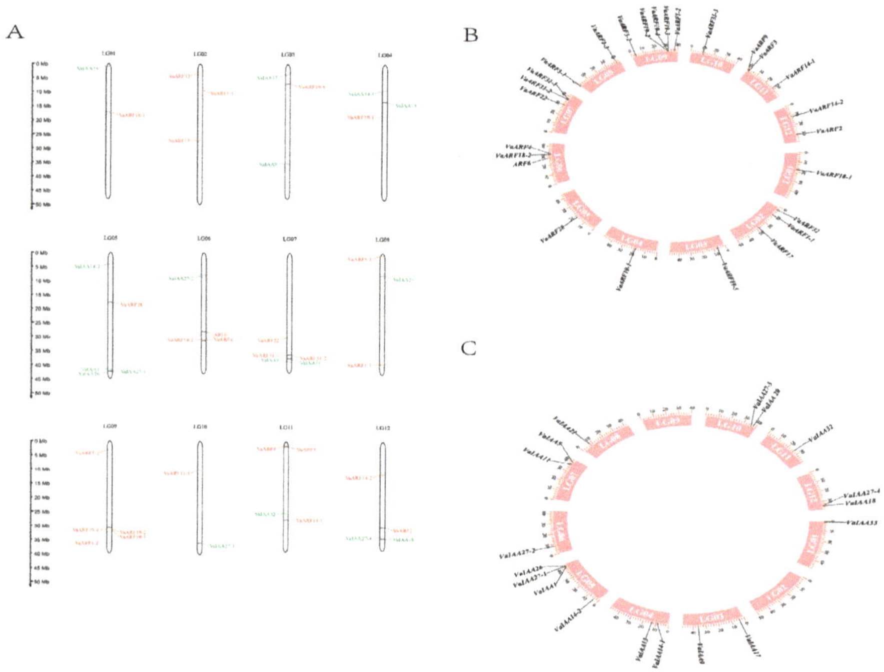

Figure 1. The distribution of *VaARF* and *VaIAA* family members on the chromosomes of *Vaccinium bracteatum*. (**A**) The chromosome mapping of *VaARF*s-IAAs and (**B**) collinearity among *VaARF* gene families within the chromosome. (**C**) collinearity among *VaIAA* gene families within the chromosome.The chromosome number is listed at the top of each vertical bar. The ruler depicts the length of the chromosome/Mb. The red color on the chromosome is the *ARF* gene, and the green color is the *IAA* gene. The gray line indicates that most of the *ARF* and *IAA* genes have a collinear relationship among species. Chromosomes 1–12 are shown in pink, with *ARF* and *IAA* gene names on the periphery.

Furthermore, the analysis also highlighted several instances of potential segmental or whole-genome duplications. For example, three pairs of *ARF* genes, two pairs of *IAA* genes, and additional clusters involving four *ARF* genes and a multi-gene family of *IAA* showed homology, suggesting their origin from more extensive duplication events. The non-synonymous to synonymous substitution ratio (Ka/Ks) for these gene pairs was calculated and found to be less than 1 for all instances, as shown in Supplementary Table S2. This indicates that the duplicated gene pairs and clusters have undergone purifying selection, which typically restricts functional diversification and preserves essential functions.

2.3. *Phylogenetic Analysis of VaARF-IAA Gene Families in Vaccinium bracteatum*

The phylogenetic relationships among auxin response factor (*ARF*) genes from *Vaccinium bracteatum* and other species were elucidated through comprehensive analyses utilizing ClustalX for multiple sequence alignment and PhyML software for phylogenetic tree construction via the maximum likelihood method. This analysis encompassed *ARF* nucleotide sequences from five species (Table S3) including *Vaccinium bracteatum*

(26 *VaARFs*), *Arabidopsis thaliana* (21 AtARFs), *Glycine max* (37 GmARFs), *Brassica napus* (27 BnARFs), and *Gossypium hirsutum* L. (37 GhARFs), totaling 146 *ARF* genes (Figure 2). The resultant phylogenetic tree distributed these 146 *ARF* genes into six distinct subgroups (I–VI), illustrating the evolutionary breadth of the *ARF* gene family across these species. Notably, *VaARF1-1* showed close evolutionary proximity to *ARF* genes from the other four examined species, while *VaARF22* shared a tight phylogenetic relationship with *AtARF3*, highlighting instances of conserved evolutionary trajectories. Additionally, 141 *IAA* genes were categorized into six subgroups (I–VI), including 18 *VaIAAs*, 24 *AtIAAs*, 45 *GmIAAs*, 31 *BnIAAs*, and 33 *GhIAAs*. Within this framework, *VaIAA1* was closely related to *GhIAA14*, and *VaIAA27-3* was near *GhIAA27*, underscoring the homology and potential functional conservation across species.

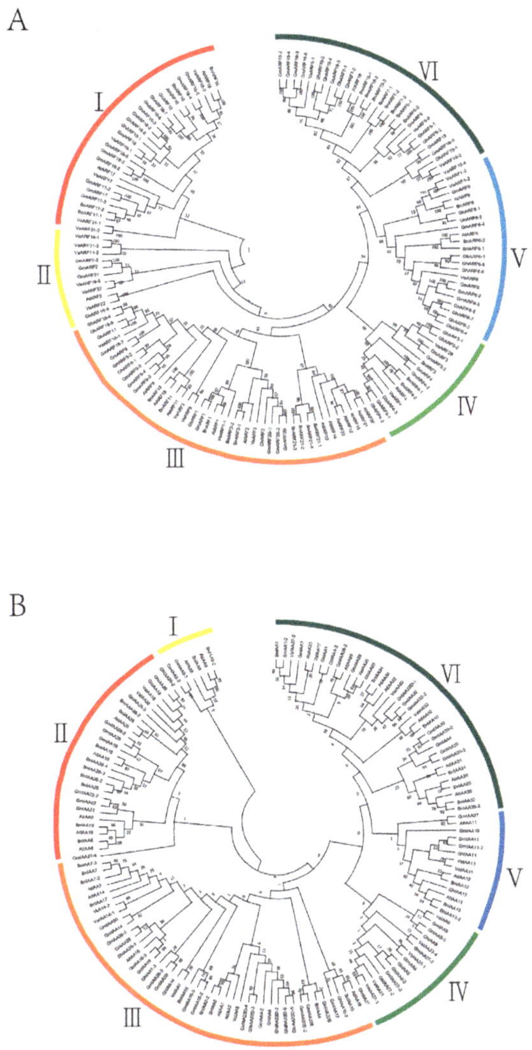

Figure 2. Phylogenetic relationships among *VaARF*, *VaIAA*, and five other families of species. (**A**) Phylogenetic analysis of *VaARF* (**B**) Phylogenetic analysis of *VaIAA*. The phylogenetic tree was established utilizing the full-length gene sequence of *VaARF-IAAs* from *Vaccinium bracteatum* and *Arabidopsis thaliana*, *Glycine max*, *Brassica napus*, and *Gossypium hirsutum* L.

2.4. Gene Structure and Conserved Domain Analysis of VaARF-IAA Gene Families in Vaccinium bracteatum

The majority of *VaARF* genes are characterized by more than five introns, highlighting the intricate architecture of these genes. Specifically, *VaARF14-1*, *VaARF14-2*, and *VaARF30-2* are unique in their structure, each comprising a single exon, devoid of any introns (Figure 3). In contrast, *VaARF17*, *VaARF19-3*, and *VaARF31-1* contain two introns each. Similarly, the majority of *VaIAA* genes possess around five introns, with *VaIAA33* and *VaIAA27-2* featuring one and two introns, respectively. This variability underscores the evolutionary diversification within the *ARF* and *IAA* gene families.

Figure 3. Genetic structure analysis of *VaARF* and *VaIAA* gene families. (**A**) Genetic structure analysis of *VaARF* (**B**) Genetic structure analysis of *VaIAA*. Exons and introns are represented by yellow boxes and black lines, respectively.

Upon domain analysis, all identified *VaARF*s were found to contain the conserved B3 domain (Figure 4A,B), and most harbored the PB1 domain, affirming the conservation of these domains within the *VaARF* family. These domains are likely crucial for the functional repertoire of the *ARF* genes, with the B3 domain involved in DNA binding and the PB1 domain involved in protein–protein interactions. Interestingly, *VaARF14* encompasses an AP2_ERF domain, and *VaARF3* uniquely possesses the KS3_2 domain, hinting at specialized functions beyond typical *ARF* gene roles. For the *VaIAA* gene family, except for *VaIAA17*, which is composed of three B3 domains, the rest predominantly feature the PB1 domain, further illustrating the functional conservation and diversification within these auxin-response gene families.

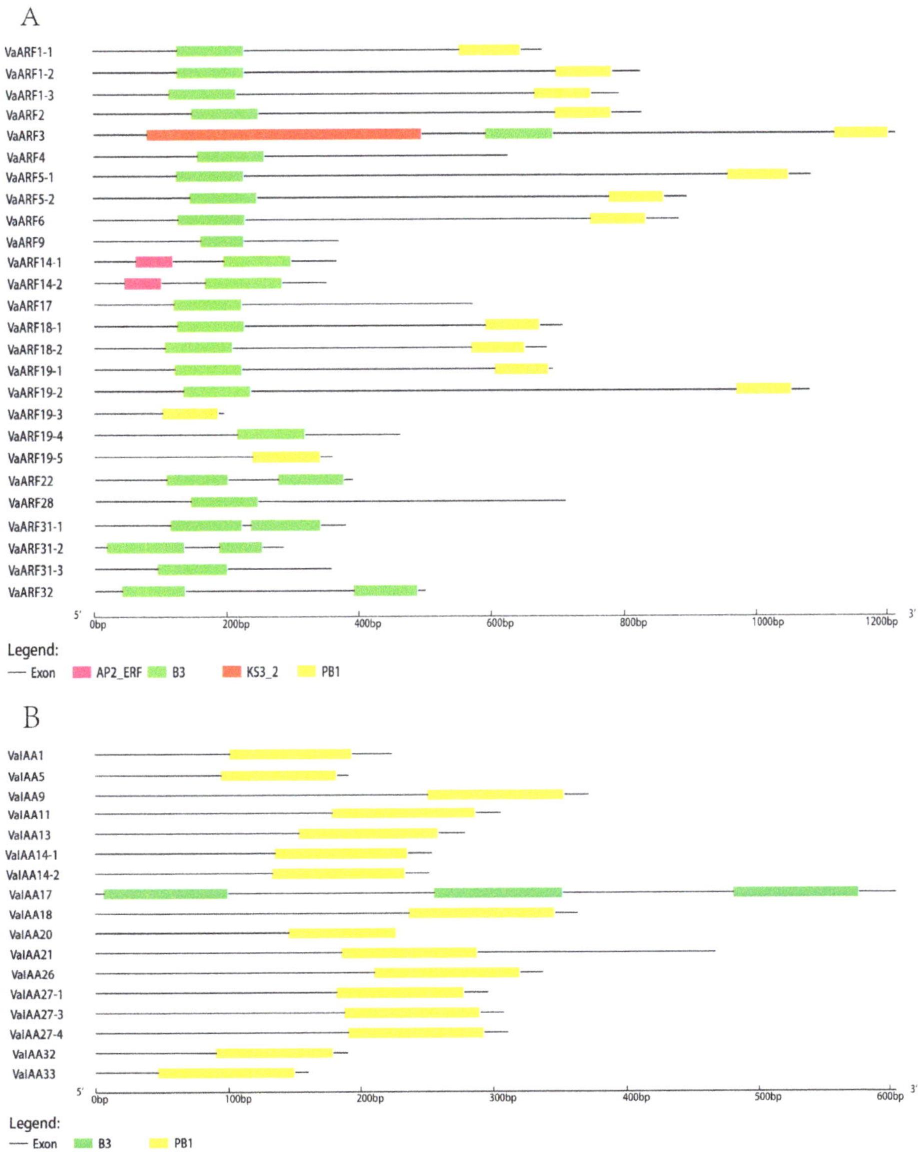

Figure 4. Domain analysis of *VaARF* and *VaIAA* gene families. (**A**) Domain analysis of *VaARFs* (**B**) Domain analysis of *VaIAAs*. Different colors represent different domains.

Based on the protein sequences of *VaARF*s and *VaIAA*s, these conserved motifs were discovered using the MEME website. The conserved motifs of VaARF and VaIAA proteins were analyzed, and 12 conserved motifs (motifs 1–12) were found in both proteins (Figure 5A,B). Their sequences are shown on the right side of Figure 5. In addition to motifs 8 and 10, the remaining 10 motifs are prevalent in most *ARF*s. *VaARF22*, *VaARF32*, *VaARF31-1*, and *VaARF31-2* contain only motif 1, and these four genes may be functionally redundant. *VaARF19-3* and *VaARF19-5* contain only motifs 8 and 10. Motif 10 is unique to *VaARF6, VaARF5-1, VaARF1-3, VaARF19-2, VaARF1-2, VaARF5-2, VaARF1-1, VaARF2, VaARF19-1, VaARF18-1*, and *VaARF18-2*. For the *IAA* gene family, all *VaIAAs* except *VaIAA27-1* contain motif 1, indicating that it plays an important and possibly structural role in *VaIAAs*.

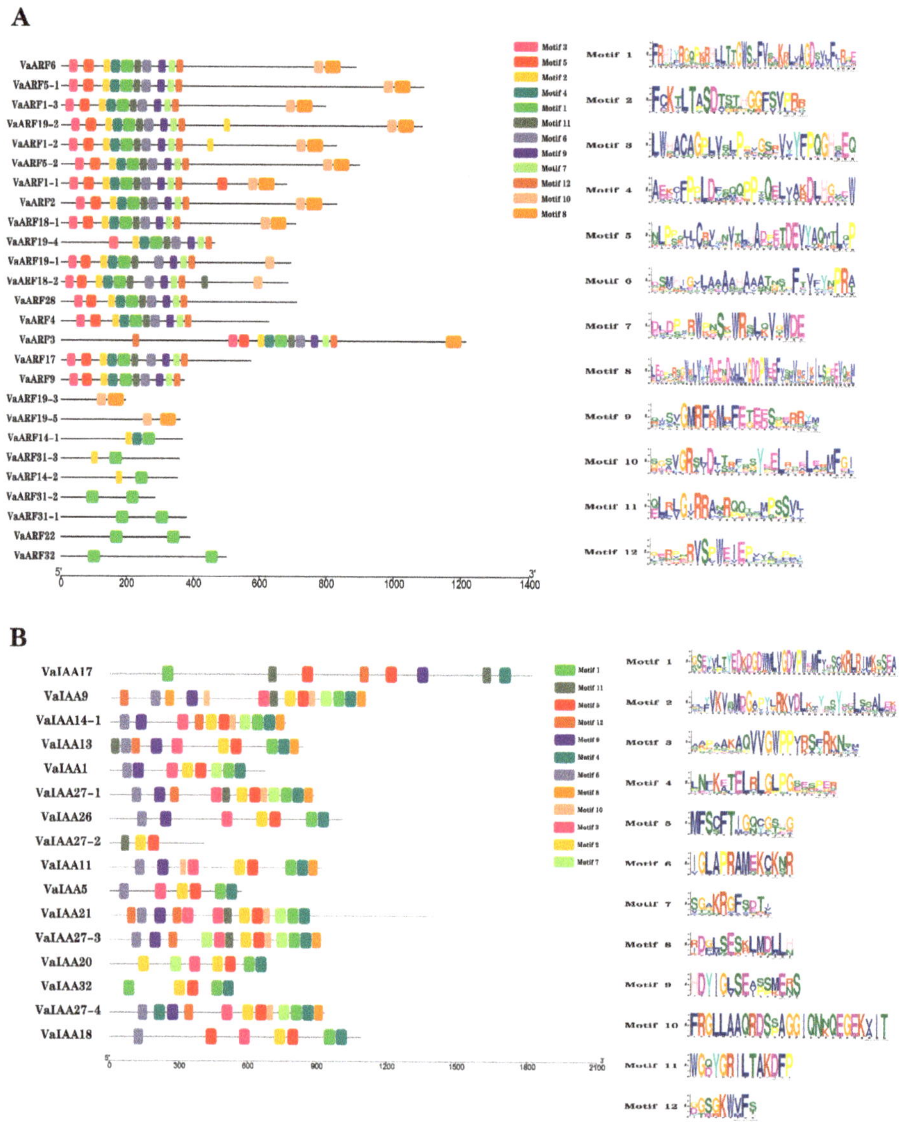

Figure 5. Conserved motifs of the *VaARF* and *VaIAA* gene families. (**A**) Conserved motifs of the *VaARFs* (**B**) Conserved motifs of the *VaIAAs*. The motif length can be calculated using the ruler at the bottom. The ten expected motifs are represented by colored boxes on the left, with the Motif LOGO on the right.

2.5. Promoter and Synteny Analysis of VaARF-IAA Gene Families in Vaccinium bracteatum

Remarkably, every member of the *VaARF-IAA* gene families was found to possess light-responsive elements (Table S4), highlighting the pivotal role of these genes in light-mediated physiological processes (Figure 6A,B). Similarly, hormone-responsive elements were ubiquitously present across all *VaARF* gene promoters, reflecting the gene family's comprehensive involvement in hormonal signaling pathways. The majority of these promoters also featured MeJA-responsiveness, MYB binding sites indicative of drought inducibility, auxin responsiveness, and gibberellin-responsive elements, among others. Additionally, elements related to low temperature, defense (TC-rich repeats), and stress (CARE) were identified, signifying the gene family's versatile role in plant growth, development, and stress response mechanisms.

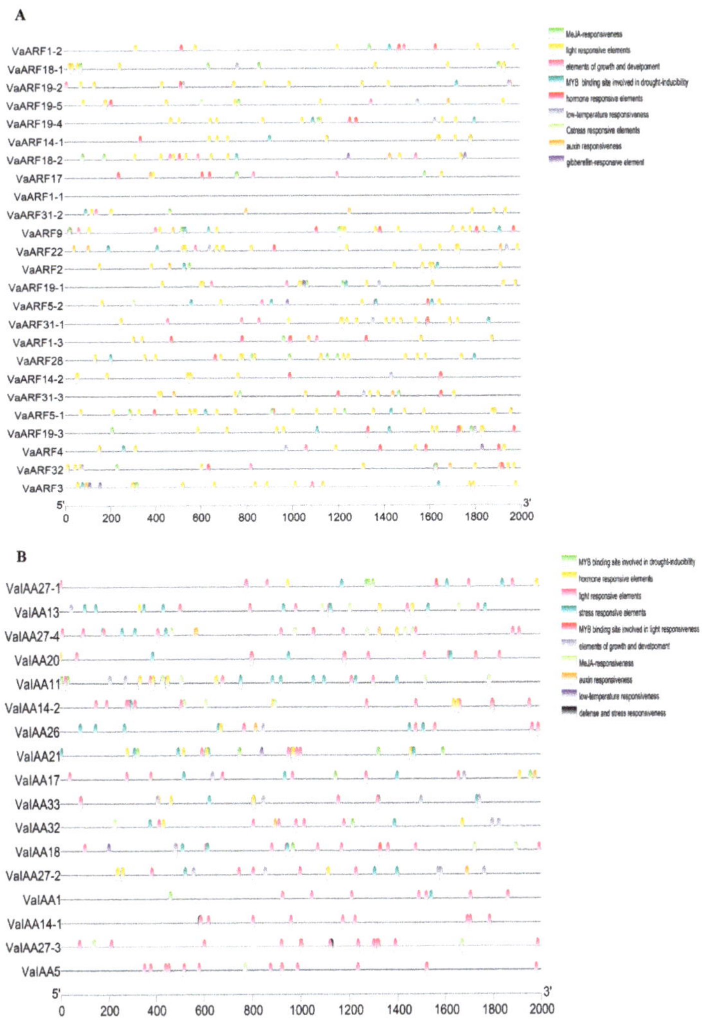

Figure 6. Analysis of cis-regulatory elements in the *VaARF* and *VaIAA* families. (**A**) Analysis of cis-regulatory elements in the *VaARFs* (**B**) Analysis of cis-regulatory elements in the *VaIAAs*. Trans-acting regulatory elements discovered in the promoter regions are represented in different blocks colored according to the type of element.

Specifically, CARE elements, associated with stress-responsive elements, were uniquely identified in the promoters of *VaARF19-5* and *VaARF5-2*, while TC-rich repeats, signifying defense and stress responsiveness, were exclusively found in *VaIAA14-1* and *VaIAA27-3*. The diversity and specificity of cis-acting elements observed in the promoters of *VaARFs* and *VaIAAs* suggest a complex regulatory network influencing plant responses to environmental cues and hormonal signals.

2.6. Expression Analysis of VaARF-IAAs during Fruit Ripening

VaARF2, *VaARF6*, *VaARF4*, *VaARF5-1*, *VaARF1-2*, *VaARF19-5*, and *VaARF28* were highly expressed during the green fruit stage, suggesting a potential correlation with fruit enlargement (Figure 7A). Conversely, *VaARF5-2*, *VaARF18-2*, and *VaARF1-1* showed heightened expression in the blue fruit stage, which may be associated with the processes involved in fruit ripening.

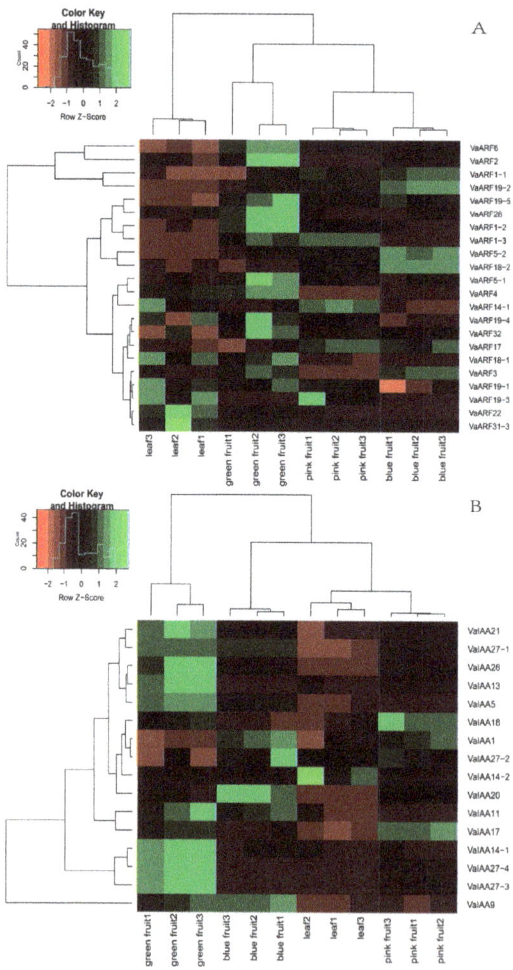

Figure 7. Expression pattern in the *VaARF* and *VaIAA* gene families during different stages of fruiting. (**A**) Expression pattern in the *VaARFs* (**B**) Expression pattern in the *VaIAAs*.

Similarly, within the *IAA* gene family, 12 genes displayed significant differential expression across the fruit development stages. *VaIAA14-1*, *VaIAA27-4*, and *VaIAA27-3* were particularly active during the green fruit stage, further associating these genes with the process of fruit enlargement. On the other hand, *VaIAA1*, *VaIAA20*, and *VaIAA27-2* demonstrated peak expression levels in blue fruit, suggesting their roles in the maturation process.

2.7. Interaction Networks of VaARF-IAAs with Functional Genes

Auxins play an integral role in orchestrating plant growth and development. They regulate gene expression through specific transcription factors and proteins, which are fine-tuned to environmental responses within the signaling cascade [39]. Key among the auxin transcription factors are the auxin response factors (*ARF*s) and the *AUX/IAA* inhibitors. In pursuit of a deeper understanding of the biological functions and regulatory networks of *ARF*s and *AUX/IAA*s, we utilized a homology-based approach to predict protein–protein interactions (PPIs). Our study revealed the presence of several PPIs within the *ARF* and *AUX/IAA* gene families (Figure 8A), with *VaIAA1* and *VaIAA20* identified as the core components within these two gene families [40].

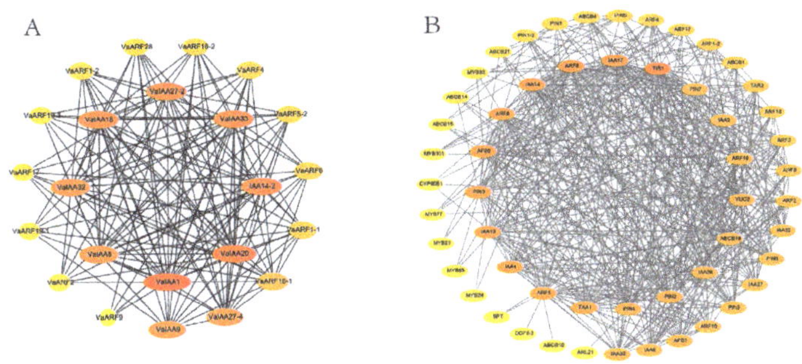

Figure 8. Interaction protein analysis of the *VaARF* and *VaIAA* gene families. (**A**) The interaction between the two gene families. (**B**) Prediction of protein–protein interactions between *ARF* protein and other functional proteins. The dark color of the circles and lines represents a high degree of interaction among proteins. The size and dark color of the circles of each protein show high interactions with other proteins.

Furthermore, we analyzed the protein–protein interactions among *ARF*s, *AUX/IAA*s, and other functional genes (Figure 8B). As anticipated, the majority of the proteins interacting with *ARF* were validated as significant constituents of the auxin response pathway [41]. For instance, *ARF* interacts with PIN-FORMED (PIN) proteins, ATP-BINDING CASSETTE subfamily B (ABCB) proteins, MYB proteins, TIR proteins, and AFB proteins [42], all of which play vital roles in plant growth and development [43].

2.8. Protein 3D Structure Analysis

Three-dimensional models were generated using the primary sequences of VaARF and VaIAA proteins, which revealed distinct structural features that may correlate with functional specificity (Figure 9A,B). The structural variability observed among the VaARF proteins, such as differences in the orientation and configuration of secondary structural elements like α-helices and β-sheets, suggests specialized interactions and functions within the plant signaling pathways. The 3D structural models of key proteins such as *VaARF1-1*, *VaARF2*, *VaARF5-1*, and *VaARF19* series show notable differences in their domain architectures and surface properties. These variations may influence their binding affinities and interactions with other molecules, possibly affecting their role in auxin signaling and response mechanisms in *Vaccinium bracteatum*. Moreover, the analysis identified specific regions within these proteins that could potentially be involved in ligand binding or protein–protein interactions, critical for their function as transcriptional regulators. For example, the predicted models suggest that the *VaARF19-5* protein might have additional binding sites, which could be crucial for its interaction with other proteins or DNA elements within the plant cell.

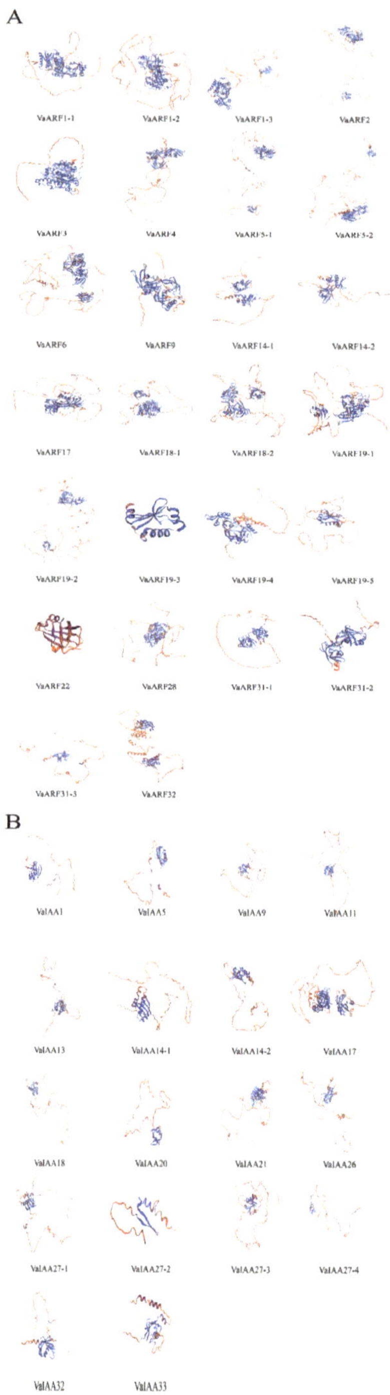

Figure 9. Three-dimensional structure model of VaARF/VaIAA proteins. (**A**) Three-dimensional structure model of VaARF proteins (**B**) Three-dimensional structure model of VaIAA proteins. Ramachandran plots were also used to validate the models of the various proteins that were obtained.

3. Discussion

3.1. Interpretation of Phylogenetic Relationships and Gene Structure Diversity

In naming these identified genes, we adhered to a systematic approach by aligning them with their closest homologs in the NCBI database, thereby ensuring that the nomenclature reflects phylogenetic relationships and facilitates cross-species comparisons [44]. This methodology not only enhances the clarity of our findings but also aligns with standard practices in genomic studies, enabling researchers to easily correlate our identified genes with known auxin response mechanisms in other species [45].

The exhaustive identification and subsequent annotation of the *VaARF* and *VaIAA* gene families underscore the complexity of auxin signaling in *Vaccinium bracteatum*. These findings lay a crucial foundation for future functional studies aimed at unraveling the roles of these genes in growth, development, and stress responses in *Vaccinium bracteatum*, potentially opening new avenues for agricultural innovation and crop improvement.

Vaccinium bracteatum, a member of the Rhododendron family, is a diploid plant species characterized by 12 distinct chromosomes, as revealed in our previous research. The distribution of *ARF* and *IAA* genes across these chromosomes is notably uneven, highlighting the complex genetic architecture of auxin response mechanisms in this species. Chromosome 9 stands out with the highest gene distribution density for *VaARF* genes, whereas chromosome 5 leads in density for *VaIAA* genes. This uneven gene distribution and evidence of gene clustering and duplication provide valuable insights into the genomic organization and evolutionary dynamics of the *ARF* and *IAA* gene families in *Vaccinium bracteatum*. Such findings not only enrich our understanding of the genetic basis of auxin response in this species but also underscore the intricate relationship between chromosomal architecture and gene function diversification in plant genomes [46].

The arrangement of *VaIAA1*, *VaIAA27-1*, and *VaIAA26* in close locations, along with *VaIAA27-4* and *VaIAA18*, paired with the clustered positioning of *VaARF* genes including *VaARF18-2* and *VaARF4*, *VaARF31-1* and *VaARF31-2*, *VaARF19-2* and *VaARF19-3*, and *VaARF9* and *VaARF3*, strongly suggests the history of gene duplication events. These clusters may indicate regions of the genome that have undergone evolutionary expansions [47], potentially contributing to the diversification of the *ARF* and *IAA* gene functions in *Vaccinium bracteatum*.

The chromosomal collinearity analysis not only elucidated the structural organization and evolutionary dynamics of the *ARF* and *IAA* gene families in *Vaccinium bracteatum* but also provided insights into the mechanisms driving gene family expansion and functional conservation. This detailed mapping and comparison help pave the way for further functional genomic studies aimed at understanding the correlations between these gene families in plant development and stress responses [48].

3.2. Role of ARF Genes in the Fruit Ripening of Vaccinium bracteatum Based on Expression Profiles

The expression analysis of auxin response factor (*ARF*) genes during fruit ripening in *Vaccinium bracteatum* was meticulously carried out using transcriptome data from four developmental stages including leaves, green fruit, red fruit, and blue fruit, with green and red fruits representing unripe stages. This comprehensive dataset, derived from triplicate samples for each stage, provided a robust foundation for exploring the dynamic expression of *ARF* genes across different phases of fruit development [49].

Expression levels were quantified using FPKM (Fragments Per Kilobase of transcript per Million mapped reads) [50], enabling a precise measurement of gene activity. Our analysis revealed that 14 *ARF* genes exhibited significant differential expression throughout the fruit development process, suggesting a pivotal role for the majority of *ARF*-related genes in fruit maturation.

This detailed expression profiling of *ARF* and *IAA* genes underscores the complex regulatory network governing fruit development in *Vaccinium bracteatum* [51]. The distinct expression patterns observed not only highlight the specific contributions of individual *ARF* and *IAA* genes to fruit enlargement and ripening but also pave the way for future studies aimed at unraveling their functional roles in these critical developmental processes [52].

Further studies, including gene knockout experiments, are necessary to determine the definitive roles of these *ARF* genes in specific fruit developmental processes, such as cell division, growth, and maturation. Additionally, exploring the links between specific auxin sources, such as *YUCCA* gene expression, and the regulation of these *ARF* genes will provide deeper insights into the hormonal control mechanisms in fruit development.

3.3. Analysis of Interaction Networks

Auxin Polar Transport (PAT) is essential for auxin's short-distance distribution, mediated by auxin carrier protein families within the cell membrane, including PIN-FORMED (PIN) proteins and ATP-BINDING CASSETTE subfamily B (ABCB) proteins [41]. The activity of these carrier proteins is influenced by the regulation of their expression and subcellular localization, which are modulated by *ARF* transcription factors in response to environmental signals.

In tomatoes, *SlPIN8* is specifically expressed in pollen, with its function possibly linked to pollen development [53] and auxin homeostasis [54]. TIR and AFB proteins are nuclear auxin-sensitive proteins containing nucleotide/leucine-rich repeats, allowing interaction with specific subsets of AUX/IAA proteins [54]. Research in *Arabidopsis* thaliana reveals that *TIR1/AFB* genes encode the primary auxin receptors that regulate plant growth, and *AFB3* mediates lateral root growth in response to nitrogen [55]. In soybeans, the *GmTIR1/AFB3* gene is a homolog of *TIR1* and *AFB3* found in *Arabidopsis thaliana*, serving to regulate soybean nodulation in response to rhizobia infection through the auxin signaling pathway [56]. In barley (*Hordeum vulgare*), the HvYUC protein is involved in mediating defense against salt stress, and a similar positive response to aluminum stress by *TAA1* through ethylene signaling has been observed [57]. The findings of this study significantly broaden our understanding of the regulatory networks of *ARF*s and *AUX/IAA*s, as well as their influence on plant growth and development [58].

3.4. Functional Implications of the Protein Structure Findings

The comprehensive phylogenetic analysis not only reveals the homologous relationships among *Vaccinium bracteatum ARF* genes, *IAA* genes, and those from other key plant species, but it also indicates a greater degree of evolutionary conservation within the *ARF* gene family compared with *IAA*s [59]. Such findings suggest that *ARF* genes may play a fundamental role in plant development and stress responses that have been preserved across divergent evolutionary paths [50]. The detailed subgrouping and cross-species comparisons provide valuable insights into the functional diversification and evolutionary history of the *ARF* and *IAA* gene families, offering a foundation for further functional characterization and exploration of their roles in plant biology [47].

The structural analysis of *VaARF* and *VaIAA* gene families in *Vaccinium bracteatum* reveals a remarkable diversity and complexity within their gene structures. A significant structural variance is observed between the *ARF* and *IAA* genes [60], where *ARF*s exhibit a higher exon/intron ratio compared with *IAA*s, suggesting a higher degree of complexity and diversity within the *ARF* family. This structural distinction could be indicative of the varied functional roles that these gene families play within plant developmental processes and stress responses. The elucidation of gene structures and domain compositions not only provides insights into the functional complexities of the *VaARF* and *VaIAA* gene families but also underscores the evolutionary strategies employed by *Vaccinium bracteatum* in modulating auxin signaling pathways. These findings pave the way for future functional characterization studies, aiming to unravel the specific roles of these gene families in plant growth, development, and adaptation to environmental cues.

In our comprehensive analysis of the promoter regions, spanning 2 kb upstream of the transcription start sites of each *VaARF* gene, we aimed to elucidate the potential regulatory mechanisms governing the *VaARF-IAA* gene family's activity in *Vaccinium bracteatum*. This analysis uncovered a plethora of cis-acting elements predominantly associated with various plant hormone responses, underscoring the intricate relationship between the *VaARF-IAA* gene family and hormonal regulation [61].

This promoter analysis not only sheds light on the potential functional diversity within the *VaARF-IAA* gene family but also provides a foundation for future studies aimed at unraveling the molecular mechanisms by which these genes modulate plant growth and stress resilience in *Vaccinium bracteatum*. The identification of unique and common regulatory elements across these promoters underscores the multifaceted roles of *ARF* and *IAA* genes in orchestrating plant development and environmental adaptation [62].

To better understand the structural basis of ARF protein function in *Vaccinium bracteatum*, we utilized the Swiss Model web-based platform to predict and analyze the 3D structures of the ARF protein family. This analysis revealed substantial structural diversity among the ARF proteins [63], reflecting their varied roles in plant physiology and development. This comprehensive structural analysis not only enhances our understanding of the functional diversity of the *ARF* gene family in *Vaccinium bracteatum* but also provides a foundation for future experimental studies aimed at validating these models and elucidating the specific roles of these proteins in plant growth and development processes [64,65].

4. Materials and Methods

4.1. Plant Resources

Samples of *Vaccinium bracteatum* were collected from the hilly subtropical region in the southern part of Anhui Province (30°51′ N, 118°23′ E). This collection enabled us to obtain essential genomic data for the species. Complete genomic data and amino acid sequences of two gene families from *Arabidopsis thaliana*, *Glycine max*, *Brassica napus*, and *Gossypium hirsutum* L. were procured respectively from the TAIR website (https://www.arabidopsis.org/) (accessed on 25 September 2023), NCBI database (https://www.ncbi.nlm.nih.gov/) (accessed on 25 September 2023) and uniport database (https://www.uniprot.org/), (accessed on 25 September 2023). Leaves and fruits were collected from three separate developmental stages including the green fruit growth phase (Stage 1), pink fruit color accumulation phase (Stage 2), and mature blue fruit phase (Stage 3). Each developmental stage was sampled from three biological replicates. The samples were flash-frozen in liquid nitrogen followed by preservation at −80 °C until RNA extraction. The resultant RNA samples were thereafter utilized for gene prediction based on the transcriptome.

4.2. Genome-Wide Identification and Classification of ARF-Aux/IAA Gene Families

Information on *Arabidopsis thaliana ARF* and *AUX/IAA* gene families was downloaded from the TAIR database (https://www.arabidopsis.org/) (accessed on 25 September 2023). The JGI Plant Site (https://phytozome.jgi.doe.gov/pz/portal.html) (accessed on 27 September 2023) was then utilized to compare the homology of amino acids linked with *Arabidopsis thaliana* ARF and AUX/IAA proteins. Furthermore, the characteristics of *ARF* and *AUX/IAA* genes were identified through the NCBI database's Blastp feature, followed by assigning specific names. The online ExPASy tool (https://www.expasy.org/) (accessed on 12 October 2023) was then utilized to obtain data on the biochemical qualities of each ARF and AUX/IAA protein, including parameters such as the number of amino acids, isoelectric point (pI), and molecular weight.

4.3. Visualization and Collinearity Analysis of Chromosomal Positioning

The chromosomal localization data for the verified *VaARF-IAA* gene family stems from the sequencing results obtained from *Vaccinium bracteatum*. Chromosome maps were illustrated using MG2C (http://mg2c.iask.in/mg2c_v2.1/) (accessed on 15 October 2023). To analyze and characterize the collinearity of the *ARF-AUX/IAA* gene families, TBtools (version 2.019) were utilized. These tools helped in identifying homologous and paralogous genes in the replication events of the *ARF* and *AUX/IAA* gene families. For the computation of the Ka/Ks ratio, which represents the rate of non-synonymous to synonymous substitutions among respective gene pairs, TBtools (version 2.019) was used. In addition, EMBL Pairwise Sequence Alignment (https://www.ebi.ac.uk/jdispatcher/psa/lalign) (accessed on 15 October 2023) was employed for carrying out a bidirectional sequence alignment, assisting in determining nucleotide level congruity within gene pairs.

4.4. The Establishment of the Phylogenetic Tree

By utilizing the tool known as "Find Best DNA/Protein Models (ML)" found in Molecular Evolutionary Genetics Analysis (version MEGA 11.0), optimal model tests were conducted on the amino acid sequences of ARF and AUX/IAA proteins. This ultimately facilitated the construction of a maximum likelihood model. Each tree node was computed utilizing 1000 bootstrap duplicates, with the remainder of the parameters set at their default values.

4.5. Analysis of Introns, Exons, and Regulatory Sequences

To illustrate the exon–intron structure diagram and the coding sequence (CDS) configuration of the *VaARF-IAA* genes, we leveraged the GENE Structure Display (GSDS 2.0) web tool (accessible at http://gsds.gao-lab.org/) (accessed on 9 January 2024). This was performed by uploading the genomic sequence and the coding sequence to create the gene structure. We analyzed the conservative base sequences of the ARF and AUX/IAA proteins in VaARF-IAAs using the online tool MEME Suite (available at https://meme-suite.org/meme/) (accessed on 20 January 2024). The base sequences found in the generated protein dataset were saved for future reference. For the visualization process, we employed TBtools software (version 2.019).

4.6. Analysis of Promoter Sequence

The promoter sequence analysis involved 2000 bp of the genome sequence upstream as the transcription initiation site. The software TBtools (version 2.019), was utilized to extract the 2000 bp upstream sequence of the *ARF* and *AUX/IAA* gene families, serving as the promoter sequence. Further analysis of the operative elements in the promoter region was conducted via the PlantCARE website (http://bioinformatics.psb.ugent.be/webtools/plantcare/html/) (accessed on 5 March 2024). Upon generating the data, it was prudently organized and simplified. The visualization process was facilitated using TBtools (version 2.019) software.

4.7. Expression Patterns within Different Fruit Stages

Total RNA was isolated from 12 samples, precisely from tissues of leaves and green, pink, and blue fruits, with each tissue represented by triplicates. The data depicted in the figure are the products of a transcriptome analysis used to examine tissue specificity and the differential expression of *ARF* and *AUX/IAA* gene families across varying regions within the plant.

The mapping of these RNA-seq reads was completed using TBtools (version 2.019), employing its default settings. Quantitative gene expression analysis was conducted using the TPM (Transcripts Per Million) algorithm. Additionally, the heatmap of the resultant TPM values for *ARF* and *AUX/IAA* from each tissue was generated using the heatmap function of TBtools (version 2.019).

4.8. Analysis of Protein–Protein Interaction Expression

Utilizing the AraNetV2 (http://www.inetbio.org/aranet/) (accessed on 12 March 2024) and STRING (http://string-db.org/) (accessed on 15 March 2024) databases, we established the interaction networks between *VaARF* and *VaIAAs*. To visualize these predicted interaction networks, we employed Cytoscape (version 3.7.2). In addition, the three-dimensional structural homology of the *VaARF-IAA* gene family's spatial protein models in Vitex agnus-castus (Chaste tree) was modeled using 3D structure models provided by ExPaSy (https://swissmodel.expasy.org/interactive) (accessed on 26 March 2024).

5. Conclusions

In our research, we identified 26 *ARF* genes and 17 *AUX-IAA* genes within the genome of *Vaccinium bracteatum*. We conducted an in-depth analysis of their phylogeny, gene structure, conserved domains, and motifs to ascertain their evolutionary affiliations with other plant species. The gene expression profiles during different stages of *Vaccinium bracteatum* fruit development revealed a spectrum of expression diversity in the *ARF* and *IAA* gene families. Furthermore, interaction analyses of *VaARF* and VaAUX-IAA with various functional genes unveiled their critical role in plant growth and development. These findings lay a solid foundation for future research into the operational characterization of *ARF* genes, *AUX-IAA* genes, and *ARF*-mediated signal transduction pathways.

Supplementary Materials: The following supporting information can be downloaded at: https://www.mdpi.com/article/10.3390/ijms25158385/s1.

Author Contributions: Method and writing—manuscript preparation, X.G. and X.L.; writing—review and editing, H.Z., L.C. and X.W.; data visualization C.Z., H.D. and H.Y.; data management Y.C., B.Z. and J.X. All authors have read and agreed to the published version of the manuscript.

Funding: This work was funded by the National Natural Science Foundation of China (NO.32371927); the Key R&D Project of Anhui Province (NO. 202204c06020029); the Science and Technology Project of Wuhu City (2023yf084); and the Excellent Innovation Research Team of Molecular Enzymology and Detection of Anhui University (2022AH010012).

Informed Consent Statement: For all the participants, written informed consent was collected by the caregivers.

Data Availability Statement: The original contributions presented in the study are included in the article/Supplementary Material, further inquiries can be directed to the corresponding authors.

Acknowledgments: Appreciate the sequencing services provided by Hefei Yuanzai Biotechnology Co., Ltd. HeFei China.

Conflicts of Interest: The authors declare no conflicts of interest.

References

1. Li, Y.; Han, S.; Qi, Y. Advances in structure and function of auxin response factor in plants. *J. Integr. Plant Biol.* **2023**, *65*, 617–632. [CrossRef]
2. Ali, S.; Wang, W.; Zhang, Z.; Xie, L.; Boer, D.R.; Khan, N. Genome-Wide Identification, Expression and Interaction Analysis of ARF and AUX/IAA Gene Family in Soybean. *Front. Biosci. (Landmark Ed.)* **2022**, *27*, 251. [CrossRef] [PubMed]
3. Li, Z.A.; Li, Y.; Liu, D.; Molloy, D.P.; Luo, Z.F.; Li, H.O.; Zhao, J.; Zhou, J.; Su, Y.; Wang, R.Z.; et al. YUCCA2 (YUC2)-Mediated 3-Indoleacetic Acid (IAA) Biosynthesis Regulates Chloroplast RNA Editing by Relieving the Auxin Response Factor 1 (ARF1)-Dependent Inhibition of Editing Factors in Arabidopsis thaliana. *Int. J. Mol. Sci.* **2023**, *24*, 16988. [CrossRef] [PubMed]
4. Luo, W.; Xiao, N.; Wu, F.; Mo, B.; Kong, W.; Yu, Y. Genome-Wide Identification and Characterization of YUCCA Gene Family in Mikania micrantha. *Int. J. Mol. Sci.* **2022**, *23*, 13037. [CrossRef]
5. Wang, Z.; Shang, Q.; Zhang, W.; Huang, D.; Pan, X. Identification of ARF genes in Juglans Sigillata Dode and analysis of their expression patterns under drought stress. *Mol. Biol. Rep.* **2024**, *51*, 539. [CrossRef] [PubMed]
6. Chandler, J.W. Auxin response factors. *Plant Cell Environ.* **2016**, *39*, 1014–1028. [CrossRef] [PubMed]
7. Zhang, J.; Li, S.; Gao, X.; Liu, Y.; Fu, B. Genome-wide identification and expression pattern analysis of the Aux/IAA (auxin/indole-3-acetic acid) gene family in alfalfa (*Medicago sativa*) and the potential functions under drought stress. *BMC Genom.* **2024**, *25*, 382. [CrossRef] [PubMed]
8. Rienstra, J.; Hernández-García, J.; Weijers, D. To bind or not to bind: How AUXIN RESPONSE FACTORs select their target genes. *J. Exp. Bot.* **2023**, *74*, 6922–6932. [CrossRef]
9. Guilfoyle, T.J. The PB1 domain in auxin response factor and Aux/IAA proteins: A versatile protein interaction module in the auxin response. *Plant Cell* **2015**, *27*, 33–43. [CrossRef]
10. Fontana, M.; Roosjen, M.; Crespo García, I.; van den Berg, W.; Malfois, M.; Boer, R.; Weijers, D.; Hohlbein, J. Cooperative action of separate interaction domains promotes high-affinity DNA binding of Arabidopsis thaliana ARF transcription factors. *Proc. Natl. Acad. Sci. USA* **2023**, *120*, e2219916120. [CrossRef]
11. Wang, H.; Jones, B.; Li, Z.; Frasse, P.; Delalande, C.; Regad, F.; Chaabouni, S.; Latché, A.; Pech, J.C.; Bouzayen, M. The tomato Aux/IAA transcription factor IAA9 is involved in fruit development and leaf morphogenesis. *Plant Cell* **2005**, *17*, 2676–2692. [CrossRef]

12. Wang, D.; Pei, K.; Fu, Y.; Sun, Z.; Li, S.; Liu, H.; Tang, K.; Han, B.; Tao, Y. Genome-wide analysis of the auxin response factors (ARF) gene family in rice (*Oryza sativa*). *Gene* **2007**, *394*, 13–24. [CrossRef]
13. Qiao, L.; Zhang, W.; Li, X.; Zhang, L.; Zhang, X.; Li, X.; Guo, H.; Ren, Y.; Zheng, J.; Chang, Z. Characterization and Expression Patterns of Auxin Response Factors in Wheat. *Front. Plant Sci.* **2018**, *9*, 1395. [CrossRef] [PubMed]
14. Wang, S.; Bai, Y.; Shen, C.; Wu, Y.; Zhang, S.; Jiang, D.; Guilfoyle, T.J.; Chen, M.; Qi, Y. Auxin-related gene families in abiotic stress response in Sorghum bicolor. *Funct. Integr. Genom.* **2010**, *10*, 533–546. [CrossRef]
15. Wang, Y.; Deng, D.; Shi, Y.; Miao, N.; Bian, Y.; Yin, Z. Diversification, phylogeny and evolution of auxin response factor (ARF) family: Insights gained from analyzing maize ARF genes. *Mol. Biol. Rep.* **2012**, *39*, 2401–2415. [CrossRef]
16. Freire-Rios, A.; Tanaka, K.; Crespo, I.; van der Wijk, E.; Sizentsova, Y.; Levitsky, V.; Lindhoud, S.; Fontana, M.; Hohlbein, J.; Boer, D.R.; et al. Architecture of DNA elements mediating ARF transcription factor binding and auxin-responsive gene expression in Arabidopsis. *Proc. Natl. Acad. Sci. USA* **2020**, *117*, 24557–24566. [CrossRef]
17. Rademacher, E.H.; Möller, B.; Lokerse, A.S.; Llavata-Peris, C.I.; van den Berg, W.; Weijers, D. A cellular expression map of the Arabidopsis AUXIN RESPONSE FACTOR gene family. *Plant J. Cell Mol. Biol.* **2011**, *68*, 597–606. [CrossRef]
18. Bargmann, B.O.; Estelle, M. Auxin perception: In the IAA of the beholder. *Physiol. Plant.* **2014**, *151*, 52–61. [CrossRef] [PubMed]
19. Ding, T.; Zhang, F.; Wang, J.; Wang, F.; Liu, J.; Xie, C.; Hu, Y.; Shani, E.; Kong, X.; Ding, Z.; et al. Cell-type action specificity of auxin on Arabidopsis root growth. *Plant J. Cell Mol. Biol.* **2021**, *106*, 928–941. [CrossRef] [PubMed]
20. Simonini, S.; Mas, P.J.; Mas, C.; Østergaard, L.; Hart, D.J. Auxin sensing is a property of an unstructured domain in the Auxin Response Factor ETTIN of Arabidopsis thaliana. *Sci. Rep.* **2018**, *8*, 13563. [CrossRef] [PubMed]
21. Song, H.; Ji, X.; Wang, M.; Li, J.; Wang, X.; Meng, L.; Wei, P.; Xu, H.; Niu, T.; Liu, A. Genome-wide identification and expression analysis of the Dof gene family reveals their involvement in hormone response and abiotic stresses in sunflower (*Helianthus annuus* L.). *Gene* **2024**, *910*, 148336. [CrossRef]
22. Marzi, D.; Brunetti, P.; Saini, S.S.; Yadav, G.; Puglia, G.D.; Dello Ioio, R. Role of transcriptional regulation in auxin-mediated response to abiotic stresses. *Front. Genet.* **2024**, *15*, 1394091. [CrossRef] [PubMed]
23. Tian, Q.; Reed, J.W. Control of auxin-regulated root development by the Arabidopsis thaliana SHY2/IAA3 gene. *Development* **1999**, *126*, 711–721. [CrossRef] [PubMed]
24. Rogg, L.E.; Lasswell, J.; Bartel, B. A gain-of-function mutation in IAA28 suppresses lateral root development. *Plant Cell* **2001**, *13*, 465–480. [CrossRef] [PubMed]
25. Uehara, T.; Okushima, Y.; Mimura, T.; Tasaka, M.; Fukaki, H. Domain II mutations in CRANE/IAA18 suppress lateral root formation and affect shoot development in Arabidopsis thaliana. *Plant Cell Physiol.* **2008**, *49*, 1025–1038. [CrossRef] [PubMed]
26. Liu, M.; Chen, Y.; Chen, Y.; Shin, J.H.; Mila, I.; Audran, C.; Zouine, M.; Pirrello, J.; Bouzayen, M. The tomato Ethylene Response Factor Sl-ERF.B3 integrates ethylene and auxin signaling via direct regulation of Sl-Aux/IAA27. *New Phytol.* **2018**, *219*, 631–640. [CrossRef] [PubMed]
27. Ebstrup, E.; Ansbøl, J.; Paez-Garcia, A.; Culp, H.; Chevalier, J.; Clemmens, P.; Coll, N.S.; Moreno-Risueno, M.A.; Rodriguez, E. NBR1-mediated selective autophagy of ARF7 modulates root branching. *EMBO Rep.* **2024**, *25*, 2571–2591. [CrossRef] [PubMed]
28. Kirolinko, C.; Hobecker, K.; Cueva, M.; Botto, F.; Christ, A.; Niebel, A.; Ariel, F.; Blanco, F.A.; Crespi, M.; Zanetti, M.E. A lateral organ boundaries domain transcription factor acts downstream of the auxin response factor 2 to control nodulation and root architecture in Medicago truncatula. *New Phytol.* **2024**, *242*, 2746–2762. [CrossRef] [PubMed]
29. Jung, H.; Lee, D.K.; Choi, Y.D.; Kim, J.K. OsIAA6, a member of the rice Aux/IAA gene family, is involved in drought tolerance and tiller outgrowth. *Plant Sci. Int. J. Exp. Plant Biol.* **2015**, *236*, 304–312. [CrossRef]
30. He, F.; Gui, L.; Zhang, Y.; Zhu, B.; Zhang, X.; Shen, M.; Wan, F.; Yang, L.; Xiao, J. Validation of reference genes for gene expression analysis in fruit development of *Vaccinium bracteatum* Thunb. using quantitative real-time PCR. *Sci. Rep.* **2022**, *12*, 16946. [CrossRef] [PubMed]
31. Wang, Y.Y.; Pan, Y.B.; Wan, Z.Y.; Li, J.J.; Bao, J.; Zhang, J.S.; Zhang, H. Anti-inflammatory Polyketides from an Endophytic Fungus Chaetomium sp. *UJN-EF006* of *Vaccinium bracteatum*. *Chem. Biodivers.* **2024**, *21*, e202400002. [CrossRef] [PubMed]
32. Oh, D.R.; Choi, C.; Kim, M.J.; Mun, B.Y.; Ko, H.; Oh, K.N.; Jo, A.; Kim, J.Y.; Bae, D. Antidepressant effects of p-coumaric acid isolated from *Vaccinium bracteatum* leaves extract on chronic restraint stress mouse model and antagonism of serotonin 6 receptor in vitro. *Phytomedicine Int. J. Phytother. Phytopharm.* **2023**, *116*, 154871. [CrossRef]
33. Verma, S.; Negi, N.P.; Pareek, S.; Mudgal, G.; Kumar, D. Auxin response factors in plant adaptation to drought and salinity stress. *Physiol. Plant.* **2022**, *174*, e13714. [CrossRef] [PubMed]
34. Chao, M.; Dong, J.; Hu, G.; Li, Y.; Huang, L.; Zhang, J.; Tang, J.; Wang, Q. Phylogeny, gene structures, and expression patterns of the auxin response factor (GhARF2) in upland cotton (*Gossypium hirsutum* L.). *Mol. Biol. Rep.* **2023**, *50*, 1089–1099. [CrossRef] [PubMed]
35. Wang, H.; Huang, H.; Shang, Y.; Song, M.; Ma, H. Identification and characterization of auxin response factor (ARF) family members involved in fig (*Ficus carica* L.) fruit development. *PeerJ* **2022**, *10*, e13798. [CrossRef] [PubMed]
36. Li, W.; Chen, F.; Wang, Y.; Zheng, H.; Yi, Q.; Ren, Y.; Gao, J. Genome-wide identification and functional analysis of ARF transcription factors in Brassica juncea var. *tumida*. *PLoS ONE* **2020**, *15*, e0232039. [CrossRef] [PubMed]
37. Xu, L.; Wang, D.; Liu, S.; Fang, Z.; Su, S.; Guo, C.; Zhao, C.; Tang, Y. Comprehensive Atlas of Wheat (*Triticum aestivum* L.) *AUXIN RESPONSE FACTOR* Expression during Male Reproductive Development and Abiotic Stress. *Front. Plant Sci.* **2020**, *11*, 586144. [CrossRef] [PubMed]

38. Nan, Q.; Qian, D.; Niu, Y.; He, Y.; Tong, S.; Niu, Z.; Ma, J.; Yang, Y.; An, L.; Wan, D.; et al. Plant Actin-Depolymerizing Factors Possess Opposing Biochemical Properties Arising from Key Amino Acid Changes throughout Evolution. *Plant Cell* **2017**, *29*, 395–408. [CrossRef]
39. Gomes, G.L.B.; Scortecci, K.C. Auxin and its role in plant development: Structure, signalling, regulation and response mechanisms. *Plant Biol.* **2021**, *23*, 894–904. [CrossRef]
40. Tiwari, S.B.; Wang, X.J.; Hagen, G.; Guilfoyle, T.J. AUX/IAA proteins are active repressors, and their stability and activity are modulated by auxin. *Plant Cell* **2001**, *13*, 2809–2822. [CrossRef]
41. Kim, J.; Harter, K.; Theologis, A. Protein-protein interactions among the Aux/IAA proteins. *Proc. Natl. Acad. Sci. USA* **1997**, *94*, 11786–11791. [CrossRef] [PubMed]
42. Xie, Q.; Frugis, G.; Colgan, D.; Chua, N.H. Arabidopsis NAC1 transduces auxin signal downstream of TIR1 to promote lateral root development. *Genes. Dev.* **2000**, *14*, 3024–3036. [CrossRef]
43. Casanova-Sáez, R.; Mateo-Bonmatí, E.; Ljung, K. Auxin Metabolism in Plants. *Cold Spring Harb. Perspect. Biol.* **2021**, *13*, a039867. [CrossRef] [PubMed]
44. Roosjen, M.; Paque, S.; Weijers, D. Auxin Response Factors: Output control in auxin biology. *J. Exp. Bot.* **2018**, *69*, 179–188. [CrossRef] [PubMed]
45. Zhang, Y.; Wu, W.; Shen, H.; Yang, L. Genome-wide identification and expression analysis of ARF gene family in embryonic development of Korean pine (*Pinus koraiensis*). *BMC Plant Biol.* **2024**, *24*, 267. [CrossRef] [PubMed]
46. Hu, W.; Zuo, J.; Hou, X.; Yan, Y.; Wei, Y.; Liu, J.; Li, M.; Xu, B.; Jin, Z. The auxin response factor gene family in banana: Genome-wide identification and expression analyses during development, ripening, and abiotic stress. *Front. Plant Sci.* **2015**, *6*, 742. [CrossRef] [PubMed]
47. Finet, C.; Berne-Dedieu, A.; Scutt, C.P.; Marlétaz, F. Evolution of the ARF gene family in land plants: Old domains, new tricks. *Mol. Biol. Evol.* **2013**, *30*, 45–56. [CrossRef] [PubMed]
48. Tinoco-Tafolla, H.A.; López-Hernández, J.; Ortiz-Castro, R.; López-Bucio, J.; Reyes de la Cruz, H.; Campos-García, J.; López-Bucio, J.S. Sucrose supplements modulate the Pseudomonas chlororaphis-Arabidopsis thaliana interaction via decreasing the production of phenazines and enhancing the root auxin response. *J. Plant Physiol.* **2024**, *297*, 154259. [CrossRef] [PubMed]
49. Wang, W.X.; Yang, C.; Xiong, W.; Chen, C.Y.; Li, N. Transcriptome-wide identification of ARF gene family in medicinal plant Polygonatum kingianum and expression analysis of PkARF members in different tissues. *Mol. Biol. Rep.* **2024**, *51*, 648. [CrossRef] [PubMed]
50. Niu, J.; Bi, Q.; Deng, S.; Chen, H.; Yu, H.; Wang, L.; Lin, S. Identification of AUXIN RESPONSE FACTOR gene family from Prunus sibirica and its expression analysis during mesocarp and kernel development. *BMC Plant Biol.* **2018**, *18*, 21. [CrossRef] [PubMed]
51. Wen, J.; Guo, P.; Ke, Y.; Liu, M.; Li, P.; Wu, Y.; Ran, F.; Wang, M.; Li, J.; Du, H. The auxin response factor gene family in allopolyploid Brassica napus. *PLoS ONE* **2019**, *14*, e0214885. [CrossRef] [PubMed]
52. Zhang, Y.; Zeng, Z.; Chen, C.; Li, C.; Xia, R.; Li, J. Genome-wide characterization of the auxin response factor (ARF) gene family of litchi (*Litchi chinensis* Sonn.): Phylogenetic analysis, miRNA regulation and expression changes during fruit abscission. *PeerJ* **2019**, *7*, e6677. [CrossRef] [PubMed]
53. Gan, Z.; Feng, Y.; Wu, T.; Wang, Y.; Xu, X.; Zhang, X.; Han, Z. Downregulation of the auxin transporter gene SlPIN8 results in pollen abortion in tomato. *Plant Mol. Biol.* **2019**, *99*, 561–573. [CrossRef]
54. Vain, T.; Raggi, S.; Ferro, N.; Barange, D.K.; Kieffer, M.; Ma, Q.; Doyle, S.M.; Thelander, M.; Pařízková, B.; Novák, O.; et al. Selective auxin agonists induce specific AUX/IAA protein degradation to modulate plant development. *Proc. Natl. Acad. Sci. USA* **2019**, *116*, 6463–6472. [CrossRef]
55. Vidal, E.A.; Moyano, T.C.; Riveras, E.; Contreras-López, O.; Gutiérrez, R.A. Systems approaches map regulatory networks downstream of the auxin receptor AFB3 in the nitrate response of Arabidopsis thaliana roots. *Proc. Natl. Acad. Sci. USA* **2013**, *110*, 12840–12845. [CrossRef] [PubMed]
56. Cai, Z.; Wang, Y.; Zhu, L.; Tian, Y.; Chen, L.; Sun, Z.; Ullah, I.; Li, X. GmTIR1/GmAFB3-based auxin perception regulated by miR393 modulates soybean nodulation. *New Phytol.* **2017**, *215*, 672–686. [CrossRef] [PubMed]
57. Vojta, P.; Kokáš, F.; Husičková, A.; Grúz, J.; Bergougnoux, V.; Marchetti, C.F.; Jiskrová, E.; Ježilová, E.; Mik, V.; Ikeda, Y.; et al. Whole transcriptome analysis of transgenic barley with altered cytokinin homeostasis and increased tolerance to drought stress. *New Biotechnol.* **2016**, *33 Pt B*, 676–691. [CrossRef]
58. Yang, Z.B.; Geng, X.; He, C.; Zhang, F.; Wang, R.; Horst, W.J.; Ding, Z. TAA1-regulated local auxin biosynthesis in the root-apex transition zone mediates the aluminum-induced inhibition of root growth in Arabidopsis. *Plant Cell* **2014**, *26*, 2889–2904. [CrossRef] [PubMed]
59. Ha, C.V.; Le, D.T.; Nishiyama, R.; Watanabe, Y.; Sulieman, S.; Tran, U.T.; Mochida, K.; Dong, N.V.; Yamaguchi-Shinozaki, K.; Shinozaki, K.; et al. The auxin response factor transcription factor family in soybean: Genome-wide identification and expression analyses during development and water stress. *DNA Res. Int. J. Rapid Publ. Rep. Genes. Genomes* **2013**, *20*, 511–524.
60. Boer, D.R.; Freire-Rios, A.; van den Berg, W.A.; Saaki, T.; Manfield, I.W.; Kepinski, S.; López-Vidrieo, I.; Franco-Zorrilla, J.M.; de Vries, S.C.; Solano, R.; et al. Structural basis for DNA binding specificity by the auxin-dependent ARF transcription factors. *Cell* **2014**, *156*, 577–589. [CrossRef] [PubMed]
61. Fan, J.; Deng, M.; Li, B.; Fan, G. Genome-Wide Identification of the Paulownia fortunei Aux/IAA Gene Family and Its Response to Witches' Broom Caused by Phytoplasma. *Int. J. Mol. Sci.* **2024**, *25*, 2260. [CrossRef] [PubMed]

62. Planchais, S.; Perennes, C.; Glab, N.; Mironov, V.; Inzé, D.; Bergounioux, C. Characterization of cis-acting element involved in cell cycle phase-independent activation of Arath;CycB1;1 transcription and identification of putative regulatory proteins. *Plant Mol. Biol.* **2002**, *50*, 111–127. [CrossRef]
63. Shen, C.; Wang, S.; Bai, Y.; Wu, Y.; Zhang, S.; Chen, M.; Guilfoyle, T.J.; Wu, P.; Qi, Y. Functional analysis of the structural domain of ARF proteins in rice (*Oryza sativa* L.). *J. Exp. Bot.* **2010**, *61*, 3971–3981. [CrossRef] [PubMed]
64. Maraschin Fdos, S.; Memelink, J.; Offringa, R. Auxin-induced, SCF(TIR1)-mediated poly-ubiquitination marks AUX/IAA proteins for degradation. *Plant J. Cell Mol. Biol.* **2009**, *59*, 100–109. [CrossRef]
65. Wang, J.L.; Wang, M.; Zhang, L.; Li, Y.X.; Li, J.J.; Li, Y.Y.; Pu, Z.X.; Li, D.Y.; Liu, X.N.; Guo, W.; et al. WAV E3 ubiquitin ligases mediate degradation of IAA32/34 in the TMK1-mediated auxin signaling pathway during apical hook development. *Proc. Natl. Acad. Sci. USA* **2024**, *121*, e2314353121. [CrossRef] [PubMed]

International Journal of *Molecular Sciences*

MDPI

Article

Predicting the Structure of Enzymes with Metal Cofactors: The Example of [FeFe] Hydrogenases

Simone Botticelli [1,2], Giovanni La Penna [2,3,*], Velia Minicozzi [1,2], Francesco Stellato [1,2], Silvia Morante [1,2], Giancarlo Rossi [1,2,4] and Cecilia Faraloni [5]

1 Department of Physics, University of Roma Tor Vergata, 00133 Rome, Italy; simone.botticelli@roma2.infn.i (S.B.); velia.minicozzi@roma2.infn.it (V.M.); francesco.stellato@roma2.infn.it (F.S.); silvia.morante@roma2.infn.it (S.M.); rossig@roma2.infn.it (G.R.)
2 Section of Roma Tor Vergata, National Institute of Nuclear Physics, 00133 Rome, Italy
3 Institute of Chemistry of Organometallic Compounds, National Research Council, 50019 Florence, Italy
4 Museo Storico della Fisica e Centro Studi e Ricerche E. Fermi, 00184 Rome, Italy
5 Institute of Bioeconomy, National Research Council, 50019 Florence, Italy
* Correspondence: giovanni.lapenna@cnr.it

Abstract: The advent of deep learning algorithms for protein folding opened a new era in the ability of predicting and optimizing the function of proteins once the sequence is known. The task is more intricate when cofactors like metal ions or small ligands are essential to functioning. In this case, the combined use of traditional simulation methods based on interatomic force fields and deep learning predictions is mandatory. We use the example of [FeFe] hydrogenases, enzymes of unicellular algae promising for biotechnology applications to illustrate this situation. [FeFe] hydrogenase is an iron–sulfur protein that catalyzes the chemical reduction of protons dissolved in liquid water into molecular hydrogen as a gas. Hydrogen production efficiency and cell sensitivity to dioxygen are important parameters to optimize the industrial applications of biological hydrogen production. Both parameters are related to the organization of iron–sulfur clusters within protein domains. In this work, we propose possible three-dimensional structures of *Chlorella vulgaris* 211/11P [FeFe] hydrogenase, the sequence of which was extracted from the recently published genome of the given strain. Initial structural models are built using: (i) the deep learning algorithm AlphaFold; (ii) the homology modeling server SwissModel; (iii) a manual construction based on the best known bacterial crystal structure. Missing iron–sulfur clusters are included and microsecond-long molecular dynamics of initial structures embedded into the water solution environment were performed. Multiple-walkers metadynamics was also used to enhance the sampling of structures encompassing both functional and non-functional organizations of iron–sulfur clusters. The resulting structural model provided by deep learning is consistent with functional [FeFe] hydrogenase characterized by peculiar interactions between cofactors and the protein matrix.

Keywords: hydrogenase; molecular modelling; structure prediction; microalgae; photobiological hydrogen production

1. Introduction

The enzyme [FeFe] hydrogenase (Hyd, hereafter) is found in bacteria and microalgae organisms [1]. The enzyme is an efficient catalyst in converting the proton of liquid water into molecular hydrogen as a gas, using either solar energy or chemical reductants [2,3]. Among different types of hydrogenases, the [FeFe] hydrogenase is more efficient in hydrogen production than the other types. Therefore, Hyd is at the core of biological production of H_2, which is a cost-effective and carbon-free strategy to produce fuels [4]. Among the organisms that express the enzyme, *Chlorella vulgaris* 211/11P *Cvu*, is an interesting microalga strain that presents a promising candidate for the production of biofuels at an industrial scale [5], due to its rapid growth in fresh water. Strain selection is fundamental to optimize the biological production of H_2 by microalgae photosynthesis at a reasonable cost [6]. The proton reduction takes place in the so-called H-cluster of Hyd, which is the active site inside the evolutionary preserved core domain of [FeFe] hydrogenases. *Cvu* microalgae were found to produce H_2 gas in relatively easy growing conditions [7]. H-cluster is arranged in a canonical cubane-type iron–sulfur cluster, $[4Fe4S]_H$, which is linked via the thiol group of a cysteine amino acid to a peculiar diiron active site, $[2Fe]_H$. The two iron atoms of $[2Fe]_H$, in proximal (Fe_p) or distal (Fe_d) positions relative to $[4Fe4S]_H$, carry biologically atypical carbon monoxide (CO) and cyanide (CN^-) ligands that stabilize the reduced Fe states. The latter ligand molecules are present in all classes of hydrogenases [8]. The two ligand molecules azadithiolate ($-(SCH_2)_2NH$, *adt* hereafter), and one of the CO molecules bridge Fe_p and Fe_d in $[2Fe]_H$, respectively. The catalytic mechanism likely involves the formation of an intermediate hydride species of the H-cluster [9]. In Figure 1, we display the structure of the Hyd H-cluster of the bacterium *Clostridium pasteurianum* [FeFe] hydrogenase I (*CpI*, hereafter), together with a schematic drawing of its chemical structure.

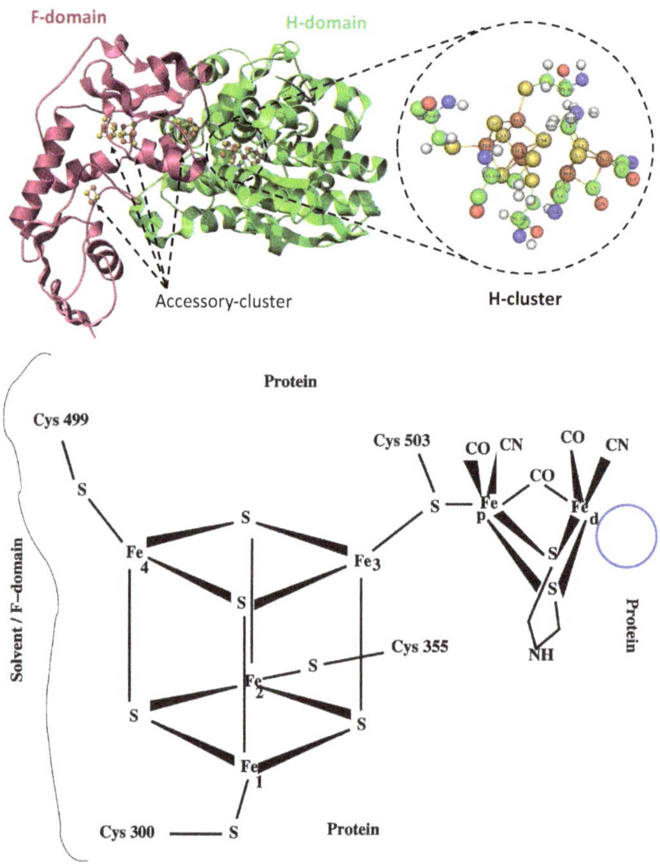

Figure 1. (**Top panel**)—[FeFe] hydrogenase crystal structure of *CpI* in bacteria *Clostridium pasteurianum* (PDB code: 6N59): H-domain and F-domain are represented as green and pink ribbons, respectively, with PyMOL [10]. H-cluster and auxiliary clusters are explicitly indicated. (**Bottom panel**)—A sketch of the chemical structure of the H-cluster. The blue circle identifies the region close to Fe_d accessible to water molecules through channels in the protein matrix.

While the role of the protein domain where the H-cluster is embedded (the H-domain) has been investigated in detail, the structural and functional role of other [FeFe] hydrogenase regions are more elusive, because limited structural information is available. Most [FeFe] hydrogenases possess additional FeS clusters, called F-clusters. The task of these auxiliary (or accessory) clusters is to transfer electrons from external donors to the catalytic site and vice versa. In the case of [FeFe] hydrogenase, where the electron external donor is often reduced ferredoxin, auxiliary clusters allow electron transfer from the protein surface where ferredoxin binds to the active site. The latter is buried into the H-domain to attract electrons, favouring protection of active reduced states from oxidation. Because of the larger solvent accessibility of F-clusters compared to the buried H-cluster, the properties of the former may influence the O_2 sensitivity more significantly than the latter, as O_2 is dissolved in the solution environment. Since the H-cluster environment is very conserved within all [FeFe] hydrogenases, it is reasonable to assume that the auxiliary clusters may differently tune O_2 sensitivity in different species [11]. From this point of view, the eventual structure of the F-domain and its interactions with the H-domain can be important features to possibly increase the resistance of Hyd to O_2 [9].

For completeness, the three-dimensional (3D) structures of three [FeFe] hydrogenases obtained by X-ray diffraction of crystals are displayed in Figure 2. The location of the H-clusters and the auxiliary [4Fe4S] F-clusters ($[4Fe4S]_F$, hereafter) are explicitly indicated.

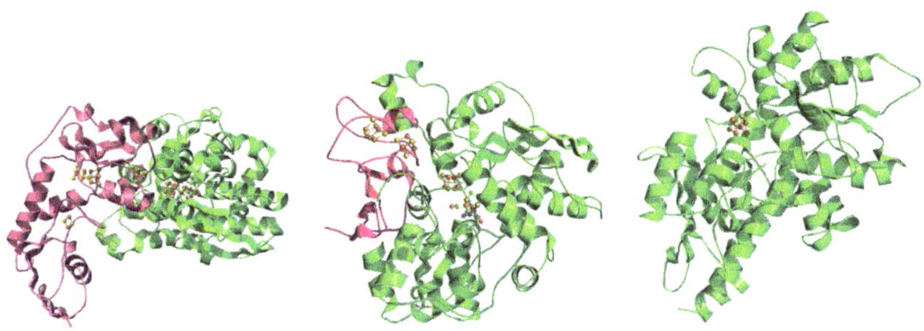

Figure 2. Three structures determined by X-ray crystallography of [FeFe] hydrogenase. (**Left**): *Cp* (HydA1 gene, *Clostridium pasteurianum* (bacteria), PDB 6N59 [12]). (**Middle**): *Dd* (HydAB gene, *Desulfovibrio desulfuricans* (bacteria), PDB 1HFE [13]). (**Right**): *Cr* (HydA1 gene, *Chlamydomonas rheinardtii* (algae), PDB 3LX4 [14]). Secondary motifs assigned with STRIDE [15] are displayed as cartoons with the use of the PyMOL program [10]. FeS clusters are displayed as spheres (S is yellow, Fe is orange). The proteins are not in scale. The size of the proteins decreases from left to right. The images are rescaled to fit into each frame. The F-domain is emphasized in pink color.

The first two structures refer to the bacteria *Clostridium pasteurianum* and *Desulfovibrio desulfuricans* (*Dd*, hereafter), respectively. They are the best characterized enzymes in the literature in terms of structure [2]. These hydrogenases are, depending on metabolites, location, and other cellular conditions, catalysts for either hydrogen uptake or hydrogen production [16]. The *Chlamydomonas rheinardtii* (*Cr*) Hyd structure is the only [FeFe] hydrogenase from microalgae, which has been crystallized, though the crystal lacks the $[2Fe]_H$ subcluster. This hydrogenase enzyme is one of the most efficient catalysts for hydrogen gas production at (temperature and pressure) room conditions. Most of the other *Cr* Hyd structures in the protein databases (see Supplementary Materials) are almost identical to 3LX4. Even the O_2 inactivated HydA1 enzyme (PDB 4R0V [17]) has a structural deviation (measured as RMSD) from 3LX4 of 0.7 Å. The overall structure of the H-domain in *Cr* Hyd is similar to that in *CpI* and *Dd* Hyd. The most interesting difference between *Cr* Hyd and the bacterial Hyd is the absence of a folded F-domain including some auxiliary clusters. *Cr* Hyd is considered as the minimal catalyst of biological H_2 production, though very sensitive to oxidation. Compared to genus *Chlamydomonas*, genus *Chlorella* microalgae can have folded F-domain and auxiliary clusters efficiently bound [11], but no crystal structures are available.

The isolation and purification of [FeFe] hydrogenase from algae is problematic [18] and this is one of the reasons why in vitro characterization of Hyd in species different from bacteria is missing, with a few exceptions though. Hyd is part of the dioxygen sensing mechanism developed by unicellular eukaryotes to adapt their metabolism to environments where the production of dioxygen by photosynthesis is not fully sustainable [16].

In this investigation, we perform the first computational study proposing a structural and functional characterization of hydrogenase in *Cvu* where also the FeS clusters required for hydrogenase functioning are included. To predict possible structures for the protein including the location and binding of iron–sulfur clusters, we start from the protein sequence employing the first two proposals produced by a routinely used tool to predict draft structures, SwissModel [19]. The first proposal is obtained by the deep learning algorithm, AlphaFold [20]. The prediction pipeline contains different moduli encompassing, in the first steps, many of the established homology modeling techniques, but in later steps builds configurations of disordered regions by using conformational maps. This structure was reproduced by us using the original AlphaFold code. The second structure is obtained by homology method (from SwissModel) based on the search of HHblits [21]. This method ignores conformational maps and therefore proposes configurations of disordered regions on the basis of the available experimental structures. Thus, a number of regions cannot be predicted and are consequently missing. Deep-learning and homology methods have been compared [22], concluding that in the absence of sufficient experimental information reliability of results depends on the specific application [23]. The two above mentioned structures are compared to a third manually constructed model, based on the highest resolution crystal structure available for [FeFe] hydrogenase. Interatomic potentials are used to set the initial structures in the environment similar to chloroplast stroma where microalgae hydrogenase is located [24].

Summarizing, we propose structures for *Cvu* Hyd according to the following protocol: (i) we analyze the annotation of the sequence as available in the literature; (ii) we assume the existence of a relation between the protein matrix and two iron–sulfur clusters, the catalytic one in the H-domain and a second accessory one in a hypothetical F-domain; (iii) we use deep-learning and homology methods to predict 3D initial structures; (iv) we perform simulations of a third molecular models where the protein, the two iron–sulfur cofactors, and a simple model of water solution environment are included.

This work extends that performed by Ordonez et al. [25], in that here we fully annotate the gene in *Cvu* which potentially expresses Hyd and we provide a reproducible pipeline designed to obtain 3D structures including functional ones. The annotated *Cvu* sequence is different from any other one previously proposed in computational modeling. The knowledge of the enzyme structure is necessary for further mechanistic studies [12]: isolation and purification of [FeFe] hydrogenase from microalgae, with yields allowing structural in vitro studies, are very difficult [18]. Because of enzyme isolation issues, even the *Cr* enzyme could be characterized in various states only in whole E. coli cells [26,27].

Our construction provides the basis for interpreting possible forthcoming experimental data coming from a possible new *Cvu* hydrogenase endowed with a single potential F-cluster on top of the H-cluster.

2. Results

2.1. Alignment and Gene Annotation

The *Cvu* Hyd sequence was determined by using the procedure described below in Methods (Section 4). In performing the alignment of the whole set of (10,723) *Cvu* nuclear genes to the eight known Hyd sequences (see Figure 3), only one gene displays both E-value smaller than 3×10^{-19} and query coverage larger than 58%. The gene in question is KAI3438965.1. It is interesting to note that for this gene the UNIPROT [28] data-base yields the following annotation: "Iron hydrogenase small subunit domain-containing protein".

The alignments reveal in all cases the potential fingerprint of the H-cluster binding domains composed by 3 conserved motifs [16,29].

```
                 224          280                465    469

P1-3        XXXCXPXXX.XXPCXXKXXX.EXMXCXXGCXXG
L1-3        XTSCCPXWX.XMPCXXKXXE.EXMACXXGCXXG
CpI         FTSCCPGWV.VMPCTSKKFE.EVMACHGGCVNG
Dd_Hyd      FTSCCPGWV.IMPCIAKKYE.EYMACPGGCVCG
Cr_Hyd      FTSCCPGWI.IMPCTRKQSE.EIMACPAGCVGG
Td_Hyd      FTSCCPAWV.VMPCVRKQGE.EVMACPGGCIGG
CspDT_Hyd   FTSCCPAWV.VMPCVRKQGE.EVMACPGGCIGG
Cf_Hyd      FTSCCPAWV.VMPCVRKQGE.EVMACPGGCIGG
Cv_Hyd      FTSCCPAWI.IMPCTAKKHE.EVMACPGGCIGG
Cv_v_Hyd    LASACPGWV.IMPCYDKKLE.EVMACPSGCLNG
Cvu_Hyd     LASACPGWV.VMPCYDKKLE.EVMACPSGCLNG

At_Nar1     LSSACPGWI.VMPCYDKKLE.EIMACPAGCLNG
Mm_Nar1     LASACPGWI.VMPCYDKKLE.EVMACPSGCLNG
Hs_Nar1     LASACPGWI.VMPCYDKKLE.EVMACPSGCLNG
Hs_Narf     LTSACPGWV.VAPCYDKKLE.EVLACAGGCLNG
Sp_Nar1     LSSSCPGWI.LMPCFDKKLE.EVMACPGGCING
Sc_Nar1     LSAVCPGFL.LMPCFDKKLE.EVNACPGACMNG

                P1/L1         P2/L2              P3/L3
```

Figure 3. Hyd fingerprint motifs (P1–3 of Ref. [16], L1–3 of Ref. [29]) in 8 different species (all species are in Supplementary Materials). The last 6 rows are the regions aligned to P1–3 and L1–3 fingerprints in Nar (see main text). Aligned Cys residues are those with numbers (top) as in the *Cvu* sequence. Abbreviations for organisms (see also Section Abbreviations) are Cf, *Chlorella fusca;* CspDT, *Chlorella* species, DT strain (Taiwan); Td, *Tetradesmus obliquus;* At, *Arabidopsis thaliana;* Sc, *Saccharomyces cerevisiae;* Sp, *Schizosaccharomyces pombe;* Hs, *Homo sapiens;* Mm, *Mus musculus. Cv_Hyd* and *Cv_v_Hyd* are the proteins expressed by gene AEA34989.1 and XP_005843810, respectively, in *Cv.*

Although in Ref. [29] only the most studied Hyd are shown and compared, the full list of well-conserved motifs is available and can be found in Refs. [16,30]. There are evident changes to sequences that have been identified in P1-3 motifs in the literature. In particular, the TSC sequence in the P1 motif of *Cvu* Hyd is not present in the known Hyd sequences. The P1 sequence change is, however, also observed in one *Cv* gene (XP_005843810.1), that is the protein sequence closest to that analyzed in this work and that bears the same annotation. The most important residues are the conserved Cys 224, 280, 465, 469 in *Cvu* Hyd sequence (red in Figure 3), all binding the H-cluster. Apart from not fully conserved residues, the alignments we have performed show that the sequence extracted from the whole *Cvu* genome represents, especially in terms of the position of Cys residues, a candidate for the FeS cluster ligand, as it is required by [FeFe] hydrogenase. In Ref. [16] (Figure 12) and first line in Figure 3 the second Cys in P1 is not always present, while it is a marker in L1 [29].

By using the available hydrogenase identifier tool [31], the identified *Cvu* gene is not classified as a hydrogenase and the same is true for the similar sequence of *Cv* (XP_005843810.1). Conversely, the *Cv* gene AEA34989.1 (see below) is identified as hydrogenase of class A. It must be noticed that in the sequence alignment of all the *Cvu* genes, no other genes display similarity with any of the existing hydrogenases, including gene AEA34989.1 of *Cv*. Since both species produce hydrogen, the whole genome search excludes that other genes can code for a possible hydrogenase in *Cvu*, differently from *Cv*.

It is interesting to notice the fairly good alignment, in fingerprint regions, between the KAI3438965.1 *Cvu* gene and the genes expressing the nuclear prelamin A recognition factor (Nar) in eukaryotic organisms [32] (see the six bottom lines in Figure 3). The Nar class of proteins has been found as a possible evidence of a "memory" of ancient Hyd in aerobic eukaryotes [16]. The sequence variability between the Hyd and Nar classes is larger in the direction of N-terminus: P1 looks more variable than P3 and indeed only Nar proteins show the ASA sequence in place of TSC. The implications of the similarity of the addressed *Cvu* gene to Nar genes concerning the ability of binding H- and F-clusters are described in more details in Section 3.

Since there are no other genes among the *Cvu* candidates for H-cluster binding while the hydrogenase activity was demonstrated in *Cvu* cultures, we are led to conclude that the hydrogenase sequence of *Cvu* is gene KAI3438965.1. Naturally, we cannot exclude that the expressed Hyd protein is eventually processed when introduced in chloroplasts and there assembled into the $[2Fe]_H$ component of H-cluster.

By inspecting the global alignments (see in Supplementary Materials, the folder Blas_Clustal_Alignments) we observe that also the Cys residues in the N-termini of *Cvu* Hyd are aligned to the species where Cys residues are present in the N-terminal region, namely *CpI, Dd* Hyd, *Cv* Hyd. In Figure 4, we display, as an example, the alignment between *Cvu* and *Cv* Hyd (gene AEA34989.1), the latter belonging to class A Hyd [31] (see above). Previous studies about the recruitment of auxiliary F-clusters along with evolutionary analysis arguments showed that Hyd in species without these aligned Cys residues in the N-terminal region (like *Cr, Cf, CspDT*, and *Td*, does not bind auxiliary F-clusters [29]. Therefore, our alignment, that is common to *CpI* and *Dd* Hyd, shows that an auxiliary conserved F-cluster can be present in *Cvu* Hyd as well as in the other Hyd sequences where an F-domain has been identified (see the crystal structures of *CpI* and *Dd* Hyd).

```
>C.Variabilis
Sequence ID: Query_106816 Length: 717
Range 1: 241 to 670

Score:146 bits(338), Expect:7e-39,
Method:Compositional matrix adjust.,
Identities:137/493(28%), Positives:164/493(33%), Gaps:82/493(16%)

|

Query  62   DEAIKVSLMDCLACSGCVT-SAEAVLLQHQSAGELLQRLADPDWTVVVSLSPQSIAALAA  120
+EA   S   C  C  C     +H    + L  L +      VV   P
Sbjct  241  EEALDHS--KCIECGQCSSVCPVGAIVEHSEWRQVLDALENKQKVMVVQTAPSVRVSIGE  298

Query  121  VHRLSSAEC-AARLTAFLRHLGVAAVFDIAAARQVALAEAAREFLQR----YRSSRGGST  175
L         + A  R LG   VFD              E   E LQR      R
Sbjct  299  ELGLAPGTVETGQMVAAQRALGFDYVFDSDFSADLTIMEEGTELLQRLGAAWRAETAAQD  358

Query  176  AEAAAAAGQDAAGGGDDMDVDRPKPSSDGHRNAVGGGSPGPLPMLASACPGWVCYAEKTH  235
A A   A     G  +             GH           PGPLPM  S CP W+    EK
Sbjct  359  AAAGSWAAAKQGHGEGE---------AHGH-------APGPLPMFISCCPAWINLVEKSY  402

Query  236  GDHVLPYIATGKSPQAVLGTLVKRQWCAAAGLAPDRVYHCSVMPCYDKKLEGARDDFWLP  295
+    P     KSPQ  +G  VK  W     GL P+ V     +MPC  KK E   R +F
Sbjct  403  PE-LIPHLSSCKSPQMMMGAVVKHYWAKKKGLKPEDVCLVGIMPCTAKKHETERKEFR-N  460

Query  296  GTSVPETDCVLATTELQELLEQRQADLHSLQGIPFDAILPAAAAAAAAPSGTRASSSGGS  355
+ D V  T E   +L +     + SL      FD  L  A  AAA       T
Sbjct  461  ENGAYDCDYVITTREFGHMLRHKKIPMPSLKPEEFDNPLGEATGAAALFGAT--------  512

Query  356  SAQHDTVQLPTSSSSGSGSGVTATASSGSGGYLEHVFRSAARELFGLELPPGPLPVKVGR  415
GG +E   R A      G  LP   L V   R
Sbjct  513  ----------------------------GGVMEAAIRTAYEIAAGEPLP--KLEVEAVR  541

Query  416  NAD-LRE--LSLEAPDGTV--------LRFAAAYGFRNIQGLMRKVKLGRCEYDYVEVMA  464
E  L L A D T              R A A G  N   L+  +  G   YD+VEVMA
Sbjct  542  GVKGVKEATLTLPANDTTLKAGVAGKEIRVAVASGIGNARHLLQRIQAGEAHYDFVEVMA  601

Query  465  CPSGCLNGGGQPKPTAKQTPAQLLEQLEELYAAGGIAA--RPEADPALQQLYMSFVQGEP  522
CP GC  GGGQPK      +  P   L  +   Y      A      +P  +  Y  F  G+P
Sbjct  602  CPGGCIGGGGQPK---THDPDAVLKRMGAIYQVDKSLALRKSHENPSIHKIYAEFL-GQP  657

Query  523  GSQAAHQLMHTHY  535
G +  H L+HTHY
Sbjct  658  GGELSHKLLHTHY  670
```

Figure 4. Sequence alignment between *Cv* Hyd (Sbjct) and *Cvu* Hyd (Query). Motifs P1, P2, and P3 (see Figure 3) are enclosed within red, green, and blue boxes, respectively. The Cys-rich motif in the N-terminus is enclosed within black boxes.

In the following analysis, we assign Cys 72, 75 and 78 in *Cvu* Hyd as ligands for an auxiliary F-cluster. To complete the binding of the auxiliary 4Fe4S cluster, we assign Cys 21 as a fourth ligand. In Nar proteins, the 4Fe4S binding pattern by four Cys residues is found to be $Cx_{38}Cx_2Cx_2C$ [33], while in *Cvu* we find $Cx_{50}Cx_2Cx_2C$ (x denotes a generic aminoacid). As shown by Nar proteins, the actual cooperation between the first Cys residue and the other three Cys residues depends on the structure of the long protein segment (38 residues) connecting the two regions. The 4-Cys binding of the auxiliary F-cluster is here adopted on the basis of the homology with the Nar proteins. The assessment of its stability is the major aim of the following study.

In summary, by using the sequence alignment of the KAI3438965.1 *Cvu* gene discussed above, we observe that the [FeFe] hydrogenase sequences feature the following three domains: the N-terminal F-domain, with at least one possible F-cluster; the central well-structured and conserved H-domain, containing the active site (H-cluster) bound by Cys residues in P1, P2, and P3 motifs; an unstructured C-terminal short segment. These domains are characterized in terms of residue numbers in Table 1.

Table 1. Summary of protein domains conserved among Hyd of different species. The numbers represent the location of the Cys residues that bind FeS clusters. In the last line, we show our hypothesis for the location of the *Cvu* F- and H-domain and C-terminus.

Species	F-domain Aux. clusters	H-domain H-cluster	C-terminus
CpI	1–209 157, 190, 193, 196 147, 150, 153, 200 94, 98, 101, 107 36, 46, 49, 64	210–568 300, 355, 499, 503	569–574
Dd Hyd	1–86 45, 66, 69, 72 35, 38, 41, 76	87–392 178, 234, 378, 372	393–421
Cr Hyd	1–64	65–487 170, 225, 417, 421	488–497
Cvu Hyd	1–88 21, 72, 75, 78	89–530 224, 280, 465, 469	531–549

In Table 1, we report the location along the amino acid sequence of the Cys residues that bind the FeS clusters in the *Cvu* F- and H-domain and C-terminus.

To confirm evidence of the *Cvu* gene KAI3438965.1 annotation to a hydrogenase function, we have performed a detailed analysis of the structural protein stability as described in the following section.

2.2. Structure Prediction

As is commonly done in similar cases, we inserted the KAI3438965.1 sequence in the SwissModel server (see Section 4 for details) and decided to continue the sequence structural study taking into further consideration the first two of the proposed structures produced in this search.

To reproduce the 3D structure of the first proposal of *Cvu* Hyd obtained by SwissModel, we used the deep learning algorithm (AlphaFold, AF hereafter, see Section 4).

Before going into the details of the 3D AF predictions, it is convenient to first describe the distribution of structural information among the sequence domains detailed in the previous section. In all available Hyd crystal structures, namely (*CpI* 6N59, *Dd* 1HFE and *Cr* 3LX4), the H-domain scaffold is made by 11 α-helical secondary domains embedded into a globule with a further small β-sheet made of 7 β strands in an antiparallel configuration. We call this arrangement of secondary motifs the "scaffold" of the H-domain. In Table S1 of Supplementary Materials, we indicate the residues belonging to the scaffold shared by *Cvu* Hyd and the three reference available crystal structures mentioned above.

We first notice that the formation of disulfide bridges between Cys residues has never been observed in hydrogenase structures obtained in anaerobic in vitro working conditions. This is confirmed by the AF predictions with no surprise as they are obtained from the available structural information. In the chloroplast, stroma hydrogenase is functional when in a reducing environment, while in the presence of dioxygen it is rapidly degraded. Thus, in the following structural refinement we shall assume that no disulfide bridges are present in the models in study.

In the AF output file "pdb_hits.hrr" (see Supplementary Materials AF_alignments/msas directory), secondary structure motifs of crystal structures coming from database search and AF predicted structures are compared using the DSSP algorithm [34]. AF then calculates the predicted alignment error (PAE) matrix [35]. The matrix provides AF's expected position error between pairs residues and it is a measure of the confidence in the relative position of two residues within the predicted structure. As displayed in Figure S2, PAE matrix for *Cvu* prediction shows large errors for three inter-domain, the hypotetical F-domain and the two extended loops visible in Figure S1 (left panel). This fact indicates that the relative positions and/or orientations of at least two segments connecting two ordered regions in the 3D structure are poorly determined. Therefore, F-domain and the disordered segments must be refined by, for instance, molecular simulations to reduce this uncertainty.

In Figure S3, we display the behavior of the structural prediction (plDDT) accuracy index used by AF along with the protein sequence in the case of the highest-score AF structure prediction (see Section 4 for details). The other predicted structures do not differ significantly in the H-domain secondary structure. The structural difference is limited to the same regions identified by PAE, in particular in the terminal regions, i.e., the N-terminus that is the possible F-domain, and the extended loops connecting secondary domains lying within the H-domain (see Figure S1, left panel).

The scaffold of the H-domain (displayed as top-right horizontal bars in the figure) is entirely made by residues with high plDDT index (blue points). The long unstructured loops that connect secondary motifs within the H-domain (residues Gly 173-Gly 212 and Ala 338-Ser 381) are constructed by AF uniquely to connect regions where structural information is high. On the other hand, the F-domain is characterized by a partially emerging secondary structure. Indeed, a short α-helix (residues Ser 81-His 89) is found by the STRIDE algorithm at the interface between the F- and the H-domain (Figure S1, left panel). Interestingly, an apparently rigid turn encompasses residues Leu 69-Asp 71 with the consequence that Cys 72, 75, 78 become close in space because of the constraint imposed by these secondary motifs. This information comes from the available crystal structures that AF elaborates, taking into account the presence of the auxiliary cluster in *CpI* that is proximal to its H-cluster. In order to better quantify the deviation of the structural prediction with respect to known available 3D structures (three crystal structures and one recent electron microscopy based model), in Table 2 we display the root-mean square deviation (RMSD, see Methods) of heavy backbone atoms belonging to different sets of aligned residues.

Table 2. RMSD (Å) of heavy backbone atoms (N, Cα, C, O), belonging to different sets of aligned residues. The left value is for the whole H-domain scaffold, middle is for the 11 helix-bundle, right is for the 7 β-strands. *Tm* is Thermotoga maritima. **1** indicates here the initial configuration generated by AlphaFold.

Reference / **Target**	***CpI***	***Dd* Hyd**	***Cr* Hyd**	***Tm* HydABC**	***Cvu* Hyd 1**
CpI	0				
Dd Hyd	1.7/1.8/0.8	0			
Cr Hyd	3.2/3.6/0.7	3.0/3.4/0.9	0		
Tm HydABC	2.2/2.3/1.6	2.3/2.4/1.6	3.7/4.2/1.5	0	
Cvu Hyd **1**	22.2/4.5/2.0	22.2/4.2/2.2	22.8/5.4/1.7	22.4/4.8/2.5	0

We measured the structural deviation among two portions of the scaffold: the 11-helices bundle and the 7 β-strands. In all cases, the deviation is measured by comparing the *Cvu* AF prediction with three reference known crystal structures and one recent cryo electron-microscopy (cEM) structure. These are *CpI*, *Dd* Hyd, *Cr* Hyd, and *Tm* HydABC. The latter is one of the chains (PDB 7P92) reported for the *Thermotoga maritima* HydABC electron bifurcating [FeFe] hydrogenase, indicated as HydABC [36]. The lower values of RMSD obtained for the β-strands compared to the values coming from the α-helices points to a more rigid structure of the H-domain portion formed by the seven β-strands. This analysis suggests that the H-domain, independently of the hydrogenase performance, can be divided in two sub-domains: one made by the α helices; the other by the two β sheets. The two domains will be used as references to describe the equilibration of the *Cvu* Hyd model into the solution environment modeled in the following simulations. As expected, after energy minimization (see below), AF predictions of the H-domain structure differ mainly in the two large extended loops.

In the following, we introduce labels that will be used in the subsequent structure refinement. Notice that this step is performed once the FeS clusters are explicitly added to initial structural models and atomistic simulations are performed. We denoted the simulation starting from the AF construction described above as simulation **1**.

A second simulation (**2**, hereafter) was started from the F-domain proposed by AF combined with the H-domain proposed by SwissModel. The H-domain is present, in the SwissModel second proposal, in the form of a partially oxidized HydA1 protein of *Cr* (PDB 4R0V [17]). This initial configuration displays the following RMSD values (in Å see Table 2) with respect to the sequence in parenthesis: 4.8/4.2/5.8 (*CpI*); 4.7/4.1/5.8 (*Dd*); 3.7/2.3/5.7 (*Cr*); 5.0/4.5/5.6 (*Tm*); 22.8/5.6/6.0 (*Cvu* **1**). It can be noticed that the deviation of the entire scaffold is smaller than in AlphaFold proposal, while the deviation of the β-sheet is larger. Indeed, the first β-strand (b1, see Table S1 in Supplementary Materials) does not get formed at the beginning of simulation **2**.

The further simulation where the F-domain has the initial backbone structure of *CpI* was denoted as simulation **3** (see Section 4 for details). In this case, the initial RMSD (see Table 2) with respect to the reference known structures is the same as in **1**.

2.3. Structure Refinement

In view of the time evolution of several structural parameters like gyration radius, solvent-accessible surface area (SASA), distance between FeS clusters (see Figure S5 and related discussion in Supplementary Materials), observed in molecular dynamics (MD) we decided to compare structural parameters averaging over the last 200 ns of the simulated MD trajectories. Simulations show that the initial configurations require several hundreds ns to settle down in the water environment. Reduction of fluctuations and the stabilization of the monitored structural parameters after 800 ns of simulation are the result of the equilibration of different initial configurations of the system.

The small structural deviation of FeS clusters in all simulations indicates that they all keep a structure functional to hydrogenase activity (see Supplementary Materials). One A crucial step underlying the hydrogenase activity is the proton reduction occurring at the Fe_d site. To probe the proton access to this site, we measured the accessibility of solvent water molecules to Fe atoms. The radial distribution function, $g(r)$ (Figure 5) of Fe-Ow pairs shows that the distal Fe_d atom is accessible to water in simulations **1** and **3**, while Fe_p is less accessible than Fe_d in all simulations (left panels). This is a consequence of the free valence left on Fe_d in all models. However, the more compact H-domain used as initial configuration in simulation **2** displays a less accessible Fe_d atom, because of the relatively smaller number of water molecules in the Fe first-coordination shell ($r < 0.4$ nm). Fe atoms of $[4Fe4S]_H$ are buried, while Fe atoms in $[4Fe4S]_F$ are more accessible to water molecules (right panels). Furthermore, Fe atoms in $[4Fe4S]_H$ clusters are significantly more accessible in the first coordination sphere in simulation **2** compared to **1** and **3**. The water molecules accessing Fe_d are exchanged by water molecules at larger distances, showing that water molecules are not trapped into protein cages because of the bias induced by the initial model construction.

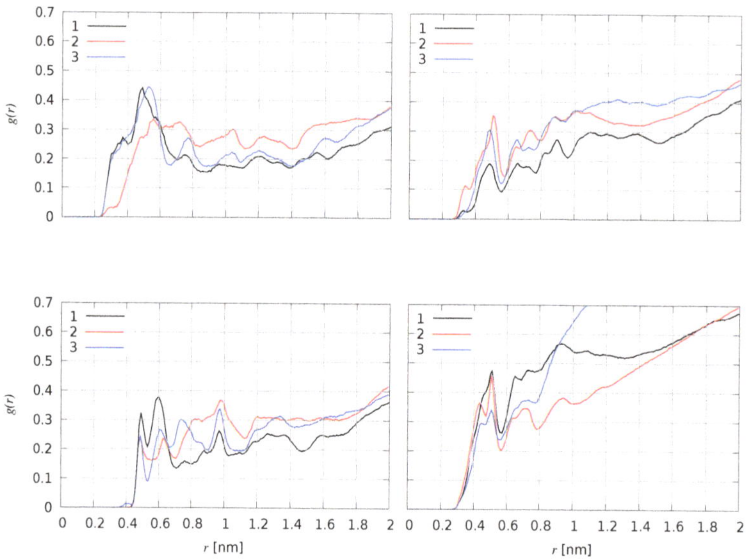

Figure 5. Radial distribution function of Fe-Ow pairs, with Ow oxygen atoms belonging to water molecules and Fe belonging to different FeS clusters. Black curves: simulation **1**. Red curves: simulation **2**. Blue curves: simulation **3**. (**Top-left**): Fe is Fe_d (distal Fe atom in $[2Fe]_H$). (**Bottom-left**): Fe is Fe_p (proximal Fe atom in $[2Fe]_H$). (**Top-right**): all Fe atoms belonging to $[4Fe4S]_H$. (**Bottom-right**): all Fe atoms belonging to $[4Fe4S]_F$. Averages are computed over the last 200 ns.

Interestingly, in all simulations we found that the H atoms of water molecules approaching Fe_d are located at a distance smaller (2.5 Å) than the corresponding O atom (3.5 Å, see the leftmost density in $g(r)$ in Figure 5). The probabilities of such water configuration are 0.07, 0.04, and 0.11 for simulations **1**, **2** and **3**, respectively. This means that the water molecule approaching Fe_d is often oriented with an H atom towards Fe_d, the site of H reduction. In Figure 6, we show a representative configuration of the F- and H-clusters after a simulation time of about 800 ns of model **1**. Figure 6 shows that the water molecule approaching Fe_d atom has the orientation expected for an efficient proton reduction [37]. In fact, one proton is at 2.1 Å from Fe_d and forms a hydrogen bond with N(*adt*) (2.2 Å). The proton of NH(*adt*) is also at a pretty small distance from Fe_d, namely 2.2 ± 0.5 Å in (**1**) and 2.5 ± 0.6 Å in (**2**).

The average distances between the $[4Fe4S]_F$ and $[4Fe4S]_H$ cluster centers of mass (see also Figure S7 in Supplementary Materials are: 17.3±0.6 Å in (**1**); 34.9±0.8 Å in (**2**); 15.1±0.9 Å in (**3**). In simulations **1** and **2**, the average distance is only slightly above the upper range of distances where electron transfer between the two clusters can occur in proteins [38,39]. However, we should consider this distance remarkably small for simulations **1** and **2** considering the fact that we started from an extended configuration of the F-domain produced by AF and SwissModel, respectively. On the other hand, the large distance observed in simulation **3** points to structural constraints in the F-domain that hinder the approach of the two clusters (see below). As discussed in a recent work [37], active [FeFe] hydrogenases, with the exception of *Cr*, show that the $[4Fe4S]_H$-$[4Fe4S]_F$ distance (measured as the distance between the centers of mass) is in the range 11–14 Å. The comparison of simulations shows that simulations **1** and **2** are in better agreement with the above observations.

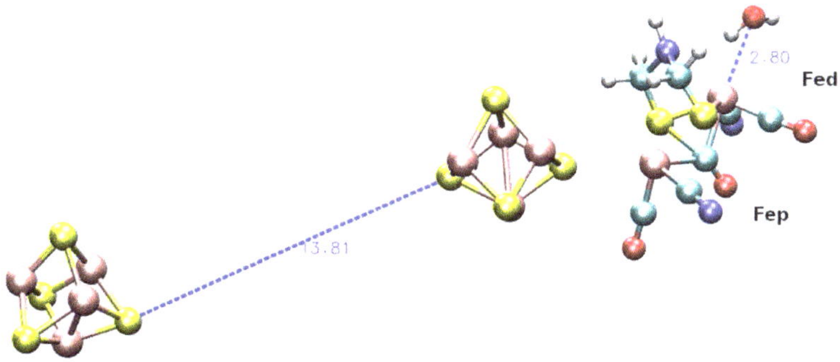

Figure 6. F-cluster and H-cluster after 800 ns in simulation **1**. FeS clusters are displayed as spheres, with Fe in pink and S in yellow. C is cyano, N blue, O red, H white. The atomic dimension is given by the van der Waals radius of each atom. Blue dashed lines display the distance vectors with the moduli as numbers in Å.

The difference between the simulations is to be ascribed to different model constructions as most of the changes in distance develop during the first 300 ns (see the time evolution of the structural parameters in Supplementary Materials).

The above observations regarding the simulation **1** can be summazized by saying that the collected simulation data are consistent with: (i) the mechanism of proton reduction at Fe_d; (ii) the protection to oxidation of $[4Fe4S]_H$; (iii) a good accessibility of the F-cluster which interacts with reductant species coming from the protein environment.

In the following, we shall describe interactions relevant to stabilize reactive configurations. In *Cvu* Hyd the first three residues of the protein motif facing Fe_d (motif P1, see Figure 3) are not conserved among the other [FeFe] hydrogenases. The ASA sequence in *Cvu* replaces TSC of all the other Hyd sequences. Ala 223 is aligned to Cys 367 in the [FeFe] hydrogenase of *Clostridium beijerinkii* (CbA5H, PDB code 6TTL [40,41]). In the crystal structure of the latter, Sγ of Cys 367 is found at a small distance from Fe_d (3.1 Å). In our simulations, the atoms contributing to hinder the access of water to Fe_d are: the NH group of *adt* ligand (in all simulations); the methyl group of Ala 223 (Cβ, in all simulations); the guanidinium group of Ala 152 (Nη) (in simulations **1** and **2**).

Cβ(Ala 223) and N(*adt*) show similar behavior in simulations **1** and **2**, as they remain stable at distances 4.3 $\pm$ 0.3 and 2.5 $\pm$ 0.1 Å from Fe_d, respectively. Nη(Arg 152), conversely, is closer to Fe_d in **2** (3.9 $\pm$ 0.8 Å) than in **1** (7.2 $\pm$ 1.0 Å), thus explaining the lower accessibility to water of Fe_d in **2** than in **1** (Figure 5, top-left). Therefore, we find that the Ala 223 and Arg 152 sidechains can act similarly to Cys 367 in the case of CbA5H in limiting water accessibility to Fe_d. As a consequence of this Fe_d hindering, the reduced state of the H-cluster is partially protected from water soluble O_2 accessibility. The latter can initiate the H-cluster oxidation, as suggested by previous experimental results [42].

The root-mean square deviation displayed in Table S3 (Supplementary Materials shows that the relaxation of the disordered loops present in all simulations (never included in RMSD calculation) affects more significantly the interactions between the H- and F-domains than the interactions between the structural elements within the H-domain. The simulations (**1** and **3**) were started from the H-domain taken from AlphaFold. They display low RMSD values for both the 11-helix bundle and the two β-sheets, including the case where the reference structure is *Cr*. Simulation **2**, started from an oxidized form of HydA1 in *Cr* (PDB 4R0V [17]), leads to structures different from the available mature crystal structures mainly in the organization of the 7 β strands. In all simulations, the H-domain scaffold does not get distorted in any significant way, as the relative RMSD relative to the initial configurations remains always smaller than 4 Å.

In agreement with the results of the SASA analysis of the different domains and sub-domains (see related discussion in Supplementary Materials), the collected RMSD values indicate that in simulations **1** and **3** structural changes of the initial structures occurs while keeping the protein scaffold almost rigid. Instead N-terminal (F-domain) and loops contribute for the largest part to structural relaxation. However, the whole protein backbone and the H-domain scaffold change more significantly when the F-domain is more structured. This occurs in simulation **3** immediately at the beginning of it and in simulation **2** more slowly during the whole time evolution.

Fluctuations are concentrated on the first few residues of the N-terminus and on the long loop connecting helices h3 and h4 (Arg 168–Cys 224). This loop encompasses a hydrophylic region and the largest fluctuation is around Ala 178, belonging to a 5-Ala segment (Ala 178–Ala 182) in the middle of the loop. Fluctuations are larger in simulation **3** than in **1**, showing a lower stability of the configuration reached after the first 800 ns by **3** compared to that reached in a similar situation by **1**. In particular, the N-terminus is fluctuating even at Cys residues involved in the binding of the F-cluster (Table 1), with a significant stress around Fe atoms. The loop around residue 351 (h7–h8) is also fluctuating. Indeed, the major difference of simulation **2** compared to simulation **1** is in the behaviour of the long h7–h8 loop. In **2**, the reduction of fluctuations in this region increases the fluctuations in the N-terminus. The different behavior of the h3–h4 loop, h7–h8 loop and N-terminus in simulations **1** and **2** appears to be an important marker of mutual interactions between the H- and F-domains. These interactions can be visualized comparing the last configurations obtained after 1 μs as shown in Figure 7.

It can be observed that in simulation **3** the atoms of the $[Fe4S4]_H$ cluster are left accessible to water (bottom panel), while in the other two simulations water accessibility is hindered by the F-domain. Focusing on the position of the long h3–h4 loop (in green) and C-terminus (in red), one notices that in simulation **1** the F-domain is wrapped around the assembly of the loop and the C-terminus (top-left). This simulation tends to organize N- and C-termini around the long h3–h4 loop. We also notice that the latter is largely embedded into the H-domain, reaching the active site, with Ala 223, that is very close to Fe_d, being the final part of this long loop. In simulation **2**, the N-terminus is more independent by the behavior of the H-domain, and it is less spread over the H-domain and C-terminus (top-right). This condition leaves the N-terminus able to better fold into a nascent β-strand (the yellow ribbons). In simulation **3**, the wrapping of the N-terminal F-domain is diverted, with respect to the case of **1** and **2**, towards other regions of the H-domain. The different disordered segments connecting the structural elements of the H-domain can act as hairpoons towards what, in our case, is the unique external body present in the H-domain environment, i.e., the F-domain.

Summarizing, the comparison between the three simulations indicate that the seven β-strands, organized in short β-sheets, constitute the most rigid portion of the H-domain scaffold. Around this scaffold, at least four disordered regions are arranged: the two terminal regions and two long loops, h3–h4 and h7–h8. We observed that the N-terminus interacts with the H-domain and that, depending on the settling of the H-domain, an F-domain either emerges as β-rich (simulation **2**), tends to release secondary motifs (simulation **1**) or keeps a helical content (simulation **3**). These features indicate inter-domain cross-talk mediated by the disordered regions in the H-domain. From simulation **3**, we can exclude that the F-domain has the structure of *CpI* because this structure provides a non-functional Hyd, with the two clusters at a large distance one from the other. In the other two cases, we observe that a possible binding of the auxiliary F-cluster is consistent with a functional hydrogenase.

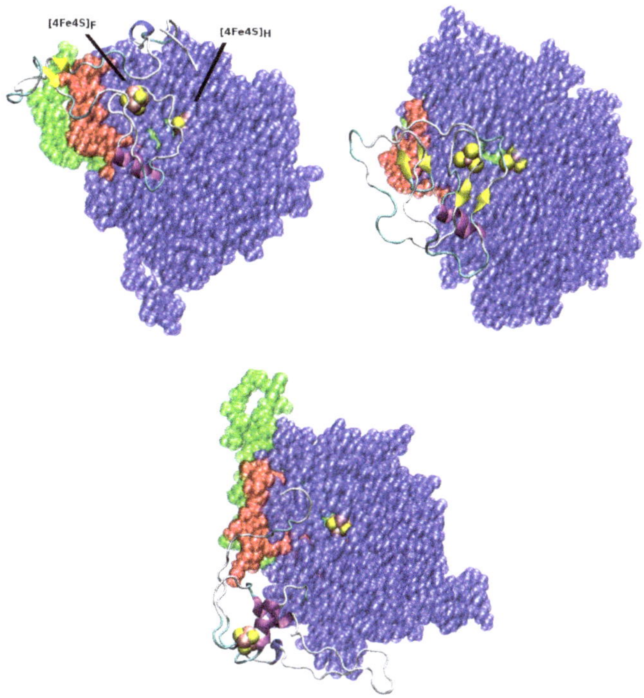

Figure 7. *Cvu* Hyd after 1 μs of MD simulation. (**Top-left**): simulation **1**. (**Top-right**): simulation **2**. (**Bottom**): simulation **3**. The H-domain is displayed as a set of blue transparent spheres. The disordered h3-h4 loop (residues Arg 168–Cys 224) is in green. The C-terminus (residues Met 531–Trp 549) is in red. The F-domain is shown as a cartoon representing the secondary motifs calculated by STRIDE. FeS clusters are displayed as spheres, with Fe in pink and S in yellow. The atomic dimension is given by the van der Waals radius of each atom. Residues His 89-Leu 530 (H-domain) have been translated and rotated in order to minimize the magnitude of the relative RMSD of heavy atom backbone of the three structures.

2.4. Enhancing Statistics

The possibility for the two FeS clusters to be pushed at short distances in simulation **1** indicates the presence of interactions eventually favouring the approach of the F-cluster towards the H-cluster. These interactions seem to emerge when the F-domain is found in a configuration different from that exhibited in the *CpI* hydrogenase.

To sample protein configurations across a broad range of values of the distance between the two FeS clusters, we subjected the system to metadynamics simulations using as an external bias a suitable function of the cluster–cluster distance. By using this method, it is possible to compare the propensity of the two FeS clusters to approach when the F-domain is initially only marginally folded by deep learning method (simulation **1**) to the situation where the F-domain was initially folded as in *CpI* (simulation **3**).

In Figure 8, we compare the free energy as a function of the d distance between the two FeS clusters (see Section 4 for details) calculated by using 100 walkers starting with AF (metadynamics **1**) to that calculated by using 100 walkers starting with *CpI*-like configurations (metadynamics **3**). The free energy as a function of the distance between N-terminus and the H-cluster of *Cr* is also shown for comparison.

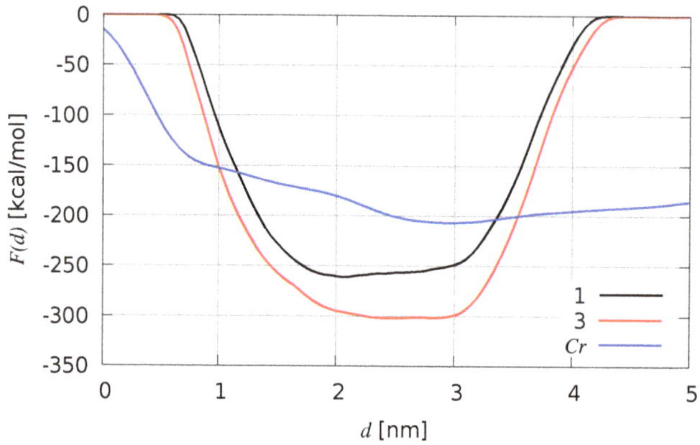

Figure 8. Free energy (F) as a function of the distance between FeS clusters' centers of mass (d) for *Cvu* Hyd, after bias build-up convergence in multiple-walkers metadynamics. The same function of the distance between N-terminus and H-cluster in *Cr* is shown for comparison. Black curve: metadynamics **1**. Red curve: metadynamics **3**; Blue curve: metadynamics of *Cr*.

The comparison between the two *Cvu* curves clearly shows that in the absence of the *CpI* folding of the F-domain (like in simulation **1**), the minimum of F is at about $d = 2$ nm, while a flat profile around d =2.5 nm is displayed by metadynamics **3**. This comparison confirms that the initial F-domain folding hinders the approach between the two clusters, requiring an extra work of about 50 kcal/mol in order to change the distance from 3 nm to the lowest d value of about 0.75 nm (obtained in both metadynamics). Since the short N-terminus of *Cr* allows a larger flexibility compared to *Cvu*, the free energy of *Cr* is flatter. But interestingly a free energy minimum at about d = 3 nm is visible and a work only $\sim$50 kcal/mol lower than metadynamics **1** of *Cvu* is measured as a result of the wrapping of the N-terminus towards the H-cluster in *Cr*.

Some of the structural parameters that have been studied in the equilibrium MD described in the previous sections are discussed below. In Table 3, we display the average values of the scaffold backbone RMSD (using the PDB structures as reference) and of the molecular compactness. Molecular compactness is measured as the ratio R between the whole protein SASA and the sum of the SASA of bare F- and H-domains. The two domains were identified in Table 1 and "bare" means that each SASA is computed ignoring the presence of all other atoms. This quantity tends to unit when the two domains do not overlap while it is lower than unit when part of the domains is excluded from the solvent by the presence of the other domain. In the case of hydrogenases, R can be slightly larger than unit because of the short independent C-terminus. Average quantities are extracted in different windows of the quantity d sampled by metadynamics. The d-windows are: $0.75 \leq d \leq 1.25$ (w1); $1.75 \leq d \leq 2.25$ (w2); $2.75 \leq d \leq 3.25$ (w3). We assume that molecules are compact when $R \leq 0.95$. The compactness is then measured as the percentage of compact configurations in each d-window. We remark that the total number of analyzed configurations is 100,000 for each metadynamics simulation.

Table 3. RMSD and compactness measured for different sets of configurations. Reference structure are 6N59 (chain A backbone) and 3LX4 (chain A backbone) for *Cvu* and *Cr*, respectively. Compact samples are the percentage of those in the given window with $R \leq 0.95$ over the whole number of collected configurations (100,000).

Configurations	⟨RMSD⟩ (Å)	Samples in Set	⟨SASA(H-Cluster)⟩	Compact Samples (%)
1/w1	3.6	31,217	1.8 ± 1.6	18
1/w2	3.5	6733	1.7 ± 1.1	4
1/w3	3.4	4220	1.3 ± 1.4	1
3/w1	4.0	43,305	1.6 ± 1.5	31
3/w2	3.9	4410	1.0 ± 0.7	2
3/w3	3.7	240	1.9 ± 0.7	0
Cr/w1	-	0	-	-
Cr/w2	2.7	14	1.7 ± 1.0	0
Cr/w3	2.8	1751	0.9 ± 0.9	1

It should be noticed that the deviation of the scaffold backbone from the reference structure slightly depends on the value of the chosen collective variable d. In *Cvu*, the larger is the distance the lower is the perturbation of the H-domain scaffold. Furthermore, in metadynamics **1** the protein scaffold is less distorted compared to the reference crystal structure than in metadynamics **3**. The *Cr* smaller deviation (RMSD lower than 3 Å) compared to *Cvu* shows that an unstructured N-terminal tail perturbs less the H-domain scaffold than the structured F-domain (metadynamics **3**). The solvent accessibility to the active H-cluster is, on average, in all cases very small, being the maximal SASA of the H-cluster 110 Å^2 for *Cr*. Most of the *Cr* configurations with H-cluster SASA larger than half the maximum display values of d larger than 4 nm. In *Cr*, the H-cluster is known to be accessible to solvent being the enzyme highly sensitive to dioxygen. In *Cvu*, conversely, the H-cluster accessibility is distributed differently. In Figure 9, we display for each configuration the sampled values of R and SASA of H-clusters.

The distribution of values shows that in *Cvu* metadynamics **1** compact configurations with low H-cluster accessibility are sampled. In *Cvu* metadynamics **3** these samples appear more rarely, while in *Cr* metadynamics most of the configurations have R larger than 0.9 and when $R \geq 0.95$ we see, as expected, the highly accessible H-clusters appearing.

In summary, the enhanced statistics metadynamics method shows for *Cvu* an interesting correlation between short distances of the FeS clusters and low accessibility of the active H-cluster to water. These properties are of interest in understanding the minimal architecture of hydrogenase that combines dihydrogen production with resistance to dioxygen. However, the dissection of assemblies between F- and H-domain in case of *Cvu* (or the N-terminal tail and the H-domain in *Cr*) requires long trajectories at constant bias and a proper reweighting of the collected sampling. This analysis will be performed in a following more detailed study.

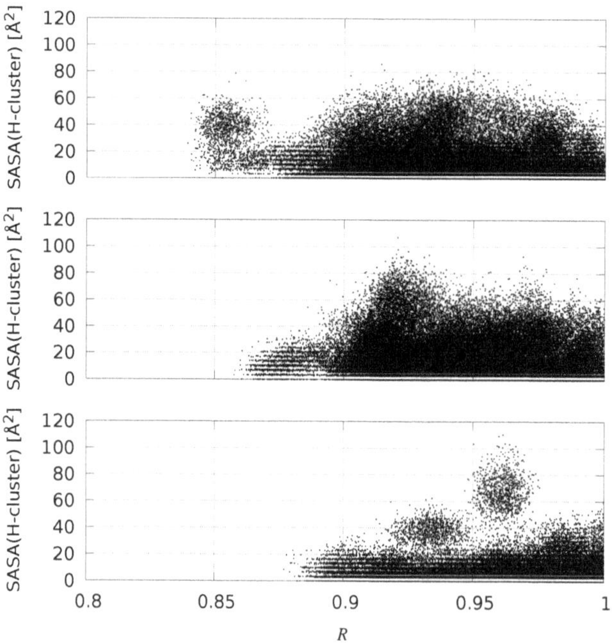

Figure 9. Sampled points in *R*-SASA(H-cluster) space for *Cvu* and *Cr* Hyd. Top–*Cvu* metadynamics **1**. Middle–*Cvu* metadynamics **3**. Bottom–*Cr* metadynamics.

3. Discussion

AlphaFold is now used via almost automated pipelines fed by sequence databases in order to both annotate and propose initial 3D structures. Structural models appear, for instance in UNIPROT, with the prefix AF, usually as proposals of the highest rank. Despite this high-throughput structural prediction available today, a detailed scrutiny of such predictions is necessary.

Therefore, once the function of [FeFe] hydrogenase is associated to the gene KAI3438965.1 of *Cvu*, the information provided by AlphaFold (version 2.2.4) or by other similar softwares needs to be carefully analysed on the basis of detailed chemical properties to be able to truly claim understanding of the possible function of the protein.

The AlphaFold alignment results (see Section 2.1 and Supplementary Materials show that the sequence encoded by gene KAI3438965.1 of *Chlorella vulgaris* has an E-value of zero, 99% query coverage and 74% identity with gene XP_005843810.1 of *Chlorella variabilis* (see Supplementary Materials). The latter *Cv* gene is annotated as region Fehyd_SSU domain-containing, but with unknown function for the coded protein. This unknown protein belongs to the so-called group C of hydrogenases, while we noticed that the known *Cv* Hyd of class A corresponds to gene AEA34989.1. Thus, the above information tells us that at least two [FeFe] hydrogenases can be coded by the *Cv* genome. It is plausible that *Chlorella vulgaris* and *Chlorella variabilis* algae strains contain different [FeFe] hydrogenases like *Clostridium pasteurianum*, that expresses three types of Hyd: *CpI*, *CpII*, and *CpIII* (see Ref. [12] and references therein).

The sequence analysis also shows that the protein encoded in KAI3438965.1 *Cvu* gene is related to proteins that have no hydrogenase function, like Nar proteins of other eukaryotic organisms [33]. However, adding the typical Hyd FeS cofactors to model structures of the protein expressed by *Cvu* KAI3438965.1 gene, we showed that one of the resulting constructs is consistent with a functional hydrogenase. It is therefore possible that the same protein turns into a hydrogenase when internalized into the chloroplast where the H-cluster is completed with the $[2Fe]_H$ component.

Different *Chlorella vulgaris* strains have been proposed as interesting candidates for sustainable biological hydrogen production [6,25,43]. Despite the phenotypical characterization available for these microalgae, only the genome of the 211/11P strain is published [5]. Moreover, the structural information about [FeFe] hydrogenases of microalgae is poor. In particular, the *Chlamydomonas rheinardtii* crystal and solution structures available in the Protein databank do not contain complete FeS clusters [14,44]. In any case, *Cr* Hyd lacks a functional F-domain and this missing domain can be responsible for the low resistance of *Cr* Hyd to oxidation [9].

Chlorella vulgaris 211/11P [FeFe] hydrogenase could be a very important system for experimental characterization of the type of cooperation between the single [4Fe4S] auxiliary cluster and the H-cluster [9]. The interactions of the iron–sulfur catalytic site (H-cluster) with the local protein environment are thought to contribute to modulate catalytic reactivity, but this has not been fully demonstrated. Some studies indicate that protein secondary interactions directly influence the relative stabilization/destabilization of different oxidation states of the metal cluster active site [2].

In some microalgae organisms, the evolutionary process has guided [FeFe]-hydrogenase to lose the F-domain and replace it with a smaller disordered domain of variable sequence [11]. Interactions between the F-domain and the H-domain modulate the accessibility to the H-cluster of O_2 and other reactive oxygen species, irrespective of their generation mechanism, thus eventually decreasing the rate of protein oxidation [42,45]. The possible protective effect depends on the mutual interactions between F- and H-domains which must favour a short distance between F- and H-clusters if it has to allow electron transfer from the exogenous reductant to Fe_d in the H-cluster. In this work, we investigated the possible protective effect of the Hyd N-terminus which is able to mimic in short microalgae Hyd the large F-domain that is present in cyanobacteria.

To investigate this possibility, we built three models of pseudo F-domains in *Cvu*. We showed that a structure of the F-domain similar to *CpI* provides a non-functional hydrogenase, because the two FeS clusters are too far apart in space to transport electrons from reductant species to the active H-cluster. Conversely, alternative structures produced by a deep learning algorithm provided a pseudo F-domain that can be functional and at the same time can hinder the H-cluster to O_2 accessibility. In another model proposal the additional FeS cluster in the F-domain is also accessible to external reductant species.

We have shown that the combination of sequence alignment, deep learning algorithms, and atomistic molecular dynamics simulations with the inclusion of inorganic cofactors, has the potential of becoming a new pipeline for model studies addressed at understanding and lately optimizing the function of [FeFe] hydrogenase.

4. Materials and Methods

For brevity, the acronyms listed at the end of the text are used in next section.

4.1. Alignment and Gene Annotation

The *Cvu* genome has been published with the code SIDB01000001.1 in Genbank [5]. To identify the protein with a candidate hydrogenase function, we aligned the whole genome of *Cvu* to the known sequences of 7 [FeFe] hydrogenase: *Clostridium pasteurianum, Desulfovibrio desulfuricans, Chlamydomonas rheinardtii, Chlorella variabilis, Chlorella* sp. DT, *Chlorella fusca, Tetradesmus obliquus*. We performed BLAST alignments [46,47] using BLOSUM90 [48] and PAM30 matrix [49,50]. Then, we singled out and extracted the sequence KAI3438965.1 that always shows the highest score in all alignments. We compare and summarize the results obtained with BLAST [51] and CLUSTAL [52] algorithms and are reported in Supplementary Materials.

4.2. Structure Prediction

In order to generate a hypothetical *Cvu* Hyd with one auxiliary [4Fe4S] cluster in the F-domain, we proceeded as explained below.

In the first case (**1**), in all the AlphaFold [20,53] (version 2.2.4) predicted structures we looked for a configuration where Cys residues 21, 72, 75, and 78 were able to bind the F-cluster (see sequence alignment in Figure 4). Then we inserted the auxiliary $[4Fe4S]_F$ cluster and minimized the energy once a suitable force-field for the system is identified (see Structure refinement below). The second model (**2**) comes from SwissModel, a fully automated protein structure homology-modeling server [19]. We choose SwissModel prediction "two", because prediction "one" was obtained from AF which is integrated into SwissModel, and it is, therefore, identical to **1**. In **2**, only a partial H-domain from residue 86 to 540 is present. Therefore, we completed the structure with the prediction made by AF for the missing residues. Once the H-domain scaffold is aligned and overlapped (see Table S1), we obtain a full model structure where residues 1–100 and 531–549 are taken from the AF prediction and residues 101–530 from the SwissModel prediction.

In the third case (model **3**), we substituted the unfolded F-domain of the AF prediction having the highest plDDT score, with the folded F-domain taken from 6N59 PDB structure of *CpI* [12]. To do so, we first replaced the coordinates of atoms in residues 1–88 of *Cvu* with the coordinates of those atoms that are shared with residues 119–206 in 6N59. We then minimized the relative root-mean square deviation (see below) of the atoms in residue 82 with respect to the same residue in *Cvu* as generated by AF. In this way, we obtain a model in which the H-domain is taken from AF and the F-domain is mapped to the coordinates of the crystal structure of *CpI*. The initial structures of the three models are displayed in Figure S1.

As a result of this procedure, we obtained three model structures with very similar H-domain scaffolds, but a variety of F-domains and disordered loops.

4.3. Structure Refinement

After the initial structure prediction (see previous sub-section), in all models we bonded all the clusters inside the protein matrix as in the H-ox state: the $[4Fe4S]_H$ cluster has the charge 2+, corresponding to the formal 2Fe(II)2Fe(III) oxidation state of iron ions; the $[2Fe]_H$ cluster has the charge -1, corresponding to the formal Fe(I)Fe(II) oxidation states of the two iron ions; the auxiliary $[4Fe4S]_F$ cluster has the same parameters of $[4Fe4S]_H$.

In order to insert the clusters and run MD simulation, we used the available force-field parameters of FeS clusters [54] that we adapted to the Charmm36 force-field for proteins [55] (see Supplementary Materials, the Force_Field folder). The FeS clusters are explicitly bonded to the protein matrix, with the bonds between Cys residues and Fe atoms described by the chosen force-field (see below). We inserted the protein (7957 atoms) in a simulation cell, adopting periodic boundary conditions, with water plus ions atoms to neutralize the system ([NaCl] = 0.15 M). All the three systems are made by $\sim 10^5$ atoms. After an initial energy minimization and equilibration in NVE statistical ensemble (20 ps), the systems were equilibrated for 100 ps in the NVT ensemble first at a temperature of $T = 150$ and than at $T = 300$ K. The configurations of the three models obtained after energy minimization are displayed with all the bonded FeS clusters in Figure S4 (see Supplementary Materials).

The equilibration was continued for another 2 ns in the NPT ensemble, at a constant $T = 300$ K with a bath relaxation time of 0.1 ps. The pressure was kept at 1 bar with a barostat relaxation time of 2 ps and compressibility of bulk water at room conditions.

Coulomb and Lennard-Jones interactions were described by means of a buffered Verlet pair list with a short-range cutoff equal to 1 nm [56]. Long-range electrostatic interactions were treated with the particle mesh Ewald method [57] with a grid spacing of 0.16 nm. The LINCS [58] and SETTLE [59] algorithms were employed to constrain, respectively, all protein bonds involving H atoms and internal degrees of freedom of water molecules, allowing to integrate the equations of motion with a 2 fs time-step. Finally, starting from the equilibrated systems, we performed 1 μs long MD simulations for each of the three different models in the NPT ensemble, initiated with different random seeds to generate the initial atomic velocities from a Maxwell distribution at 300 K. The simulation orthogonal cell's sides (L_x, L_y, and L_z) were, on average after 800 ns of equilibration for simulation **1**: $\langle L_x \rangle =$ 11.330 ± 0.006 nm; $\langle L_y \rangle = 7.352 \pm 0.004$ nm; $\langle L_z \rangle = 12.496 \pm 0.006$ nm, for simulation **3**: $\langle L_x \rangle = 10.858 \pm 0.006$ nm; $\langle L_y \rangle = 7.9678 \pm 0.004$ nm; $\langle L_z \rangle = 10.437 \pm 0.006$ nm.

4.4. Enhancing Statistics with Well-Tempered Metadynamics

Many computational methods have been devised to address the well-known limitations of equilibrium molecular dynamics, that is the insufficient sampling of configurations of macromolecules when keeping the conditions of thermal equilibrium. Among the many techniques used to enhance sampling, we decided to use altruistic multiple-walkers well-tempered metadynamics [60,61]. The method defines a collective variable (CV), that we denoted as ξ, function of atomic positions, q. The values, s, taken by ξ can be used to label the system macrostates. The set of coordinates q labels the system microstates, with each set of q yielding one of the possible values of s. If ergodicity holds, infinitely long simulations of a trajectory $q(t)$ in a given statistical ensemble would correctly sample the statistical weight of s. However, because of the huge number of ways in which certain values of s of ξ are encountered, compared to others, actual numerical simulations in practice only sample the maximally degenerate values of ξ. This is the case where ξ is the CV associated to folding/unfolding events. Generalized statistical ensembles try to address this problem by biasing the trajectory to spend more time where ξ has a low degeneracy and less time where ξ has a large degeneracy. The sampling of configurations obtained with the bias of the inverse probability of ξ is called metastatistics.

In our application, the enhanced statistics (metastatistics, hereafter) is obtained from a swarm of trajectories (called walkers) provided by metadynamics while building up a suitable external bias, which is then kept fixed when collecting configurations in the final step of the NpT simulation. The time evolution of each of the walkers starts from an initial configurations built either on the basis of AF predictions, (metadynamics **1**) or manual construction (metadynamics **3**). The initial configurations of each walker are different mainly because of the different interactions with the macromolecular environment. The initial differences are then amplified by the different construction of the external bias so that a variety of configurations among the many possible ones labelled by the values of the collective variable ξ are sampled.

The external bias that induces ξ broadening is progressively built adding repulsive gaussian functions of the collective variable term in the potential. In well-tempered metadynamics [62], as the simulation proceeds, the width of the added Gaussian remains constant but its height steadly decreases. The bias magnitude, which increases monotonically, eventually changes very little with time. As we said, at the beginning the space of CV is flooded by gaussians of height w. With the progress of flooding, the height of the newly added Gaussians decrease. This behavior is very important in highly complex biological systems, where the biasing potential should never reach any excessively large value.

In this work, we used a biasing factor $\frac{T+\Delta T}{\Delta T}$ =16 (ΔT =20 K) for *Cvu*, and 8 for *Cr* molecules. In order to sample both active and inactive structures for *Cvu*, we choose as CV the distance between the center of mass of the two [4Fe4S] clusters, one in the H-cluster and the other in the F-cluster. In particular, there is an effective catalytic function when the CV is inside a specific range (11–14 Å).

In the case of *Cr*, where the F-cluster is not present, we take as CV the distance between the center of mass of the single [4Fe4S] cluster and the N-terminus of the molecule.

While only the highest-score AlphaFold prediction was used in conventional simulation **1**, in metadynamics **1** we used the five highest-score AlphaFold predictions. Each of the 5 configurations was randomly rotated 19 times and then inserted into an orthorhombic simulation cell. We thus obtained 5×20 configurations, each used as starting configuration of a walker. In metadynamics **3**, we used the single manual construction randomly rotated

100 times. The bias construction, providing the free energy as a function of the chosen collective variable d, lasted 28 ns. Gaussian functions were added to the bias every 10 ps. The bias exchange among the 100 walkers was performed every 2 ns. The free energy displayed in Figure 8 is the opposite of the converged bias. Data reported in Table 3 are obtained averaging over the last 10 ns simulation collected at constant bias after convergence and over all 100 walkers. Averages are not reweighted to cancel the bias because at this stage we want to describe the behavior of configurations sampled at different d values.

4.5. Analysis

In order to compute average quantities in conventional MD **1–3**, we used the last 200 ns of each simulations. Structural quantities of interest are described below. RMSD measures the average distance between structure at time t_2 with respect to a reference structure at time t_1. The definition of RMSD is:

$$\mathrm{RMSD}(t_2, t_1) = \sqrt{\frac{1}{M}\sum_{i=1}^{N_{atoms}} m_i |r_i(t_1) - r_i(t_2)|^2}\,, \tag{1}$$

where M is the total mass of the molecule, m_i and r_i are the mass and position of atom i. The structure at time t_2 is standardly translated and rotated with respect to the reference one so as to minimize the RMSD. RMSD is useful to compare two structures and look for the differences along the simulation.

Root-mean square fluctuation (RMSF) instead measures the average distance between atoms belonging to the same residue i, averaged over time. In our analysis we computed such quantity as emerging from a sub-set of eigenvectors of the covariance matrix, as it is routinely made in principal component analysis (PCA) [63]. To simplify the analysis we used the first 4 eigenvectors (descending order of eigenvalues) of the covariance matrix. Also, we used the backbone heavy atoms to compute the covariance matrix, to focus the study on major structural changes of the protein

The gyration radius (R_g) is defined by the formula

$$R_g(t) = \sqrt{\frac{1}{M}\sum_{i=1}^{N_{atoms}} m_i |r_i(t) - r_{CM}|^2}\,, \tag{2}$$

where CM identifies the center of mass of the molecule.

In order to characterize the accessibility of the solvent to protein surface and clusters atoms, we calculated the solvent accessible surface areas (SASA) [64] and the radial distribution function ($g(r)$) of water molecules, respectively. The SASA is measured with a spherical probe mimicking a water molecule (with a radius of 1.4 Å). Standard atomic radii [65] are used for protein atoms. In SASA calculations Fe is assumed to have zero radius since it is buried by ligand atoms. The function $g(r)$ is defined as:

$$g(r) = \frac{1}{\langle\rho_b\rangle_{local}} \frac{1}{N_a} \sum_{i\in a}^{N_a} \sum_{j\in b}^{N_b} \frac{\delta(r_{ij} - r)}{4\pi r^2}\,, \tag{3}$$

where δ is the Dirac function and $\langle\rho_b\rangle_{local}$ the density of atoms of type b averaged over all spheres of radius r around atoms of type a.

Finally, salt bridges are the number of contacts between two groups of atoms, a and b with opposite charge. Choosing a as Cδ(Glu), Cγ(Asp), C-terminus and b as Nζ(Lys), Nη(Arg) and N-terminus, we define

$$\mathrm{SB} = \sum_{i\in a}\sum_{j\in b} S_{ij} \tag{4}$$

where S_{ij} is the step function, which is equal to 1 if the contact (distance ≤ 4 Å) between atoms i and j is formed and zero otherwise.

All MD simulations were performed with the NAMD 2.13-14 [66] and GROMACS 2021.5 [67,68] codes. Analysis was performed with GROMACS 2021.5, VMD 1.9.3 [69] and PLUMED 2.8.1 [70] softwares. The TopoGromacs Tool [71] was used to convert Charmm topology and parameters into Gromacs input data-sets.

5. Conclusions

In this work, we have presented a computational study of the *Chlorella vulgaris* 211/11P putative [FeFe] hydrogenase. The unicellular algal strain produces, under suitable conditions, pure hydrogen gas by using sunlight as a primary energy source and water as a hydrogen source. The system we are studying is challenging even for advanced annotation and structure prediction methods, because of the need for also including non-protein cofactors correctly bound to the protein.

We found that sequence alignment analysis, AlphaFold prediction models, and simulations including the required metal clusters, are consistent with the data published in the literature [2,9,12–14,37]. Our analysis led us to conjecture that the *C. vulgaris* 211/11P strain [FeFe] hydrogenase belongs to a novel family of [FeFe] hydrogenases. The main points on which we base our conclusion that the above represents a new hydrogenase are: (i) it consists of a functional H-domain, with the peculiar H-cluster binding motif ASACPGW replacing the conserved TSCCPGW motif of most of all the other known [FeFe] hydrogenases; (ii) it can contain a single auxiliary 4Fe4S cluster bound to a non-conserved F-domain; (iii) depending on the F-domain degree of structural arrangement, significant effects on the structural parameters are observed; (iv) we can exclude that the potential F-domain has the structure of [FeFe] hydrogenase I in *Chlostridium pasteurianum*.

Taken together, the models we investigated reveal important properties of the [FeFe] hydrogenase of *Chlorella vulgaris* 211/11P strain, highlighting the need for further investigation of the interactions between F- and H-domains. Indeed, understanding the cross-talk between the two domains and, in general, between disordered and ordered regions, is necessary to optimize the hydrogen production efficiency of these enzymes.

Supplementary Materials: The following supporting information can be downloaded at: https://www.mdpi.com/article/10.3390/ijms25073663/s1.

Author Contributions: Conceptualization, G.L.P., V.M. and C.F.; Data curation, S.B. and G.L.P.; Funding acquisition, V.M. and S.M.; Investigation, S.B., G.L.P. and C.F.; Methodology, S.B. and G.L.P.; Project administration, V.M.; Resources, C.F.; Validation, G.L.P., F.S., S.M. and C.F.; Writing—original draft, S.B., G.L.P., G.R. and C.F.; Writing revision, G.L.P. All authors have read and agreed to the published version of the manuscript.

Funding: This work was supported by PRIN-201744NR8S (MIUR, Italy) and PON "Ricerca e innovazione" 2014–2020 (Italy).

Institutional Review Board Statement: Not applicable.

Informed Consent Statement: Not applicable.

Data Availability Statement: Part of the data to reproduce the reported calculations are available online at: https://doi.org/10.3390/ijms25073663. A document, SM.pdf, in Supplementary Materials explains how to use the provided files deposited in Supplementary Materials. Any further data useful to reproduce the reported calculations are available on request to the corresponding author.

Acknowledgments: We thank Matteo Ballottari's group for providing us the draft of annotated genome of *Chlorella vulgaris* 211/11P, before UNIPROT elaboration of data deposited in GenBank. Numerical calculations have been made possible through: the ISCRA B project HydGen on Leonardo high-performance infrastructure (CINECA, Italy); the CINECA-INFN agreement, on Marconi-100 and Leonardo HPC infrastructures (CINECA, Italy).

Conflicts of Interest: The authors declare no conflicts of interest.

Abbreviations

Hyd	[FeFe] hydrogenase
Cp	Clostridium pasteurianum
CpI	[FeFe] hydrogenase of Cp, gene CpI
Cv	Chlorella variabilis
Cv Hyd	[FeFe] hydrogenase of Cv, gene HydA1
Cvu	Chlorella vulgaris 211/11P strain
Cvu Hyd	[FeFe] hydrogenase of Cvu, gene KAI3438965.1
Cr	Chlamydomonas rheinardtii
Cr Hyd	[FeFe] hydrogenase of Cr, gene HydA1
Dd	Desulfovibrio desulfuricans
Dd Hyd	[FeFe] hydrogenase of Dd HydAB
MD	molecular dynamics
MtD	molecular well-tempered metadynamics
AF	AlphaFold deep learning
SM	Supplementary Materials

References

1. Seibert, M.; Torzillo, G. (Eds.) *Microalgal Hydrogen Production*; Comprehensive Series in Photochemical & Photobiological Sciences; The Royal Society of Chemistry: Piccadilly, UK, 2018; pp. P001–P496. [CrossRef]
2. Lubitz, W.; Ogata, H.; RÃ¼diger, O.; Reijerse, E. Hydrogenases. *Chem. Rev.* **2014**, *114*, 4081–4148. [CrossRef]
3. Morra, S. Fantastic [FeFe]-Hydrogenases and Where to Find Them. *Front. Microbiol.* **2022**, *13*. [CrossRef] [PubMed]
4. Yin, Y.; Wang, J. Chapter 8—Production of biohydrogen. In *Biofuels and Biorefining*; Gómez Castro, F.I., Gutiérrez-Antonio, C., Eds.; Elsevier: Amsterdam, The Netherlands, 2022; pp. 283–337. [CrossRef]
5. Cecchin, M.; Marcolungo, L.; Rossato, M.; Girolomoni, L.; Cosentino, E.; Cuine, S.; Li-Beisson, Y.; Delledonne, M.; Ballottari, M. *Chlorella vulgaris* genome assembly and annotation reveals the molecular basis for metabolic acclimation to high light conditions. *Plant J.* **2019**, *100*, 1289–1305. [CrossRef] [PubMed]
6. Masojídek, J.; Karolína, R.; Rearte, T.A.; Celis Plá, P.S.; Torzillo, G.; Silva Benavides, A.M.; Neori, A.; Gómez, C.; Álvarez-Gómez, F.; Lukeš, M.; et al. Changes in photosynthesis, growth and biomass composition in outdoor *Chlorella* g120 culture during the metabolic shift from heterotrophic to phototrophic cultivation regime. *Algal Res.* **2021**, *56*, 102303. [CrossRef]
7. Touloupakis, E.; Faraloni, C.; Silva Benavides, A.M.; Masojídek, J.; Torzillo, G. Sustained photobiological hydrogen production by Chlorella vulgaris without nutrient starvation. *Int. J. Hydrogen Energy* **2021**, *46*, 3684–3694. [CrossRef]
8. Lubitz, W.; Ogata, H. Hydrogenases, Structure and Function. In *Encyclopedia of Biological Chemistry*, 2nd ed.; Lennarz, W.J., Lane, M.D., Eds.; Academic Press: Waltham, MA, USA, 2013; pp. 562–567. [CrossRef]
9. Birrell, J.A.; Rodríguez-Maciá, P.; Reijerse, E.J.; Martini, M.A.; Lubitz, W. The catalytic cycle of [FeFe] hydrogenase: A tale of two sites. *Coord. Chem. Rev.* **2021**, *449*, 214191. [CrossRef]
10. Schrödinger, L. *The PyMOL Molecular Graphics System, Version 1.8*; Schrödinger, LLC: New York, NY, USA, 2015.
11. Meuser, J.E.; Boyd, E.S.; Ananyev, G.; Karns, D.; Radakovits, R.; Narayana Murthy, U.M.; Ghirardi, M.L.; Dismukes, G.C.; Peters, J.W.; Posewitz, M.C. Evolutionary significance of an algal gene encoding an [FeFe]-hydrogenase with F-domain homology and hydrogenase activity in *Chlorella variabilis* NC64A. *Planta* **2011**, *234*, 829–843. [CrossRef] [PubMed]
12. Artz, J.H.; Zadvornyy, O.A.; Mulder, D.W.; Keable, S.M.; Cohen, A.E.; Ratzloff, M.W.; Williams, S.G.; Ginovska, B.; Kumar, N.; Song, J.; et al. Tuning Catalytic Bias of Hydrogen Gas Producing Hydrogenases. *J. Am. Chem. Soc.* **2020**, *142*, 1227–1235. [CrossRef]
13. Nicolet, Y.; Piras, C.; Legrand, P.; Hatchikian, C.E.; Fontecilla-Camps, J.C. Desulfovibrio desulfuricans iron hydrogenase: The structure shows unusual coordination to an active site Fe binuclear center. *Structure* **1999**, *7*, 13–23. [CrossRef]
14. Mulder, D.W.; Boyd, E.S.; Sarma, R.; Lange, R.K.; Endrizzi, J.A.; Broderick, J.B.; Peters, J.W. Stepwise [FeFe]-hydrogenase H-cluster assembly revealed in the structure of HydA Delta EFG. *Nature* **2010**, *465*, 248–251. [CrossRef]
15. Heinig, M.; Frishman, D. STRIDE: A web server for secondary structure assignment from known atomic coordinates of proteins. *Nucleic Acids Res.* **2004**, *32*, W500–W502. [CrossRef] [PubMed]
16. Vignais, P.M.; Billoud, B. Occurrence, Classification, and Biological Function of Hydrogenases: An Overview. *Chem. Rev.* **2007**, *107*, 4206–4272. [CrossRef] [PubMed]
17. Swanson, K.D.; Ratzloff, M.W.; Mulder, D.W.; Artz, J.H.; Ghose, S.; Hoffman, A.; White, S.; Zadvornyy, O.A.; Broderick, J.B.; Bothner, B.; et al. [FeFe]-Hydrogenase Oxygen Inactivation Is Initiated at the H Cluster 2Fe Subcluster. *J. Am. Chem. Soc.* **2015**, *137*, 1809–1816. [CrossRef]
18. Kamp, C.; Silakov, A.; Winkler, M.; Reijerse, E.J.; Lubitz, W.; Happe, T. Isolation and first EPR characterization of the [FeFe]-hydrogenases from green algae. *Biochim. Biophys. Acta Bioenerg.* **2008**, *1777*, 410–416. [CrossRef] [PubMed]

19. Waterhouse, A.; Bertoni, M.; Bienert, S.; Studer, G.; Tauriello, G.; Gumienny, R.; Heer, F.T.; de Beer, T.A.; Rempfer, C.; Bordoli, L.; et al. SWISS-MODEL: Homology modelling of protein structures and complexes. *Nucleic Acids Res.* **2018**, *46*, W296–W303. [CrossRef] [PubMed]
20. Jumper, J.; Evans, R.; Pritzel, A.; Green, T.; Figurnov, M.; Ronneberger, O.; Tunyasuvunakool, K.; Bates, R.; Žídek, A.; Hassabis, D.; et al. Highly accurate protein structure prediction with AlphaFold. *Nature* **2021**, *596*, 583–589. [CrossRef] [PubMed]
21. Zimmermann, L.; Stephens, A.; Nam, S.Z.; Rau, D.; Kübler, J.; Lozajic, M.; Gabler, F.; Söding, J.; Lupas, A.N.; Alva, V. A Completely Reimplemented MPI Bioinformatics Toolkit with a New HHpred Server at its Core. *J. Mol. Biol.* **2018**, *430*, 2237–2243. [CrossRef]
22. Binbay, F.A.; Rathod, D.C.; George, A.A.P.; Imhof, D. Quality Assessment of Selected Protein Structures Derived from Homology Modeling and AlphaFold. *Pharmaceuticals* **2023**, *16*, 1662. [CrossRef]
23. Liu, D.; Yuan, C.; Guo, C.; Huang, M.; Lin, D. Structural and Functional Insights into the Stealth Protein CpsY of Mycobacterium tuberculosis. *Biomolecules* **2023**, *13*, 1611. [CrossRef]
24. Nagy, V.; Podmaniczki, A.; Vidal-Meireles, A.; Tengölics, R.; Kovács, L.; Rákhely, G.; Scoma, A.; Tóth, S.Z. Water-splitting-based, sustainable and efficient H2 production in green algae as achieved by substrate limitation of the Calvin-Benson-Bassham cycle. *Biotechnol. Biofuels* **2018**, *11*, 69. [CrossRef]
25. Tamayo-Ordoñez, Y.d.J.; Ayil-Gutiérrez, B.A.; Moreno-Davila, I.M.M.; Tamayo-Ordoñez, F.A.; Córdova-Quiroz, A.V.; Poot-Poot, W.A.; Damas-Damas, S.; Villanueva-Alonzo, H.d.J.; Tamayo-Ordoñez, M.C. Bioinformatic analysis and relative expression of hyd and fdx during H2 production in microalgae. *Phycol. Res.* **2023**, *71*, 37–55. [CrossRef]
26. Mészáros, L.S.; Ceccaldi, P.; Lorenzi, M.; Redman, H.J.; Pfitzner, E.; Heberle, J.; Senger, M.; Stripp, S.T.; Berggren, G. Spectroscopic investigations under whole-cell conditions provide new insight into the metal hydride chemistry of [FeFe]-hydrogenase. *Chem. Sci.* **2020**, *11*, 4608–4617. [CrossRef] [PubMed]
27. Lorenzi, M.; Ceccaldi, P.; Rodríguez-Maciá, P.; Redman, H.J.; Zamader, A.; Birrell, J.A.; Mészáros, L.S.; Berggren, G. Stability of the H-cluster under whole-cell conditions—Formation of an Htrans-like state and its reactivity towards oxygen. *J. Biol. Inorg. Chem.* **2022**, *27*, 345–355. [CrossRef] [PubMed]
28. Consortium, T.U. UniProt: The Universal Protein Knowledgebase in 2023. *Nucleic Acids Res.* **2022**, *51*, D523–D531. [CrossRef] [PubMed]
29. Chien, L.F.; Kuo, T.T.; Liu, B.H.; Lin, H.D.; Feng, T.Y.; Huang, C.C. Solar-to-bioH2 production enhanced by homologous overexpression of hydrogenase in green alga *Chlorella* sp. DT. *Int. J. Hydrogen Energy* **2012**, *37*, 17738–17748. [CrossRef]
30. Greening, C.; Biswas, A.; Carere, C.R.; Jackson, C.J.; Taylor, M.C.; Stott, M.B.; Cook, G.M.; Morales, S.E. Genomic and metagenomic surveys of hydrogenase distribution indicate H2 is a widely utilised energy source for microbial growth and survival. *ISME J.* **2016**, *10*, 761–777. [CrossRef] [PubMed]
31. Søndergaard, D.; Pedersen, C.N.S.; Greening, C. HydDB: A web tool for hydrogenase classification and analysis. *Sci. Rep.* **2016**, *6*, 34212–34218. [CrossRef]
32. Balk, J.; Pierik, A.J.; Netz, D.J.A.; Mühlenhoff, U.; Lill, R. The hydrogenase-like Nar1p is essential for maturation of cytosolic and nuclear iron–sulphur proteins. *EMBO J.* **2004**, *23*, 2105–2115. [CrossRef]
33. Urzica, E.; Pierik, A.J.; Mühlenhoff, U.; Lill, R. Crucial Role of Conserved Cysteine Residues in the Assembly of Two Iron-Sulfur Clusters on the CIA Protein Nar1. *Biochemistry* **2009**, *48*, 4946–4958. [CrossRef]
34. Kabsch, W.; Sander, C. Dictionary of protein secondary structure—Pattern-recognition of hydrogen-bonded and geometrical features. *Biopolymers* **1983**, *22*, 2577–2637. [CrossRef]
35. Varadi, M.; Anyango, S.; Deshpande, M.; Nair, S.; Natassia, C.; Yordanova, G.; Yuan, D.; Stroe, O.; Wood, G.; Laydon, A.; et al. AlphaFold Protein Structure Database: Massively expanding the structural coverage of protein-sequence space with high-accuracy models. *Nucleic Acids Res.* **2021**, *50*, D439–D444. [CrossRef] [PubMed]
36. Furlan, C.; Chongdar, N.; Gupta, P.; Lubitz, W.; Ogata, H.; Blaza, J.N.; Birrell, J.A. Structural insight on the mechanism of an electron-bifurcating [FeFe] hydrogenase. *eLife* **2022**, *11*, e79361. [CrossRef] [PubMed]
37. Puthenkalathil, R.C.; Ensing, B. Fast Proton Transport in FeFe Hydrogenase via a Flexible Channel and a Proton Hole Mechanism. *J. Phys. Chem. B* **2022**, *126*, 403–411. [CrossRef] [PubMed]
38. Beratan, D.N.; Onuchic, J.N.; Winkler, J.R.; Gray, H.B. Electron-Tunneling Pathways in Proteins. *Science* **1992**, *258*, 1740–1741. [CrossRef]
39. Kaila, V.R.I. Long-range proton-coupled electron transfer in biological energy conversion: Towards mechanistic understanding of respiratory complex I. *J. R. Soc. Interface* **2018**, *15*, 20170916. [CrossRef]
40. Winkler, M.; Duan, J.; Rutz, A.; Felbek, C.; Scholtysek, L.; Lampret, O.; Jaenecke, J.; Apfel, U.P.; Gilardi, G.; Valetti, F.; et al. A safety cap protects hydrogenase from oxygen attack. *Nat. Commun.* **2021**, *12*, 756. [CrossRef]
41. Rutz, A.; Das, C.K.; Fasano, A.; Jaenecke, J.; Yadav, S.; Apfel, U.P.; Engelbrecht, V.; Fourmond, V.; Léger, C.; Schäfer, L.V.; et al. Increasing the O_2 Resistance of the [FeFe]-Hydrogenase CbA5H through Enhanced Protein Flexibility. *ACS Catal.* **2023**, *13*, 856–865. [CrossRef]
42. Stripp, S.T.; Goldet, G.; Brandmayr, C.; Sanganas, O.; Vincent, K.A.; Haumann, M.; Armstrong, F.A.; Happe, T. How oxygen attacks [FeFe] hydrogenases from photosynthetic organisms. *Proc. Natl. Acad. Sci. USA* **2009**, *106*, 17331–17336. [CrossRef]

43. Hwang, J.H.; Kim, H.C.; Choi, J.A.; Abou-Shanab, R.; Dempsey, B.A.; Regan, J.M.; Kim, J.R.; Song, H.; Nam, I.H.; Kim, S.N.; et al. Photoautotrophic hydrogen production by eukaryotic microalgae under aerobic conditions. *Nat. Commun.* **2014**, *5*, 3234. [CrossRef]
44. Duan, J.; Senger, M.; Esselborn, J.; Engelbrecht, V.; Wittkamp, F.; Apfel, U.P.; Hofmann, E.; Stripp, S.T.; Happe, T.; Winkler, M. Crystallographic and spectroscopic assignment of the proton transfer pathway in [FeFe]-hydrogenases. *Nat. Commun.* **2018**, *9*, 4726–4736. [CrossRef]
45. Outten, F.W. Iron-sulfur clusters as oxygen-responsive molecular switches. *Nat. Chem. Biol.* **2007**, *3*, 206–207. [CrossRef] [PubMed]
46. Sayers, E.W.; Bolton, E.E.; Brister, J.R.; Canese, K.; Chan, J.; Comeau, D.; Connor, R.; Funk, K.; Kelly, C.; Kim, S.; et al. Database resources of the national center for biotechnology information. *Nucleic Acids Res.* **2021**, *50*, D20–D26. [CrossRef] [PubMed]
47. Altschul, S.F.; Gish, W.; Miller, W.; Myers, E.W.; Lipman, D.J. Basic local alignment search tool. *J. Mol. Biol.* **1990**, *215*, 403–410. [CrossRef] [PubMed]
48. Henikoff, S.; Henikoff, J.G. Amino acid substitution matrices from protein blocks. *Proc. Natl. Acad. Sci. USA* **1992**, *89*, 10915–10919. [CrossRef] [PubMed]
49. Dayhoff, M.; Schwartz, R.M.; Orcutt, B. Chapter: A Model of Evolutionary Change in Proteins. In *Atlas of Protein Sequence and Structure*; National Biomedical Research Foundation: Washington, DC, USA, 1978; Volume 5, pp. 345–352.
50. Schwarz, R.; Dayhoff, M. Chapter: Matrices for Detecting Distant Relationships. In *Atlas of Protein Sequence and Structure*; National Biomedical Research Foundation: Washington, DC, USA, 1979; pp. 353–358.
51. Madeira, F.; Pearce, M.; Tivey, A.R.N.; Basutkar, P.; Lee, J.; Edbali, O.; Madhusoodanan, N.; Kolesnikov, A.; Lopez, R. Search and sequence analysis tools services from EMBL-EBI in 2022. *Nucleic Acids Res.* **2022**, *50*, W276–W279. [CrossRef]
52. Sievers, F.; Higgins, D.G. Clustal Omega for making accurate alignments of many protein sequences. *Protein Sci.* **2018**, *27*, 135–145. [CrossRef] [PubMed]
53. Baek, M.; DiMaio, F.; Anishchenko, I.; Dauparas, J.; Ovchinnikov, S.; Lee, G.R.; Wang, J.; Cong, Q.; Kinch, L.N.; Schaeffer, R.D.; et al. Accurate prediction of protein structures and interactions using a three-track neural network. *Science* **2021**, *373*, 871–876. [CrossRef] [PubMed]
54. Chang, C.H.; Kim, K. Density Functional Theory Calculation of Bonding and Charge Parameters for Molecular Dynamics Studies on [FeFe] Hydrogenases. *J. Chem. Theory Comput.* **2009**, *5*, 1137–1145. [CrossRef]
55. Best, R.B.; Zhu, X.; Shim, J.; Lopes, P.E.M.; Mittal, J.; Feig, M.; MacKerell, A.D. Optimization of the Additive CHARMM All-Atom Protein Force Field Targeting Improved Sampling of the Backbone phi, psi and Side-Chain chi1 and chi2 Dihedral Angles. *J. Chem. Theory Comput.* **2012**, *8*, 3257–3273. [CrossRef]
56. Páll, S.; Hess, B. A flexible algorithm for calculating pair interactions on SIMD architectures. *Comput. Phys. Commun.* **2013**, *184*, 2641–2650. [CrossRef]
57. Darden, T.; York, D.; Pedersen, L. Particle Mesh Ewald: An N-log(N) Method for Ewald Sums in Large Systems. *J. Chem. Phys.* **1993**, *98*, 10089–10092. [CrossRef]
58. Miyamoto, S.; Kollman, P.A. Settle: An analytical version of the SHAKE and RATTLE algorithm for rigid water models. *J. Comput. Chem.* **1992**, *13*, 952–962. [CrossRef]
59. Hess, B.; Bekker, H.; Berendsen, H.J.C.; Fraaije, J.G.E.M. LINCS: A linear constraint solver for molecular simulations. *J. Comput. Chem.* **1997**, *18*, 1463–1472. [CrossRef]
60. Hošek, P.; Kříž, P.; Toulcová, D.; Spiwok, V. Multisystem altruistic metadynamics- Well-tempered variant. *J. Chem. Phys.* **2017**, *146*, 125103. [CrossRef] [PubMed]
61. Nobili, G.; Botticelli, S.; La Penna, G.; Morante, S.; Rossi, G.; Salina, G. Probing protein stability: Towards a computational atomistic, reliable, affordable, and improvable model. *Front. Mol. Biosci.* **2023**, *10*, 1122269. [CrossRef] [PubMed]
62. Bussi, G.; Laio, A.; Tiwary, P. Metadynamics: A Unified Framework for Accelerating Rare Events and Sampling Thermodynamics and Kinetics. In *Handbook of Materials Modeling: Methods: Theory and Modeling*; Springer International Publishing: Cham, Switzerland, 2018; pp. 1–31. [CrossRef]
63. Hess, B. Convergence of sampling in protein simulations. *Phys. Rev. E* **2002**, *65*, 031910-10. [CrossRef] [PubMed]
64. Eisenhaber, F.; Lijnzaad, P.; Argos, P.; Sander, C.; Scharf, M. The Double Cubic Lattice Method: Efficient Approaches to Numerical Integration of Surface Area and Volume and to Dot Surface Contouring of Molecular Assemblies. *J. Comput. Chem.* **1995**, *16*, 273–284. [CrossRef]
65. Bondi, A. van der Waals Volumes and Radii. *J. Phys. Chem.* **1964**, *68*, 441–451. [CrossRef]
66. Phillips, J.C.; Hardy, D.J.; Maia, J.D.C.; Stone, J.E.; Ribeiro, J.V.; Bernardi, R.C.; Buch, R.; Fiorin, G.; Hénin, J.; Jiang, W.; et al. Scalable molecular dynamics on CPU and GPU architectures with NAMD. *J. Chem. Phys.* **2020**, *153*, 044130. [CrossRef]
67. Berendsen, H.J.C.; van der Spoel, D.; van Drunen, R. GROMACS: A message-passing parallel molecular dynamics implementation. *Comput. Phys. Commun.* **1995**, *91*, 43–56. [CrossRef]
68. Abraham, M.J.; Murtola, T.; Schulz, R.; Páll, S.; Smith, J.C.; Hess, B.; Lindahl, E. GROMACS: High performance molecular simulations through multi-level parallelism from laptops to supercomputers. *SoftwareX* **2015**, *1–2*, 19–25. [CrossRef]
69. Humphrey, W.; Dalke, A.; Schulten, K. VMD visual molecular dynamics. *J. Mol. Graph.* **1996**, *14*, 33–38 [CrossRef]

70. Bonomi, M.; Branduardi, D.; Bussi, G.; Camilloni, C.; Provasi, D.; Raiteri, P.; Donadio, D.; Marinelli, F.; Pietrucci, F.; Broglia, R.A.; et al. PLUMED: A Portable Plugin for Free-Energy Calculations with Molecular Dynamics. *J. Comput. Phys.* **2009**, *180*, 1961–1972. [CrossRef]
71. Vermaas, J.V.; Hardy, D.J.; Stone, J.E.; Tajkhorshid, E.; Kohlmeyer, A. TopoGromacs: Automated Topology Conversion from CHARMM to GROMACS within VMD. *J. Chem. Inform. Model.* **2016**, *56*, 1112–1116. [CrossRef]

International Journal of
Molecular Sciences

MDPI

Article

The Human Mutation K237_V238del in a Putative Lipid Binding Motif within the V-ATPase a2 Isoform Suggests a Molecular Mechanism Underlying Cutis Laxa

Anh Chu [1], Yeqi Yao [1], Miroslawa Glibowicka [2], Charles M. Deber [2,3] and Morris F. Manolson [1,3,*]

1 Faculty of Dentistry, University of Toronto, Toronto M5G 1G6, ON, Canada; anhnt.chu@mail.utoronto.ca (A.C.); yeqiyao@gmail.com (Y.Y.)
2 Division of Molecular Medicine, Research Institute, Hospital for Sick Children, Toronto M5G 0A4, ON, Canada; miragl@ymail.com (M.G.); deber@sickkids.ca (C.M.D.)
3 Department of Biochemistry, Faculty of Medicine, University of Toronto, Toronto M5S 1A8, ON, Canada
* Correspondence: m.manolson@utoronto.ca

Abstract: Vacuolar ATPases (V-ATPases), proton pumps composed of 16 subunits, are necessary for a variety of cellular functions. Subunit "a" has four isoforms, a1–a4, each with a distinct cellular location. We identified a phosphoinositide (PIP) interaction motif, $KX_nK(R)IK(R)$, conserved in all four isoforms, and hypothesize that a/PIP interactions regulate V-ATPase recruitment/retention to different organelles. Among the four isoforms, a2 is enriched on Golgi with a2 mutations in the PIP motif resulting in cutis laxa. We hypothesize that the hydrophilic N-terminal (NT) domain of a2 contains a lipid-binding domain, and mutations in this domain prevent interaction with Golgi-enriched PIPs, resulting in cutis laxa. We recreated the cutis laxa-causing mutation K237_V238del, and a double mutation in the PIP-binding motif, K237A/V238A. Circular dichroism confirmed that there were no protein structure alterations. Pull-down assays with PIP-enriched liposomes revealed that wildtype a2NT preferentially binds phosphatidylinositol 4-phosphate (PI(4)P), while mutants decreased binding to PI(4)P. In HEK293 cells, wildtype a2NT was localized to Golgi and co-purified with microsomal membranes. Mutants reduced Golgi localization and membrane association. Rapamycin depletion of PI(4)P diminished a2NT-Golgi localization. We conclude that a2NT is sufficient for Golgi retention, suggesting the lipid-binding motif is involved in V-ATPase targeting and/or retention. Mutational analyses suggest a molecular mechanism underlying how a2 mutations result in cutis laxa.

Keywords: V-ATPases; V-ATPase a2 isoforms; protein–lipid interaction; phosphoinositides; PI(4)P; cutis laxa

1. Introduction

Vacuolar H^+-ATPases (V-ATPases) are conserved ATP-dependent proton pumps responsible for the maintenance of organelle luminal pH in eukaryotic cells [1–5]. They are multi-subunit complexes comprised of a cytosolic V_1 sector responsible for ATP hydrolysis coupled to proton translocation through a membrane-bound V_o sector [6]. Several subunits have tissue- or organelle-specific isoforms thought to mediate V-ATPase localization to their various functional cellular destinations [7,8]. In mammals, there are four isoforms of the V_o-*a* subunit, a1–a4, which play key roles in V-ATPase localization [9–13]. The a1–a3 isoforms are expressed ubiquitously in different tissues, in which a1 and a2-containing V-ATPases are mainly found on intracellular membranes [14–20], while a3 are found in both intracellular compartments and the plasma membrane [21–23]. The a4 isoform is restricted to the kidney [24,25], epididymis [26], and inner ear [27], where a4-containing V-ATPases are targeted to the plasma membrane. Mutations in "a" subunit isoforms are linked to various diseases. Mutations within a1 have been linked to epileptic encephalopathy [14,16].

The autosomal recessive disease, cutis laxa, is characterized by a glycosylation malfunction in the Golgi and is associated with mutations in the a2 isoform [28–31]. Mutations in a3 and a4 are linked to osteopetrosis [32–35] and distal renal tubular acidosis [25,36,37], respectively.

The 90kDa *a*-subunit contains a cytosolic N-terminal domain and a membrane-bound C-terminal domain consisting of eight transmembrane helices (TMs) [10,38,39]. In *Saccharomyces cerevisiae*, the sorting signal for V-ATPases within the N-terminal half of the *a*-subunit homolog, Stv1p, is well characterized [10,11]. The sorting information, $W^{83}KY$, present in the Stv1p isoform dictates the retention of Stv1p-containing V-ATPases within the Golgi network [10]. However, little is known about the targeting signals for the mammalian Golgi-specific a2 isoform. Evidence in yeast suggests the interaction of the *a* subunits with membrane phosphoinositides can account for the membrane retention of the V-ATPases at specific locations [40,41].

Phosphoinositides (PIPs) are generated by reversible phosphorylation of the precursor phosphatidylinositol (PI) at the inositol headgroup, mediated by organelle-specific kinases and phosphatases [42–44]. Organelle-specific distribution and composition of the seven different PIPs play important roles in membrane protein trafficking [42,45]. Spatial or temporal enrichment of PIPs at microdomains within organelle membranes is involved in regulating the activity of membrane-bound ion channels and transporters [45,46]. The cytosolic N-terminal domain of the yeast *a* isoform, Vph1p, was recruited to the vacuolar membrane when the level of endosome/vacuole-specific $PI(3,5)P_2$ was elevated [41]. In contrast, Stv1p was shown to bind directly to a Golgi-specific PI(4)P, and the interaction was attributed to its cytosolic N-terminal domain [40].

In previous work, we showed that the plasma membrane-specific a4 isoform preferentially binds to $PI(4,5)P_2$ and that the a4NT-$PI(4,5)P_2$ interaction is responsible for membrane retention [47]. We proposed a putative binding domain within the N-terminal half of the *a* subunit, containing a conserved basic motif (K/R)X(K/R)(K/R). In the present study, we found that the a2 isoform interacts in vitro with PI(4)P, a Golgi-enriched PIP, and show that PI(4)P at the Golgi helps to retain the cytosolic N-terminal domain of a2 at the Golgi membrane. Additionally, we show that the cutis laxa causing mutation K237_V238del within the critical a2 isoform $K^{237}VKK^{240}$ binding motif not only reduced interactions with PI(4)P but also disrupted protein membrane retention, suggesting the molecular mechanism underlying the disease.

2. Results

2.1. Mutations within the Putative Binding Motif reduced Interaction of a2 with PI(4)P-Enriched Liposomes In Vitro

In our previous work, we proposed a putative lipid binding domain located at the distal lobe of the N-terminal half of a4. We further suggested that a conserved basic motif, (K/R)X(K/R)(K/R), is critical for the interaction with the acidic headgroup of PIPs [47]. In *a*2, we propose this critical lipid binding motif is $K^{237}VKK^{240}$ (Figure 1). Using mutagenesis to verify the basic motif, we generated two mutations in the *a*2 N-terminal domain (a2NT), K237A/V238A and K237_V238del, both within the putative binding motif. The K237_V238del is a mutation found within humans that results in cutis laxa [48].

We first assessed whether either mutation affected protein folding using circular dichroism (CD) spectroscopy. WT a2NT contains a mainly helical structure, exhibited by two negative minima at 222 nm and 208 nm. The CD spectra of the mutants and WT, all in aqueous buffer, overlap at the characteristic wavelengths, indicating that the mutations did not alter protein structure (Figure 2A). To mimic the membrane environment and determine whether the presence of micelles enhances the helicity of the proteins, we added detergent with an SDS-to-peptide ratio of 370:1. The spectra of WT and mutants behaved similarly in the presence of SDS micelles with increased negative ellipticity at 208 nm and 222 nm, suggesting the membrane-bound behavior of the proteins (Figure 2B). Variation in the signal of the positive peak at 190–200 nm is likely due to different trace amounts of imidazole in the buffer.

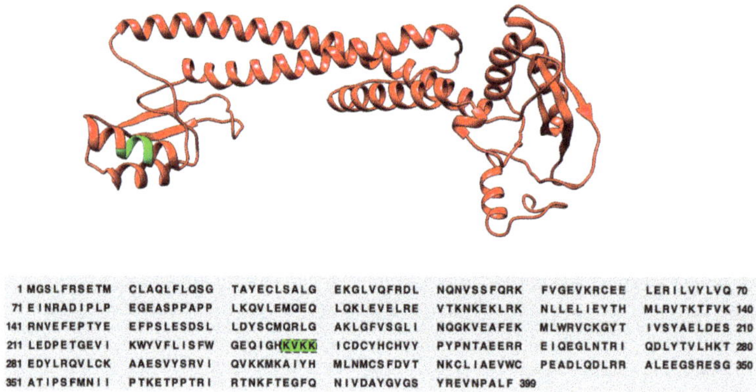

Figure 1. a2NT model generated by Phyre2.0 with the critical basic $K^{237}VKK^{240}$ motif in the distal lobe highlighted in green.

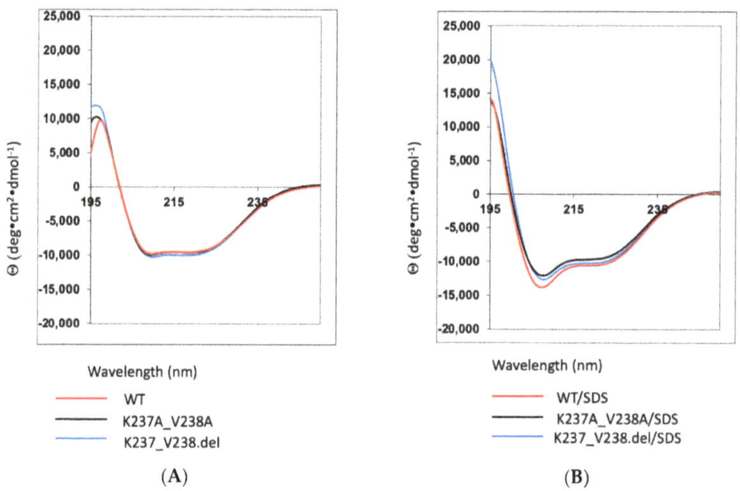

Figure 2. Mutations K237A/V238A and K237_V238del do not affect protein folding. (**A**) Circular dichroism spectra of a2NT wild-type (red), mutant K237A/V238A (black), and mutant K237_V238.del (blue) in 50 mM Tris pH8.0 in the absence and (**B**) in the presence of 10mM SDS.

We tested in vitro PIP interactions of wildtype and mutants with PIP-enriched liposomes using a liposome pull-down assay. Wildtype and mutants were incubated with PolyPIPosome liposomes enriched with different PIPs, including the Golgi-specific PI(4)P, and its derivatives PI(3,4)P_2,PI(4,5)P_2 and PI(3,4,5)P_3 (labelled as PIP3). The protein–liposome complexes are collected via high-speed centrifugation and resolved by Western blot. WT a2NT showed significantly higher binding to PI(4)P compared to PI(3,4)P_2 and PIP3 (Figure ??A,B). While both mutations visually appear to reduce association with all polyPIPosome liposomes (Figure ??B left panel), only the liposomes enriched with PI(4)P resulted in a significant reduction (Figure ??B right panel). The binding differences between WT and mutants to I(3,4)P_2, PI(4,5)P_2, and PIP3 were not statistically significant (Figure ??C). This result is consistent with the fact that the a2 isoform is localized in Golgi and that PI(4)P is primarily found in Golgi.

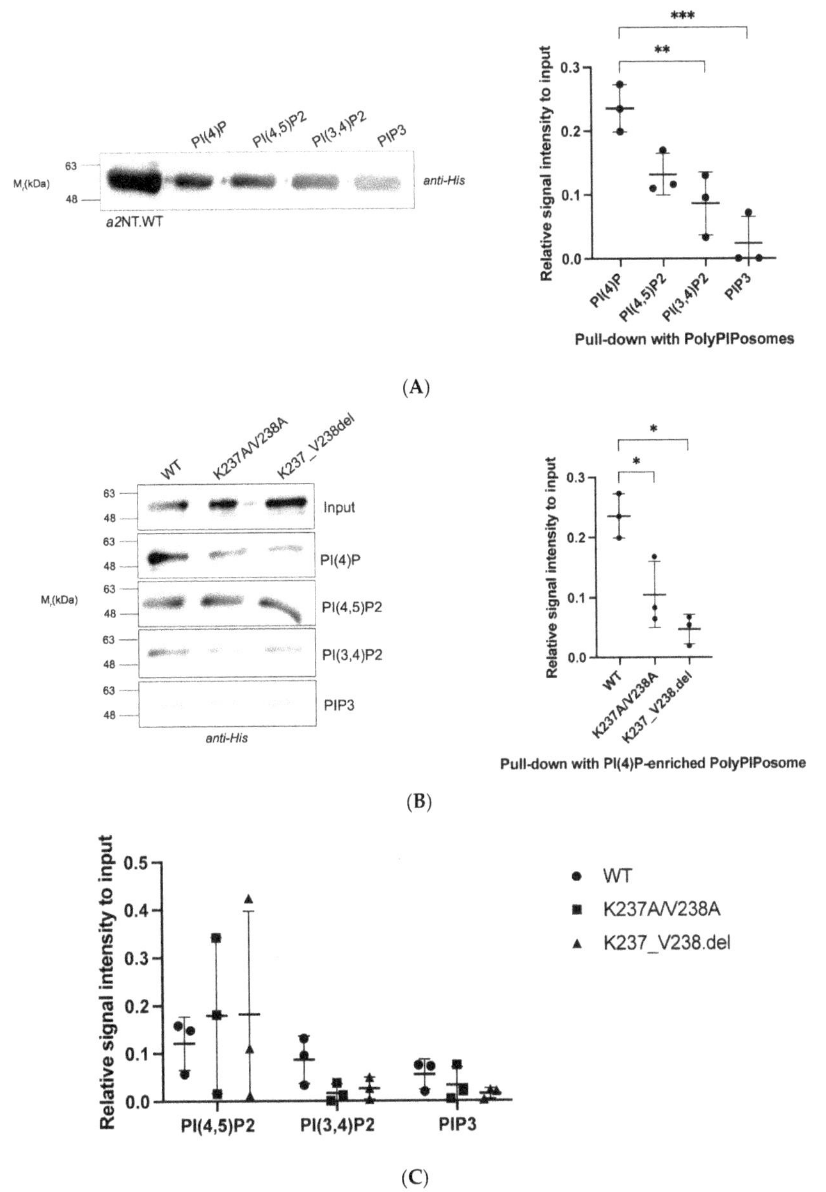

Figure 3. Mutations within the putative K[237]VKK[240] binding motif reduced interaction of a2NT with PI(4)P-enriched liposomes in vitro. (**A**) Liposome pull-down assay with PolyPIPosomes (Echelon) enriched with the indicated PIPs and HIS-tagged a2WT. In total, 20 µg of protein was incubated for 1 h at room temperature with 20 µL of 1 mM PolyPIPosomes containing 5% of the indicated PIPs in binding buffer (50 mM Tris, pH 7.5, 150 mM NaCl, 0.05% Nonidet P-40). Additionally, 5 µg of purified HIS-tagged a2WT was used as the input for loading control. (Right) Quantification was performed by measuring the intensity ratio of protein pulled down with the liposomes relative to input (n = 3). (**B**) Liposome pull-down assay of WT and mutant proteins with PolyPIPosomes (Echelon) enriched with indicated PIPs (PI(4)P, PI(4,5)P_2, PI(3,4)P_2, and PIP3. (Right) Quantification

by intensity ratio of WT and mutants pulled down with PI(4)P-enriched liposomes with respect to input. (C) Quantification by intensity ratio of WT and mutants pulled down with liposomes enriched with PI(4,5)P_2, PI(3,4)P_2, and PIP_3 with respect to input. n = 3 for all figures. Error bars indicate ± S.D. Statistical significance was analyzed by one-way ANOVA with Dunnett's multiple comparisons test comparing mutants to WT. * indicates $p < 0.05$. ** indicates $p < 0.01$, *** indicates $p < 0.001$.

These data suggest that the basic motif $K^{237}VKK^{240}$ in a2 is required for PI(4)P interaction, and mutations within this motif negatively impact the interaction. This further suggests that disruption of the a2-PI(4)P interaction could be the molecular mechanism underlying the disease-causing mutation, K237_V238del.

2.2. Mutations K237A/V238A and K237_V238del Reduce a2NT Golgi Localization

To assess cytosolic a2NT membrane association, we expressed FLAG-tagged wildtype a2NT and mutants a2NT K237A/V238A and K237_V238del in HEK293 cells and performed subcellular fractionation. Although the a2NT.FLAG lacks a transmembrane domain, it was still detected in microsomal fractions, suggesting that the membrane-bound C-terminal domain is not essential for membrane retention and that the N-terminal domain is sufficient to bring the protein to the membrane (Figure 4A). There was a significant decrease in the amount of both mutants in the microsomal fraction compared to the wildtype (Figure 4B). These results indicate that both mutations disrupt PI(4)P binding in vitro and reduce membrane retention in vivo.

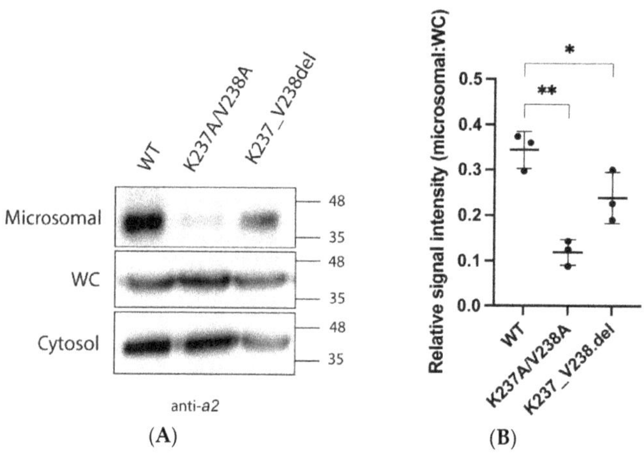

Figure 4. a2NT-WT co-purified with microsomal membranes, and mutants a2NT-K237A/V238A and K237_V238.del reduced membrane retention. (**A**) Plasmids containing a2NT wildtype (WT) and mutants K237A/V238A, K237_V238.del were transfected into HEK293 cells. Cellular fractionation performed to obtain cytosolic (cytosol) and microsomal fractions (microsomal). (**B**) Quantification was assessed by comparing the relative pixel intensity of microsomal fraction to whole cell extracts. (n = 3). Statistical significance was analyzed by one-way ANOVA with Dunnett's multiple comparisons test comparing mutants to WT. * indicates $p < 0.05$, ** indicates $p < 0.01$.

Immunofluorescence microscopy was used to visualize the localization of wildtype and mutant a2NT in the HEK293 cells. Wildtype a2NT.FLAG (red) were enriched at the Golgi, visualized with the Golgi specific marker, Tgn38-CFP (Figure 5A, top row). As a2NT was recruited to the Golgi in the absence of a membrane-bound C-terminal domain, this suggests that Golgi sorting information lies within the cytosolic N-terminal half. In contrast, there was a significant reduction in the Golgi localization of a2.K237A/V238A (Figure 5A, middle row) and a2.K237_V238del (Figure 5A, bottom row) (Figure 5B). This result aligns with the diminished membrane retention of the two mutants in microsomal fractions, supporting the hypothesis that the PI(4)P binding motif in a2 is, in part, responsible for Golgi membrane targeting/retention.

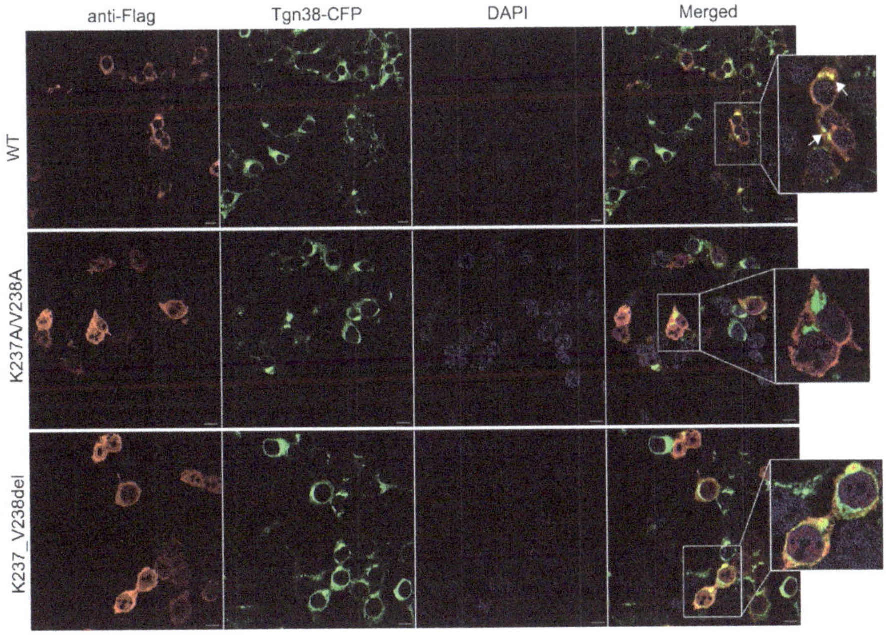

(A)

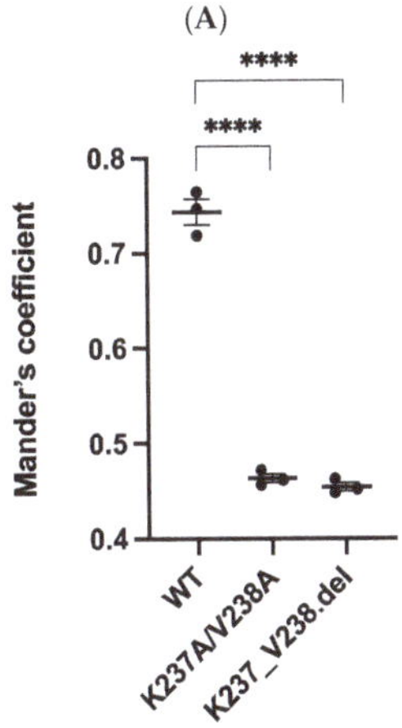

(B)

Figure 5. Mutations affect a2NT recruitment to Golgi. White arrows indicate the presence of a2NT at the Golgi. (**A**) Plasmids containing FLAG-tagged a2NT wildtype (WT), mutants K237A/V238A and K237_V238del were co-transfected with Tgn38-CFP in HEK293 cells. Cells were fixed, permeabilized, and stained for FLAG-tagged proteins (red) and DAPI (blue). (**B**) Quantification. A minimum of 30 cells from each cell line were measured for the intensity of the red signal at the vicinity of the Golgi membrane (green). Data represent mean value ± SEM from three independent experiments. Statistical significance was analyzed by one-way ANOVA with Dunnett's multiple comparisons test comparing mutants to WT. **** indicates $p < 0.0001$, scare bar = 10 μm.

2.3. Depletion of Golgi PI(4)P Impairs a2NT Recruitment to Golgi

We next tested whether the depletion of Golgi PI(4)P impairs a2NT Golgi recruitment. Sac1 is a PI(4)P phosphatase converting PI(4)P to PI [49,50]. We recruited Sac1 phosphatase to Golgi using the rapamycin-induced dimerization method [51–53]. Sac1 phosphatase coupled to FK506 binding protein FKBP (Sac1-FKBP) and Golgi membrane anchor Tgn38 coupled to FKBP-rapamycin binding domain FRB (Tgn38-FRB-CFP) were dimerized by the addition of rapamycin (Figure 6A). Sac1-PJ phosphatase is recruited to the Golgi (Figure 6B), where it converts PI(4)P to PI, resulting in the depletion of the Golgi pool of PI(4)P.

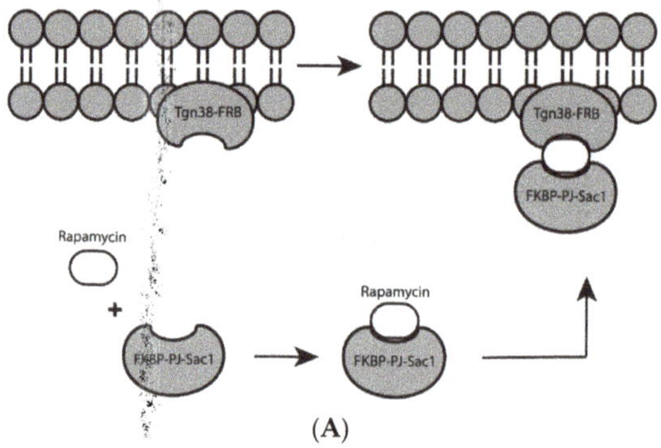

(**A**)

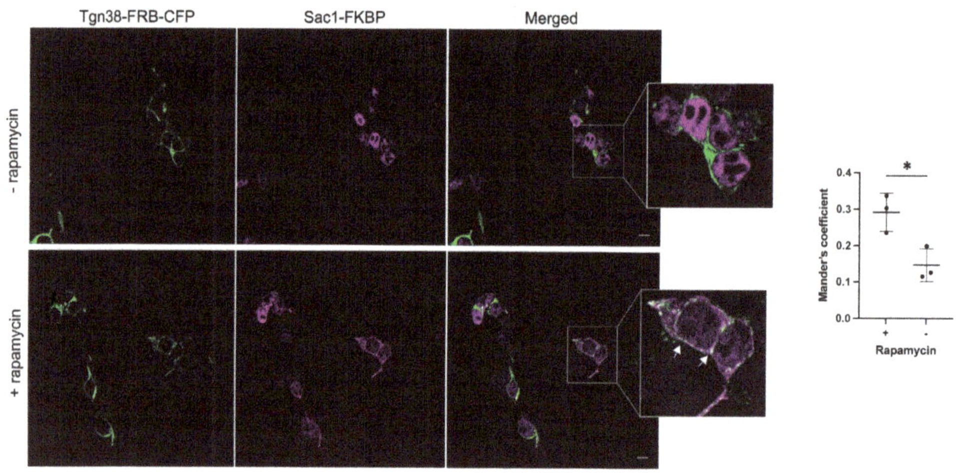

(**B**)

Figure 6. Rapamycin recruits Sac1-FKBP to the Golgi in the presence Tgn38-FRB. (**A**) Schematic illustration of rapamycin-induced dimerization. Sac1 phosphatase coupled to FK506 binding protein FKBP (Sac1-FKBP) and Golgi membrane anchor Tgn38 coupled to FKBP-rapamycin binding domain FRB (Tgn38-FRB-CFP) were dimerized by the addition of rapamycin. (**B**) Sac1-FKBP (magenta) recruitment to Golgi (white arrow), labeled by Tgn38 (green), upon treatment with 100 nM of rapamycin for 15 min at room temperature before fixing. Quantification: A minimum of 30 cells from each cell line. Data represent mean value SEM from three independent experiments. A paired t-test was run to analyze the significance in mean difference. * indicates $p < 0.05$, scare bar = 10 µm.

HEK293 cells were co-transfected with plasmids containing a2NT.FLAG, Sac1-PJ-FKBP, and Tgn38-FRB-CFP. a2NT co-localized with Tgn38-CFP, indicating recruitment to the Golgi (Figure 7A, top row). Additionally, 100 nM rapamycin decreased the a2NT intensity at the Golgi (Figure 7A, bottom row), suggesting that depletion of Golgi PI(4)P impairs a2NT's localization/retention at the Golgi.

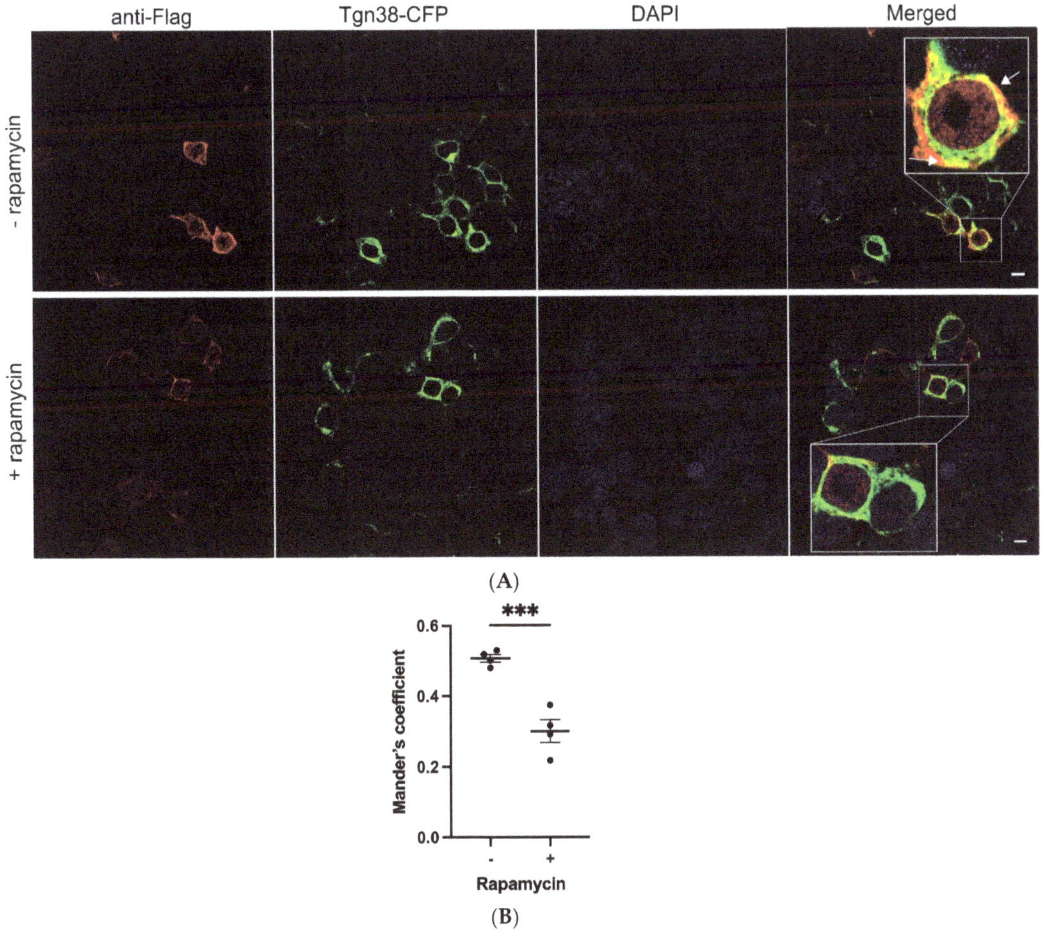

Figure 7. Depletion of Golgi PI(4)P impairs a2NT recruitment to Golgi. White arrows indicate the localization of a2NT at Golgi (**A**) Plasmids containing FLAG-tagged a2NT wildtype (WT) were co-transfected with Tgn38-FRB-CFP (green) and Sac1-FKBP in HEK293 cells. Cells were treated with 100 nM of rapamycin for 15 min at 30 h post-transfection. Cells were then fixed, permeabilized, and stained for FLAG-tagged proteins (red) and DAPI (blue). (**B**) Quantification was performed with a minimum of 30 cells from each cell line measured for the intensity of the red signal in the vicinity of Golgi (green). Data represent mean value SEM from three independent experiments. A paired *t*-test was run to analyze the significance in mean difference. *** indicates $p < 0.001$. scare bar = 10 µm.

3. Discussion

The cytosolic N-terminal domain of the *a* subunit serves as a connector for V_1 and V_o assembly [54] and is an important target for multiple V-ATPase regulators [15,17,20,21]. Experiments with chimeras of the two yeast orthologs, Vph1p and Stv1p, showed that the *a*NT contains information for both the localization and regulation of V-ATPase by reversible assembly [13]. Phosphoinositides regulate transmembrane channels and transporters [42,45,46]. Interactions between *a* subunit isoforms and different PIPs may impact both functional regulation and localization, which, in turn, could account for the differences in functional destinations among *a* subunit isoforms.

Previously, we proposed a putative lipid binding domain in the *a* subunit and the possible involvement of this domain in V-ATPase regulation [47]. Sequence alignment of the four mammalian V-ATPase *a* subunit isoforms with the two yeast orthologs revealed a conserved basic motif $KX_nK(R)IK(R)$ required for the PIP's interaction in the N-terminal domain of a4. Here, we provide evidence that the Golgi-specific a2 isoform directly interacts with PI(4)P, a Golgi-enriched PIP. Our results with a2 mutations, K237A/V238A and K237_V238del, indicate that mutations within the binding motif ($K^{237}VKK^{240}$) compromised the membrane association and Golgi localization, further suggesting that this basic motif is essential for PI(4)P binding. The depletion of Golgi PI(4)P with rapamycin had similar effects on a2 localization, indicating that a2-PI(4)P interaction is important for the Golgi localization of a2-containing V-ATPases. A recent study identified another PI(4)P interaction sequence $K^{221}WY$ within the vicinity of the putative binding domain, and the mutation of the K221 residue compromised PI(4)P binding [55]. PIPs are expressed as lipid rafts within the membrane [56,57] so that multiple basic residues are exposed to the membrane, which would strengthen protein–lipid interactions. We hypothesize that the $K^{221}WY$ sequence could help to strengthen the PIP/protein interaction or potentially define the PIP's specificity of the a2 isoform.

PIP binding is often associated with protein conformational changes [58–61]. Structural analyses of yeast V-ATPases suggest that there are conformational changes between Vph1pNT in holoenzyme V_1-V_o and in free V_o, as well as movement of the N-terminal domain in different states of the active enzyme [6,62]. Our putative lipid binding motif is within the distal lobe of the *a*NT, which rotates between the different V-TPase active states. Similar to the stabilization of K^+ channel Kir2.2 upon binding to $PI(4,5)P_2$ [59], it is possible that binding to the Golgi PI(4)P traps the a2 isoform at the Golgi membrane as well as in a conformational state that promotes V-ATPase assembly and/or activity. Studies in yeast indicate that mutations resulting in the loss of PI(4)P binding compromise Stv1p-containing V-ATPases' function, resulting in growth defects at alkaline pH [40]. Structural studies of conformational changes induced by PIP binding can provide mechanistic insights for the functional role of this interaction. Nevertheless, *a*NT is the target of multiple V-ATPase regulators [15,17,20,41,63], and PIP interaction is only one mode of V-ATPase regulation at specific membranes.

The a2 K237_V238del mutation has been identified in patients with cutis laxa [48]. Here, we show that this mutation disrupted PI(4)P interaction and compromised Golgi localization. The characterization of conserved residues implicated in diseases has been successfully used to determine functional domains and has informed the discovery of novel therapeutic targets [64–66]. Understanding *a*-PIP interactions and their impact on V-ATPase localization and regulation could similarly inform the development of a therapeutic control of V-ATPase subpopulations, enabling the inhibition, specifically, of V-ATPases involved in osteoporosis [7,21], and metastatic cancer [67,68].

4. Materials and Methods

4.1. Expression and Purification of Human a2NT Wildtype and Mutants K237A/V238A, K237_V238.del from E. coli

pET32a+ :: human V-ATPase a2NT (MM1115). N-terminal domain of human a2 from ATG to T400 was obtained by PCR with primers MO501: 5′-ACGTGGTACCA TGGGCTCCATGTTCCGGAG and MO502: 5′-ACGTGAATTCACAGATCTCCGCCGGTGT AGGGAGCGGGGTTGAC. The PCR product was cloned into pET32a(+) plasmids between *KpnI* and *EcoRV* sites, resulting in MM1111. pcDNA3 :: human V-ATPase a2NT (MM1127) *KpnI* and *EcoRV* insert of a2NT in MM1115 was moved into pcDNA3.1+; the new construct was named MM1127. pET32a+ :: human V-ATPase a2NT K237A/V238A (MM1121) Q5 Site-Directed Mutagenesis Kit (NEB E0554S) was used on MM1115 to make human V-ATPase a2NT.K237A/V238A mutant with primers MO523: 5′-GCAACATCGACGTCACCCAGCAG and MO524: 5′-ACGTGAATTCTTAGGTGTAGGGGGCTGGGTTTATC, sequencing verified. pcDNA3 :: human V-ATPase a2NT K237A/V238A (MM1128) *KpnI* and *EcoRV* fragment of MM1121 was ligated into pcDNA3.1+, and the new construct was named MM1128. pET32a+ :: human V-ATPase a2NT K237_V238.del (MM1122) Q5 Site-Directed Mutagenesis Kit (NEB E0554S) was used on MM1115 to make human V-ATPase a2NT.K238_V238.del mutant. pcDNA3 :: human V-ATPase a2NT K237_V238.del (MM1129) The *KpnI* and *EcoRV* insert of MM1122 was ligated into pcDNA3.1+; the new construct was named MM1129.

The N-terminal domain of human a2 (amino acid 1–400) with a 6X His tag at the carboxyl end was expressed in E.coli Rosetta (DE3) via pET32a plasmid and purified as described in a previous study [47].

4.2. PolyPIPosome Pull-Down Assay

In total, 20 µg of purified proteins, a2NT wildtype, K237A/V238A, and K237_V238.del, was incubated with 20 µL 1 mM PolyPIPosomes (Echelon, US) and 200 µL of binding buffer (50 mM Tris pH8.0, 150 mM NaCl, and 0.05% Nonidet P-40). Pull-down assay was performed as in [47] as well as a Western blot with mouse anti-His antibody (Sigma SAB1305538) and goat anti-mouse IgG secondary antibody (Invitrogen 31430).

4.3. HEK293 Transfection and Cellular Fractionation

HEK293 cells (ATCC, US) were cultured on 10 cm culture dishes in Dulbecco's modified Eagle's medium (DMEM) (Gibco, US) containing 10% fetal bovine serum (FBS) and 0.5% antibiotics, and grown in a 95% air, 5% CO_2 humidified environment at 37 °C. pcDNA3 plasmids of human a2 N-terminal domain (amino acid 1–400) wildtype and mutants K237A/V238A, K237_V238.del, 5 µg of plasmid/dish, were transfected into HEK293 cells using PolyJet Reagent (SignaGen, US) in accordance with the procedure recommended by the manufacturer. Cellular fractionation was as described in [47]. The fractions were analyzed by Western blot with anti-atp6v0a2 (Abcam, UK ab96803).

4.4. Immunofluorescence

In the rapamycin treatment experiment, cells were treated with 100 nM of rapamycin for 15 min at room temperature before fixing. Images were acquired with a confocal microscope (Leica Confocal SP8, Germany) using a 63x oil objective. Colocalizations were measured with Mander's coefficient M1 [69]. Antibodies used were as follows: mouse anti-DDDDK tag (Abcam ab18230), goat anti-mouse IgG Alexa Fluor ™ 647 (Invitrogen A21235), and DAPI stain (Invitrogen D1306).

4.5. Statistical Analysis

GraphPad Prism 9.4.1 software was used for statistical analysis and statistical graph production. One-way ANOVA followed by Dunnett's multiple comparison test or Student's *t*-test were used as indicated in figure legends. In figures, asterisks are used as follows: * indicates $p < 0.05$, ** indicates $p < 0.01$, and *** indicates $p < 0.001$. The experimental results are expressed as the mean ± SEM.

Author Contributions: A.C. performed all experiments except those represented in Figure 2, which were performed by M.G. and Y.Y. prepared constructs used in the study and provided technical expertise. C.M.D. helped with the initial design of CD experiments and provided expertise in the data interpretation of the experiments. M.F.M. supervised the project, participated in data interpretation, and revised the draft of the manuscript. All authors have read and agreed to the published version of the manuscript.

Funding: This work was supported by Operating Grants PJT-14850 from the Canadian Institutes of Health Research and RGPIN-2022-05169 from the Natural Science and Engineering Research Council of Canada to M.F.M, and by Operating Grant #376666 from the Canadian Institutes of Health Research to C.M.D.

Institutional Review Board Statement: The HEK293 cell line was obtained from ATCC. Human a2NT plasmid constructs were cloned in the lab as described above.

Informed Consent Statement: Not applicable.

Data Availability Statement: The original contributions presented in the study are included in the article material, further inquiries can be directed to the corresponding author.

Acknowledgments: The protein constructs Tgn38-FRB-CFP and Sac1-FKBP are kind gifts from Sergio Grinstein's lab.

Conflicts of Interest: The authors declare no conflicts of interest. The funders had no role in the design of the study; in the collection, analyses, or interpretation of data; in the writing of the manuscript; or in the decision to publish the results.

References

1. Kibak, H.; Taiz, L.; Starke, T.; Bernasconi, P.; Gogarten, J.P. Evolution of structure and function of V-ATPases. *J. Bioenerg. Biomembr.* **1992**, *24*, 415–424. [CrossRef] [PubMed]
2. Futai, M.; Sun-Wada, G.H.; Wada, Y.; Matsumoto, N.; Nakanishi-Matsui, M. Vacuolar-type ATPase: A proton pump to lysosomal trafficking. *Proc. Jpn. Acad. Ser. B Phys. Biol. Sci.* **2019**, *95*, 17. [CrossRef]
3. Nelson, N. Structure and function of V-ATPases in endocytic and secretory organelles. *J. Exp. Biol.* **1992**, *172*, 149–153. [CrossRef]
4. Forgac, M. Vacuolar ATPases: Rotary proton pumps in physiology and pathophysiology. *Nat. Rev. Mol. Cell Biol.* **2007**, *8*, 917–929. [CrossRef] [PubMed]
5. Futai, M.; Oka, T.; Sun-Wada, G.; Moriyama, Y.; Kanazawa, H.; Wada, Y. Luminal acidification of diverse organelles by V-ATPase in animal cells. *J. Exp. Biol.* **2000**, *203 Pt 1*, 107–116. [CrossRef]
6. Zhao, J.; Benlekbir, S.; Rubinstein, J.L. Electron cryomicroscopy observation of rotational states in a eukaryotic V-ATPase. *Nature* **2015**, *521*, 241–245. [CrossRef] [PubMed]
7. Kartner, N.; Manolson, M.F. V-ATPase subunit interactions: The long road to therapeutic targeting. *Curr. Protein Pept. Sci.* **2012**, *13*, 164–179. [CrossRef]
8. Wada, Y.; Sun-Wada, G.H.; Tabata, H.; Kawamura, N. Vacuolar-type proton ATPase as regulator of membrane dynamics in multicellular organisms. *J. Bioenerg. Biomembr.* **2008**, *40*, 53–57. [CrossRef]
9. Perzov, N.; Padler-Karavani, V.; Nelson, H.; Nelson, N. Characterization of yeast V-ATPase mutants lacking Vph1p or Stv1p and the effect on endocytosis. *J. Exp. Biol.* **2002**, *205 Pt 9*, 1209–1219. [CrossRef]
10. Finnigan, G.C.; Cronan, G.E.; Park, H.J.; Srinivasan, S.; Quiocho, F.A.; Stevens, T.H. Sorting of the yeast vacuolar-type, proton-translocating ATPase enzyme complex (V-ATPase): Identification of a necessary and sufficient Golgi/endosomal retention signal in Stv1p. *J. Biol. Chem.* **2012**, *287*, 19487–19500. [CrossRef]
11. Manolson, M.F.; Wu, B.; Proteau, D.; Taillon, B.E.; Roberts, B.T.; Hoyt, M.A.; Jones, E.W. STV1 gene encodes functional homologue of 95-kDa yeast vacuolar H(+)-ATPase subunit Vph1p. *J. Biol. Chem.* **1994**, *269*, 14064–14074. [CrossRef]
12. Hinton, A.; Sennoune, S.R.; Bond, S.; Fang, M.; Reuveni, M.; Sahagian, G.G.; Jay, D.; Martinez-Zaguilan, R.; Forgac, M. Function of a subunit isoforms of the V-ATPase in pH homeostasis and in vitro invasion of MDA-MB231 human breast cancer cells. *J. Biol. Chem.* **2009**, *284*, 16400–16408. [CrossRef] [PubMed]
13. Kawasaki-Nishi, S.; Bowers, K.; Nishi, T.; Forgac, M.; Stevens, T.H. The amino-terminal domain of the vacuolar proton-translocating ATPase a subunit controls targeting and in vivo dissociation, and the carboxyl-terminal domain affects coupling of proton transport and ATP hydrolysis. *J. Biol. Chem.* **2001**, *276*, 47411–47420. [CrossRef] [PubMed]
14. Aoto, K.; Kato, M.; Akita, T.; Nakashima, M.; Mutoh, H.; Akasaka, N.; Tohyama, J.; Nomura, Y.; Hoshino, K.; Ago, Y.; et al. ATP6V0A1 encoding the a1-subunit of the V0 domain of vacuolar H(+)-ATPases is essential for brain development in humans and mice. *Nat. Commun.* **2021**, *12*, 2107. [CrossRef] [PubMed]
15. Wallings, R.; Connor-Robson, N.; Wade-Martins, R. LRRK2 interacts with the vacuolar-type H+-ATPase pump a1 subunit to regulate lysosomal function. *Hum. Mol. Genet.* **2019**, *28*, 2696–2710. [CrossRef] [PubMed]
16. Morel, N.; Dedieu, J.C.; Philippe, J.M. Specific sorting of the a1 isoform of the V-H+ATPase a subunit to nerve terminals where it associates with both synaptic vesicles and the presynaptic plasma membrane. *J. Cell Sci.* **2003**, *116 Pt 23*, 4751–4762. [CrossRef] [PubMed]
17. Zhang, W.; Wang, D.; Volk, E.; Bellen, H.J.; Hiesinger, P.R.; Quiocho, F.A. V-ATPase V0 sector subunit a1 in neurons is a target of calmodulin. *J. Biol. Chem.* **2008**, *283*, 294–300. [CrossRef] [PubMed]
18. Marshansky, V. The V-ATPase a2-subunit as a putative endosomal pH-sensor. *Biochem. Soc. Trans.* **2007**, *35 Pt 5*, 1092–1099. [CrossRef] [PubMed]
19. Merkulova, M.; McKee, M.; Dip, P.V.; Gruber, G.; Marshansky, V. N-terminal domain of the V-ATPase a2-subunit displays integral membrane protein properties. *Protein Sci.* **2010**, *19*, 1850–1862. [CrossRef]
20. Marshansky, V.; Hosokawa, H.; Merkulova, M.; Bakulina, A.; Dip, P.V.; Thaker, Y.R.; Bjargava, A.; Tonra, J.R.; Ausiello, D.A.; Gruber, G. Structural model of a2-subunit N-terminus and its binding interface for Arf-GEF CTH2: Implication for regulation of V-ATPase, CTH2 function and rational drug design. *Curr. Top. Membr.* **2019**, *83*, 77–106.

21. Chu, A.; Zirngibl, R.A.; Manolson, M.F. The V-ATPase a3 Subunit: Structure, Function and Therapeutic Potential of an Essential Biomolecule in Osteoclastic Bone Resorption. *Int. J. Mol. Sci.* **2021**, *22*, 6934. [CrossRef]
22. Matsumoto, N.; Sekiya, M.; Sun-Wada, G.H.; Wada, Y.; Nakanishi-Matsui, M. The lysosomal V-ATPase a3 subunit is involved in localization of Mon1-Ccz1, the GEF for Rab7, to secretory lysosomes in osteoclasts. *Sci. Rep.* **2022**, *12*, 8455. [CrossRef]
23. Manolson, M.F.; Yu, H.; Chen, W.; Yao, Y.; Li, K.; Lees, R.L.; Heersche, J.N. The a3 isoform of the 100-kDa V-ATPase subunit is highly but differentially expressed in large (>or=10 nuclei) and small (<or= nuclei) osteoclasts. *J. Biol. Chem.* **2003**, *278*, 49271–49278.
24. Brown, D.; Sabolic, I.; Gluck, S. Polarized targeting of V-ATPase in kidney epithelial cells. *J. Exp. Biol.* **1992**, *172*, 231–243. [CrossRef]
25. Oka, T.; Murata, Y.; Namba, M.; Yoshimizu, T.; Toyomura, T.; Yamamoto, A.; Sun-Wada, G.H.; Hamasaki, N.; Wada, Y.; Futai, M. a4, a unique kidney-specific isoform of mouse vacuolar H+-ATPase subunit a. *J. Biol. Chem.* **2001**, *276*, 40050–40054. [CrossRef]
26. Pietrement, C.; Sun-Wada, G.H.; Silva, N.D.; McKee, M.; Marshansky, V.; Brown, D.; Futai, M.; Breton, S. Distinct expression patterns of different subunit isoforms of the V-ATPase in the rat epididymis. *Biol. Reprod.* **2006**, *74*, 185–194. [CrossRef]
27. Golder, Z.J.; Karet Frankl, F.E. Extra-renal locations of the a4 subunit of H(+)ATPase. *BMC Cell Biol.* **2016**, *17*, 27. [CrossRef] [PubMed]
28. Kornak, U.; Reynders, E.; Dimopoulou, A.; van Reeuwijk, J.; Fischer, B.; Rajab, A.; Budde, B.; Nurnberg, P.; Foulquier, F.; ARCL Debré-type Study Group; et al. Impaired glycosylation and cutis laxa caused by mutations in the vesicular H+-ATPase subunit ATP6V0A2. *Nat. Genet.* **2008**, *40*, 32–34. [CrossRef] [PubMed]
29. Van Maldergem, L.; Dobyns, W.; Kornak, U. ATP6V0A2-Related Cutis Laxa. In *GeneReviews®*; Adam, M.P., Mirzaa, G.M., Pagon, R.A., Wallace, S.E., Bean, L.J.H., Gripp, K.W., Amemiya, A., Eds.; University of Washington: Seattle, WA, USA, 1993.
30. Beyens, A.; Moreno-Artero, E.; Bodemer, C.; Cox, H.; Gezdirici, A.; Yilmaz Gulec, E.; Kahloul, N.; Khau Van Kien, P.; Ogur, G.; Harroche, A.; et al. ATP6V0A2-related cutis laxa in 10 novel patients: Focus on clinical variability and expansion of the phenotype. *Exp. Dermatol.* **2019**, *28*, 1142–1145. [CrossRef]
31. Hucthagowder, V.; Morava, E.; Kornak, U.; Lefeber, D.J.; Fischer, B.; Dimopoulou, A.; Aldinger, A.; Choi, J.; Davis, E.C.; Abuelo, D.N.; et al. Loss-of-function mutations in ATP6V0A2 impair vesicular trafficking, tropoelastin secretion and cell survival. *Hum. Mol. Genet.* **2009**, *18*, 2149–2165. [CrossRef] [PubMed]
32. Matsumoto, N.; Matsukawa, R.; Takahashi, S.; Kudo, K.; Sun-Wada, G.H.; Wada, Y.; Nakanishi-Matsui, M. V-ATPase a3 isoform mutations identified in osteopetrosis patients abolish its expression and disrupt osteoclast function. *Exp. Cell Res.* **2020**, *389*, 111901. [CrossRef]
33. Ochotny, N.; Voronov, I.; Owen, C.; Aubin, J.E.; Manolson, M.F. The R740S mutation in the V-ATPase a3 subunit results in osteoclast apoptosis and defective early-stage autophagy. *J. Cell. Biochem.* **2013**, *114*, 2823–2833. [CrossRef]
34. Ochotny, N.; Flenniken, A.M.; Owen, C.; Voronov, I.; Zirngibl, R.A.; Osborne, L.R.; Henderson, J.E.; Adamson, S.L.; Rossant, J.; Manolson, M.F.; et al. The V-ATPase a3 subunit mutation R740S is dominant negative and results in osteopetrosis in mice. *J. Bone Miner. Res.* **2011**, *26*, 1484–1493. [CrossRef]
35. Pangrazio, A.; Caldana, M.E.; Lo Iacono, N.; Mantero, S.; Vezzoni, P.; Villa, A.; Sobacchi, C. Autosomal recessive osteopetrosis: Report of 41 novel mutations in the TCIRG1 gene and diagnostic implications. *Osteoporos. Int.* **2012**, *23*, 2713–2718. [CrossRef]
36. Stehberger, P.A.; Schulz, N.; Finberg, K.E.; Karet, F.E.; Giebisch, G.; Lifton, R.P.; Geibel, J.P.; Wagner, C.A. Localization and regulation of the ATP6V0A4 (a4) vacuolar H+-ATPase subunit defective in an inherited form of distal renal tubular acidosis. *J. Am. Soc. Nephrol.* **2003**, *14*, 3027–3038. [CrossRef]
37. Stover, E.H.; Borthwick, K.J.; Bavalia, C.; Eady, N.; Fritz, D.M.; Rungroj, N.; Giersch, A.B.; Morton, C.C.; Axon, P.R.; Akil, I.; et al. Novel ATP6V1B1 and ATP6V0A4 mutations in autosomal recessive distal renal tubular acidosis with new evidence for hearing loss. *J. Med. Genet.* **2002**, *39*, 796–803. [CrossRef]
38. Srinivasan, S.; Vyas, N.K.; Baker, M.L.; Quiocho, F.A. Crystal structure of the cytoplasmic N-terminal domain of subunit I, a homolog of subunit a, of V-ATPase. *J. Mol. Biol.* **2011**, *412*, 14–21. [CrossRef] [PubMed]
39. Kartner, N.; Yao, Y.; Bhargava, A.; Manolson, M.F. Topology, glycosylation and conformational changes in the membrane domain of the vacuolar H+-ATPase a subunit. *J. Cell. Biochem.* **2013**, *114*, 1474–1487. [CrossRef] [PubMed]
40. Banerjee, S.; Kane, P.M. Direct interaction of the Golgi V-ATPase a-subunit isoform with PI(4)P drives localization of Golgi V-ATPases in yeast. *Mol. Biol. Cell* **2017**, *28*, 2518–2530. [CrossRef] [PubMed]
41. Li, S.C.; Diakov, T.T.; Xu, T.; Tarsio, M.; Zhu, W.; Couoh-Cardel, S.; Weisman, L.S.; Kane, P.M. The signaling lipid PI(3,5)P(2) stabilizes V(1)-V(o) sector interactions and activates the V-ATPase. *Mol. Biol. Cell* **2014**, *25*, 1251–1262. [CrossRef] [PubMed]
42. Roth, M.G. Phosphoinositides in constitutive membrane traffic. *Physiol. Rev.* **2004**, *84*, 699–730. [CrossRef] [PubMed]
43. Payrastre, B. Phosphoinositides: Lipid kinases and phosphatases. *Methods Mol. Biol.* **2004**, *273*, 201–212.
44. Kanaho, Y.; Suzuki, T. Phosphoinositide kinases as enzymes that produce versatile signaling lipids, phosphoinositides. *J. Biochem.* **2002**, *131*, 503–509. [CrossRef]
45. Posor, Y.; Jang, W.; Haucke, V. Phosphoinositides as membrane organizers. *Nat. Rev. Mol. Cell Biol.* **2022**, *23*, 797–816. [CrossRef] [PubMed]
46. Larijani, B.; Pytowski, L.; Vaux, D.J. The enigma of phosphoinositides and their derivatives: Their role in regulation of subcellular compartment morphology. *Biochim. Biophys. Acta Biomembr.* **2022**, *1864*, 183780. [CrossRef] [PubMed]

47. Chu, A.; Yao, Y.; Saffi, G.T.; Chung, J.H.; Botelho, R.J.; Glibowicka, M.; Deber, C.M.; Manolson, M.F. Characterization of a PIP Binding Site in the N-Terminal Domain of V-ATPase a4 and Its Role in Plasma Membrane Association. *Int. J. Mol. Sci.* **2023**, *24*, 4867. [CrossRef]
48. Fischer, B.; Dimopoulou, A.; Egerer, J.; Gardeitchik, T.; Kidd, A.; Jost, D.; Kayserili, H.; Alanay, Y.; Tantcheva-Poor, I.; Mangold, E.; et al. Further characterization of ATP6V0A2-related autosomal recessive cutis laxa. *Hum. Genet.* **2012**, *131*, 1761–1773. [CrossRef]
49. Zewe, J.P.; Wills, R.C.; Sangappa, S.; Goulden, B.D.; Hammond, G.R. SAC1 degrades its lipid substrate PtdIns4P in the endoplasmic reticulum to maintain a steep chemical gradient with donor membranes. *eLife* **2018**, *7*, e35588. [CrossRef]
50. Del Bel, L.M.; Brill, J.A. Sac1, a lipid phosphatase at the interface of vesicular and nonvesicular transport. *Traffic* **2018**, *19*, 301–318. [CrossRef]
51. Kim, S.; Park, J.; Jeon, B.W.; Hwang, G.; Kang, N.Y.; We, Y.; Park, W.Y.; Oh, E.; Kim, J. Chemical control of receptor kinase signaling by rapamycin-induced dimerization. *Mol. Plant* **2021**, *14*, 1379–1390. [CrossRef]
52. Wang, Y.; Barnett, S.F.H.; Le, S.; Guo, Z.; Zhong, X.; Kanchanawong, P.; Yan, J. Label-free Single-Molecule Quantification of Rapamycin-induced FKBP-FRB Dimerization for Direct Control of Cellular Mechanotransduction. *Nano Lett.* **2019**, *19*, 7514–7525. [CrossRef] [PubMed]
53. Dickson, E.J.; Jensen, J.B.; Hille, B. Golgi and plasma membrane pools of PI(4)P contribute to plasma membrane PI(4,5)P2 and maintenance of KCNQ2/3 ion channel current. *Proc. Natl. Acad. Sci. USA* **2014**, *111*, E2281–E2290. [CrossRef] [PubMed]
54. Couoh-Cardel, S.; Milgrom, E.; Wilkens, S. Affinity Purification and Structural Features of the Yeast Vacuolar ATPase Vo Membrane Sector. *J. Biol. Chem.* **2015**, *290*, 27959–27971. [CrossRef] [PubMed]
55. Mitra, C.; Winkley, S.; Kane, P.M. Human V-ATPase a-subunit isoforms bind specifically to distinct phosphoinositide phospholipids. *J. Biol. Chem.* **2023**, *299*, 105473. [CrossRef] [PubMed]
56. Helms, J.B.; Zurzolo, C. Lipids as targeting signals: Lipid rafts and intracellular trafficking. *Traffic* **2004**, *5*, 247–254. [CrossRef] [PubMed]
57. Golub, T.; Wacha, S.; Caroni, P. Spatial and temporal control of signaling through lipid rafts. *Curr. Opin. Neurobiol.* **2004**, *14*, 542–550. [CrossRef] [PubMed]
58. Huang, C.L.; Feng, S.; Hilgemann, D.W. Direct activation of inward rectifier potassium channels by PIP2 and its stabilization by Gbetagamma. *Nature* **1998**, *391*, 803–806. [CrossRef]
59. Zhang, H.; He, C.; Yan, X.; Mirshahi, T.; Logothetis, D.E. Activation of inwardly rectifying K+ channels by distinct PtdIns(4,5)P2 interactions. *Nat. Cell Biol.* **1999**, *1*, 183–188. [CrossRef]
60. Lopes, C.M.; Zhang, H.; Rohacs, T.; Jin, T.; Yang, J.; Logothetis, D.E. Alterations in conserved Kir channel-PIP2 interactions underlie channelopathies. *Neuron* **2002**, *34*, 933–944. [CrossRef]
61. Milburn, C.C.; Deak, M.; Kelly, S.M.; Price, N.C.; Alessi, D.R.; Van Aalten, D.M. Binding of phosphatidylinositol 3,4,5-trisphosphate to the pleckstrin homology domain of protein kinase B induces a conformational change. *Biochem. J.* **2003**, *375 Pt 3*, 531–538. [CrossRef]
62. Mazhab-Jafari, M.T.; Rohou, A.; Schmidt, C.; Bueler, S.A.; Benlekbir, S.; Robinson, C.V.; Rubinstein, J.L. Atomic model for the membrane-embedded VO motor of a eukaryotic V-ATPase. *Nature* **2016**, *539*, 118–122. [CrossRef]
63. Sautin, Y.Y.; Lu, M.; Gaugler, A.; Zhang, L.; Gluck, S.L. Phosphatidylinositol 3-kinase-mediated effects of glucose on vacuolar H+-ATPase assembly, translocation, and acidification of intracellular compartments in renal epithelial cells. *Mol. Cell. Biol.* **2005**, *25*, 575–589. [CrossRef] [PubMed]
64. Guevara-Coto, J.; Schwartz, C.E.; Wang, L. Protein sector analysis for the clustering of disease-associated mutations. *BMC Genom.* **2014**, *15* (Suppl. 11), S4. [CrossRef] [PubMed]
65. Valastyan, J.S.; Lindquist, S. Mechanisms of protein-folding diseases at a glance. *Dis. Model. Mech.* **2014**, *7*, 9–14. [CrossRef] [PubMed]
66. Okiyoneda, T.; Veit, G.; Dekkers, J.F.; Bagdany, M.; Soya, N.; Xu, H.; Roldan, A.; Verkman, A.S.; Kurth, M.; Simon, A.; et al. Mechanism-based corrector combination restores DeltaF508-CFTR folding and function. *Nat. Chem. Biol.* **2013**, *9*, 444–454. [CrossRef] [PubMed]
67. Stransky, L.; Cotter, K.; Forgac, M. The Function of V-ATPases in Cancer. *Physiol. Rev.* **2016**, *96*, 1071–1091. [CrossRef]
68. Alves, M.G.O.; Garcia-Garcia, A.; Perez-Sayans, M. V-ATPases and Their Implication in Oral Cancer. In *Regulation of Ca2^{+}-ATPases, V-ATPases and F-ATPases*, 1st ed.; Chakraborti, S., Dhalla, N.S., Eds.; Springer: Cham, Switzerland, 2016; Volume 14, pp. 393–405.
69. Pike, J.A.; Styles, I.B.; Rappoport, J.Z.; Heath, J.K. Quantifying receptor trafficking and colocalization with confocal microscopy. *Methods* **2017**, *115*, 42–54. [CrossRef]

International Journal of
Molecular Sciences

MDPI

Article

Structural Insights into the Penicillin-Binding Protein 4 (DacB) from *Mycobacterium tuberculosis*

Sung-Min Kang [1] and Do-Hee Kim [2,3,*]

[1] College of Pharmacy, Duksung Women's University, Seoul 01369, Republic of Korea
[2] Jeju Research Institute of Pharmaceutical Sciences, College of Pharmacy, Jeju National University, Jeju 63243, Republic of Korea
[3] Interdisciplinary Graduate Program in Advanced Convergence Technology & Science, Jeju National University, Jeju 63243, Republic of Korea
* Correspondence: doheekim@jejunu.ac.kr; Tel.: +82-64-754-8195

Abstract: *Mycobacterium tuberculosis*, a major cause of mortality from a single infectious agent, possesses a remarkable mycobacterial cell envelope. Penicillin-Binding Proteins (PBPs) are a family of bacterial enzymes involved in the biosynthesis of peptidoglycan. PBP4 (DacB) from *M. tuberculosis* (MtbPBP4) has been known to function as a carboxypeptidase, and the role and significance of carboxypeptidases as targets for anti-tuberculosis drugs or antibiotics have been extensively investigated over the past decade. However, their precise involvement remains incompletely understood. In this study, we employed predictive modeling and analyzed the three-dimensional structure of MtbPBP4. Interestingly, MtbPBP4 displayed a distinct domain structure compared to its homologs. Docking studies with meropenem verified the presence of active site residues conserved in PBPs. These findings establish a structural foundation for comprehending the molecular function of MtbPBP4 and offer a platform for the exploration of novel antibiotics.

Keywords: *Mycobacterium tuberculosis*; antibiotics; penicillin-binding protein

1. Introduction

Mycobacterium tuberculosis is a dangerous bacterium responsible for tuberculosis (TB), which is a significant global health concern, especially in developing countries [1]. *M. tuberculosis* has a thick, lipid-rich cell wall, which acts as a barrier and restricts the entry of many drugs, including β-lactam antibiotics. This impermeable cell wall hampers the penetration of antibiotics into the bacterial cell, reducing their efficacy. Additionally, it is crucial to acknowledge the significant role of efflux pumps in contributing to the beta-lactam tolerance of *M. tuberculosis* [2,3]. Therefore, β-lactam antibiotics are not typically used as first-line drugs for TB treatment due to their inherent resistance to *M. tuberculosis*. As a result, standard β-lactam antibiotic treatment for TB involves a combination of drugs [4,5]. Recently, multidrug-resistant (MDR) TB has become a critical issue in the treatment of *M. tuberculosis* infections [6]. MDR-TB occurs when the bacterium becomes resistant to two of the most potent first-line anti-tuberculosis drugs: isoniazid and rifampicin [7]. Patients with MDR-TB can experience treatment failure, the prolonged transmission of drug-resistant strains, and increased mortality rates [8]. Moreover, the emergence of extensively drug-resistant (XDR)TB has further complicated the management of TB infections [9]. Consequently, there is an urgent need to discover new antibiotics to effectively combat multidrug-resistant tuberculosis [10].

In bacteria, the peptidoglycan layer is essential for maintaining the shape and stability of the bacterial cell wall [11]. Penicillin-Binding Proteins (PBPs) are a family of bacterial enzymes involved in the biosynthesis of peptidoglycan. PBPs catalyze the polymerization of the glycan strand (transglycosylation) and the formation of cross-links between glycan chains (transpeptidation). Additionally, certain PBPs hydrolyze the terminal D-alanine of stem pentapeptides (DD-carboxypeptidation) or the peptide bond that links two glycan strands (endopeptidation) [12,13]. PBPs are classified into two main categories based on their molecular weight: high-molecular-weight (HMW) PBPs and low-molecular-weight (LMW) PBPs [14]. Class A PBPs are HMW PBPs with molecular weights typically exceeding 100 kDa [15]. They are responsible for crosslinking peptidoglycans in the cell wall and exhibit both transglycosylase and transpeptidase activities [16]. Class B PBPs have an N-terminal domain of unknown function and a C-terminal domain containing transpeptidase activity [17]. They are also HMW PBPs but are smaller in size compared to Class A PBPs [18]. Class C PBPs are LMW PBPs, usually having molecular weights below 50 kDa [19]. Although not directly involved in cell wall synthesis, they possess transpeptidase activity and participate in the remodeling and maintenance of the cell wall [13].

In *M. tuberculosis*, there are seven putative PBPs, comprising DacB, DacB1, DacB2, PbpA, PbpB, PonA1, and PonA2 [20]. Rv3627c is PBP4 (DacB) from *M. tuberculosis* (MtbPBP4) belonging to the Class C PBP group [21]. MtbPBP4 exhibits the characteristics of a novel carboxypeptidase [22], such as specific serine-type carboxypeptidase activity, and it cleaves the cross-links involving terminal alanine [23]. The mode of action of β-lactam antibiotics on MtbPBP4 involves inhibiting its carboxypeptidase activity. This activity is crucial for cleaving peptide bonds in peptidoglycan precursors and assembling the bacterial cell wall. By covalently binding to the active site serine of MtbPBP4, β-lactam antibiotics disrupt cell wall synthesis, leading to defects in the peptidoglycan structure and, eventually, bacterial cell lysis. This inhibition of cell wall assembly contributes to the anti-bacterial effects of β-lactam antibiotics against *M. tuberculosis* [22,24].

In this study, we aimed to gain structural insights into MtbPBP4 through a combination of bioinformatics, size-exclusion chromatography (SEC), modeling, molecular docking, and molecular dynamics simulations. The sequence analysis based on bioinformatics indicates that MtbPBP4 belongs to the peptidase family S13 as a D-Ala-D-Ala carboxypeptidase C and has conserved active site residues. The SEC results verified the state of MtbPBP4 as a monomer, and three-dimensional structural modeling exhibited that MtbPBP4 contains the conserved penicillin-binding (PB) domain with distinct differences with its homologs. Additionally, molecular docking with meropenem confirmed the putative interaction residues of MtbPBP4. This information can provide valuable insights to aid in the design of specific inhibitors that effectively disrupt the activity of MtbPBP4, assisting the development of novel and potent anti-tuberculosis drugs to combat MDR-TB effectively.

2. Results and Discussion

2.1. Sequence Analysis

According to a sequence-based analysis using InterPro [25], a transmembrane region at the N-terminus of MtbPBP4 (residues 1–28) is predicted to be a signal peptide, suggesting its involvement in membrane targeting or secretion. The remaining region (residues 29–461) is expected to be located outside the membrane, indicating that PBP4 functions in the extracellular region. InterPro analysis identified this region as belonging to peptidase family S13, specifically the D-Ala-D-Ala carboxypeptidase C (InterPro000667, residues 73–459) family, which possesses the conserved motif SXXK as the active site residue. In a previous study by Sauvage et al., PBPs from selected bacteria were classified, and Rv3627c (MtbPBP4) was assigned to class C PBP4 [13]. Except for MtbPBP4, the three-dimensional structures of several PBP4 homologs have been determined, including PBP4 from *Escherichia coli* (PDB ID: 2EX2) [26], PBP4a from *Bacillus subtilis* (PDB ID: 1W5D) [27], R39 from Actinomycetes (PDB ID: 1W79) [28], and PBP4 from *Haemophilus influenzae* (PDB ID: 3A3D) [29].

The sequence of MtbPBP4 was compared with other PBP4 homologs using MultAlin [30] and ESPript 3.0 [31], as depicted in Figure 1. In MtbPBP4, like its homologs, the SXXK motif is present, specifically Ser114-Thr115-Asn116-Lys117, which is conserved across PBP4 proteins [20]. Additionally, two other conserved motifs typical of LMW-PBPs were identified in MtbPBP4 as Ser295-Asp296-Asn297 (SXN) and Lys408-Thr409-Gly410 (KTG).

These highly conserved residues are anticipated to play a critical role as a carboxypeptidase in the mycobacterial cell wall. However, PBP4 homologs exhibited a notable distinction in their sequence, which inserts approximately 70 amino acids near Arg241, adjacent to the β7 strand of MtbPBP4. The region anticipated to be the β7 in the secondary structure appears to deviate from a perfect β strand due to the lack of sufficient predictive data. This distinction in prediction results is a consequence of incomplete data in the prediction.

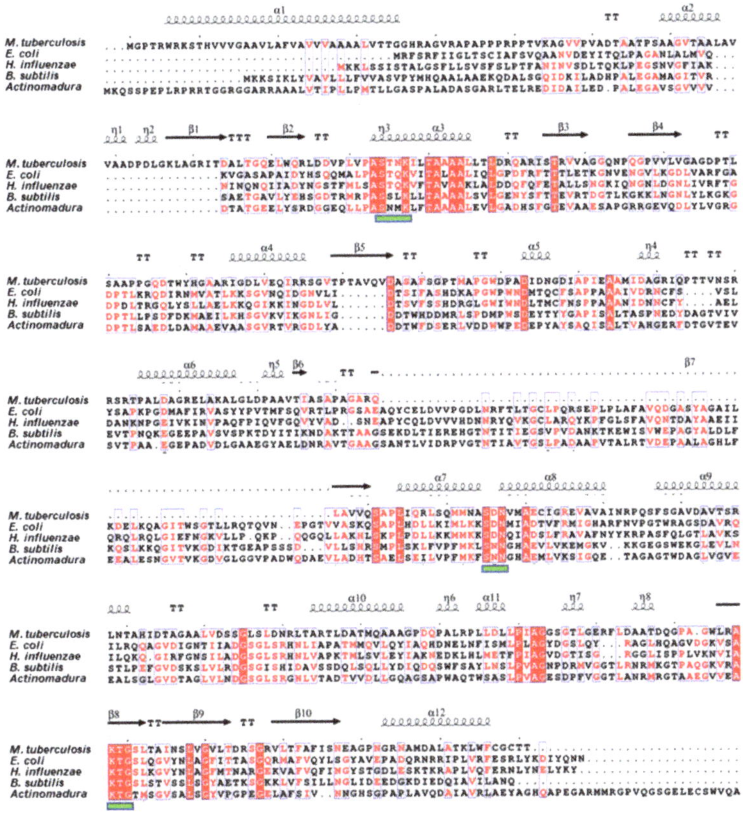

Figure 1. Sequence alignment of MtbPBP4 with its homologs. The alignment results of PBP4 from *M. tuberculosis* with PBP4 from *E. coli* (PDB ID: 2EX2), PBP4 from *H. influenzae* (PDB ID: 3A3D), PBP4a from *B. subtilis* (PDB ID: 1W5D), and R39 from Actinomycetes (PDB ID: 1W79) were analyzed using MultAlin [30] and ESPript 3.0 [31]. Conserved and similar residues are shown in red and blue boxes, respectively. The conserved motifs among PBP4s are marked by the green boxes under the sequences.

2.2. Oligomeric State of MtbPBP4 in Solution

To verify the oligomeric state of MtbPBP4 in solution, we compared the elution time after gel filtration using four references (ribonuclease A 13.7 kDa, carbonic anhydrase 29.0 kDa, ovalbumin 44.0 kDa, and conalbumin 75.0 kDa) from the Gel Filtration Calibration kits (Cytiva, Marlborough, MA, USA). It was calculated that the elution time corresponded to a molecular weight of 36.2 kDa (Figure 2a). To further confirm this, an additional experiment using SEC-MALS was conducted, and the resulting peak corresponded in size to 37.6 kDa (Figure 2b). When considering the inclusion of the N-terminal 6xHis-tag, the theoretical molecular weight should be 42.3 kDa. However, our experimental results consistently indicated values of 37.6 kDa and 36.2 kDa. Despite a slight deviation from the theoretical value, this outcome confirms that MtbPBP4 primarily exists as a monomer in solution, and consequently, a monomer model was employed to construct a 3D model of MtbPBP4 to predict its structure. However, it should be acknowledged that only the soluble part of the protein was cloned and purified, so the transmembrane domain may contribute to a different oligomeric state of the protein.

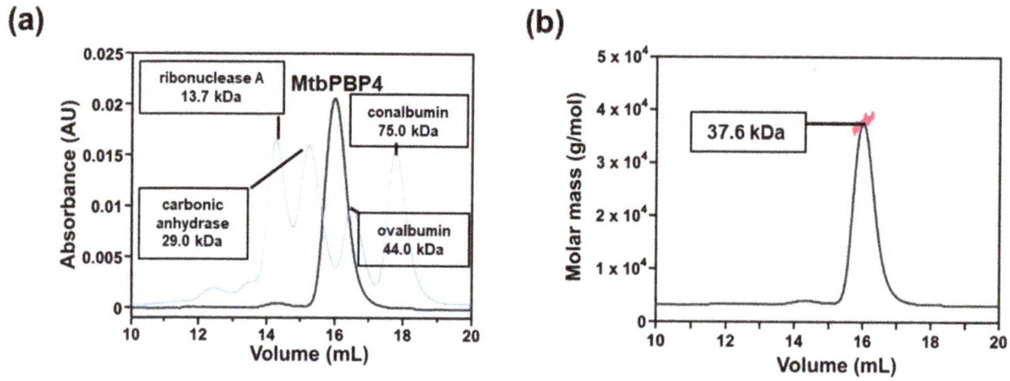

Figure 2. Oligomeric state analysis of MtbPBP4. (**a**) SEC chromatograms of MtbPBP4 and reference proteins and SEC-MALS chromatogram of MtbPBP4 showing the oligomeric state in solution. The UV absorption of MtbPBP4 (gray) and reference proteins (blue) at 280 nm are plotted as functions of the elution volume. (**b**) The calculated molecular mass in SEC-MALS is indicated as red dots. This result indicates that MtbPBP4 exists as a monomer in solution.

2.3. Overall Structure of MtbPBP4

To obtain structural insights into the role of MtbPBP4, a 3D model of MtbPBP4 was constructed using ColabFold: AlphaFold2 with the assistance of MMseqs2 [32] (pLDDT confidence 92.41). The N-terminal signal peptides (Met1–Ala28) and a long unstructured loop were excluded during the modeling process. As a result, the predicted region aligns with the region that was cloned and purified. The overall structure of MtbPBP4 consists of 11 α-helices and 10 β-sheets (Figure 1). The structure of MtbPBP4 can be primarily divided into two domains: domain I (part I: Pro67–Asp127, part II: Leu284–Thr461) and domain II (Asp128–Pro283) (Figure 3a). Domain I, also known as the penicillin-binding (PB) domain, is composed of a five-β-stranded sheet sandwiched between eight α-helices. On the other hand, domain II contains the topology of half a Rossmann fold, comprising a four-β-stranded sheet and two α-helices. Interestingly, domain II is inserted between domain I (part I) at the N-terminus and domain I (part II) at the C-terminus, creating a unique structural arrangement.

To assess the stability and flexibility of the predicted structure, MD simulations were performed. The RMSD of the Cα atoms was monitored over the course of 100 ns of simulation (Figure 4). The Cα RMSD values remained relatively constant, fluctuating between 0.1 and 0.2 nm (approximately 0.15 nm), indicating the structural stability of MtbPBP4 during the simulation. This result suggests that MtbPBP4 maintained its overall structure and did not undergo significant conformational changes throughout the simulation, indicating its stability under the simulated conditions.

According to the Dali search results [33], the overall fold of MtbPBP4 exhibits similarity to the structures of PBP4 homologs, with Z-scores exceeding 37 (Table 1). The primary structural difference between MtbPBP4 and its homologs lies in the presence of an additional domain, referred to as domain III. In the homologs of MtbPBP4, an extra domain III, consisting of approximately 70 additional amino acids, can be found between domain II, resembling the arrangement of Matryoshka (Russian dolls) (Figure 3b). However, despite the presence of domain III, domains I and II still exhibit a good superimposition between MtbPBP4 and its homologs (Table 1).

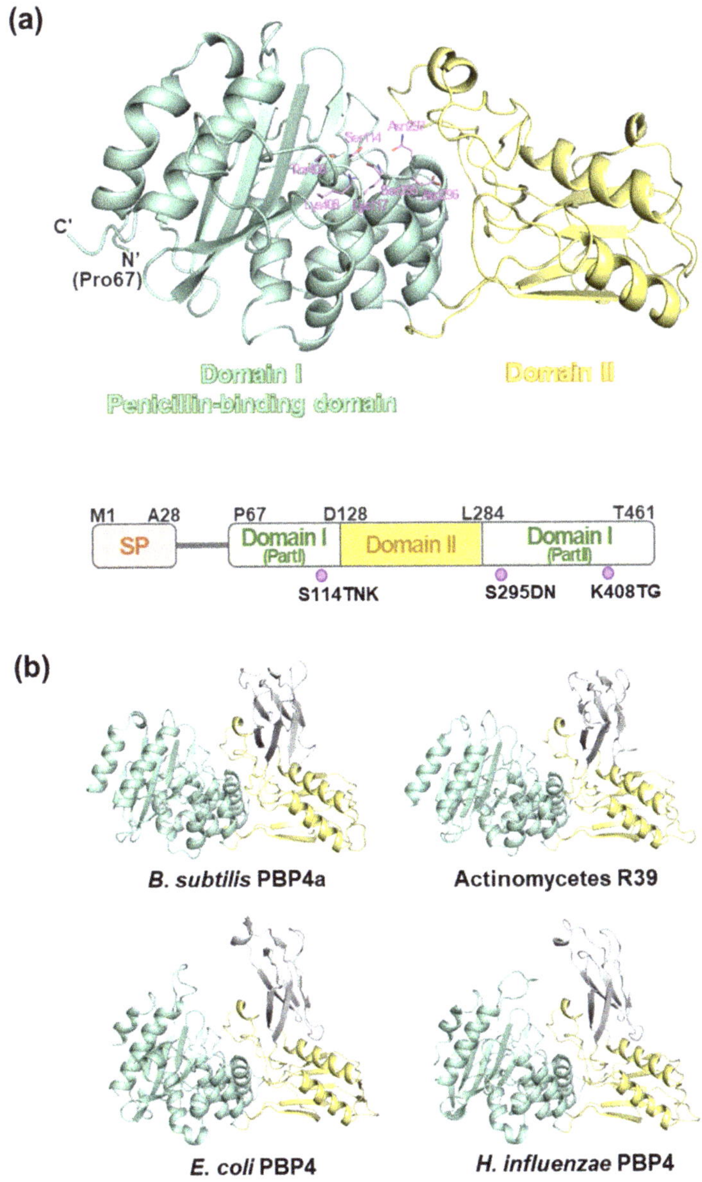

Figure 3. Overall predicted structure of MtbPBP4. (**a**) The overall predicted MtbPBP4 structure was divided into two domains: domain I, as the penicillin-binding domain, is colored in light green, and domain II is colored in light yellow. The residues of conserved motifs are represented in sticks and colored in light purple. The sequence diagram of MtbPBP4 is also denoted in the lower part. The signal peptide region (Met1- Ala28) is colored in orange. (**b**) Structures of PBP4 homologs. The domain II region is located between part I and part II of domain I. Domains I, II, and III are colored in light green, light yellow, and gray, respectively.

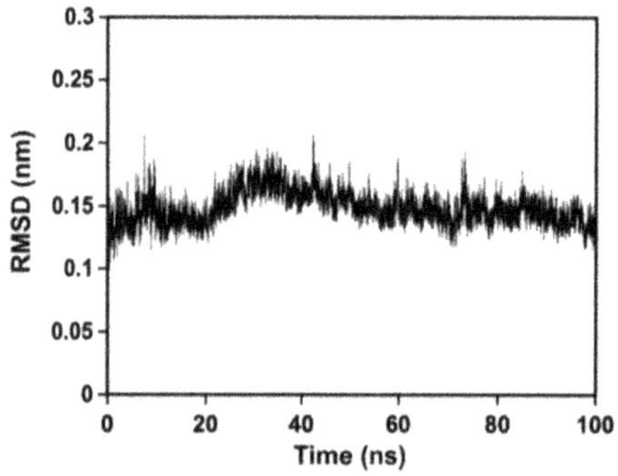

Figure 4. Structural dynamics analysis of MtbPBP4. The RMSD values of the Cα atoms from MtbPBP4 are plotted from 100 ns of MD simulations.

Table 1. Comparison of the homologs of PBP4.

Organism	PDB Code	Chain	No. of Cα Aligned	RMSD (Å)	Z-Score	% Identity
B. subtilis	1W5D	A	345 (361/458:DALI)	1.370 (1.8: DALI)	40.8	27
Actinomycetes	1W79	A	335 (363/466: DALI)	1.190 (1.8:DALI)	40.9	30
E. coli	2EX2	A	330 (355/456:DALI)	2.518 (2.3:DALI)	37.7	28
H. influenzae	3A3D	A	326 (354/453:DALI)	1.250 (2.2: DALI)	37.8	26

2.4. Active Site Cleft of MtbPBP4

Our analysis of the putative active site in MtbPBP4, conducted using CASTp [34], identified a cleft in the binding pocket with a volume of 468.099 Å^3 and a surface area of 400.493 Å^2. This binding pocket is located between domains I and II (Figure 5a) and includes functionally important conserved residues such as Ser114, Ser295, Asp296, Asn297, and Thr409. The electrostatic potential (ESP) surface of MtbPBP4 displays a different charge distribution when compared to its homologs (Figure 5b,c). Notably, the presence of the additional β-barrel domain III results in varying pocket sizes among PBP4s. These differences between MtbPBP4 and its homologs may contribute to substrate specificity, possibly affecting the interactions with ligands or substrates at the binding pocket.

Previous results have indicated that MtPBP4 functions as a carboxypeptidase, cleaving the N-terminal D-alanine from mycobacterial peptidoglycan [34]. Other D,D-carboxypeptidases identified in *M. tuberculosis* include DacB1 and DacB2, and the carboxypeptidase activity of DacB2 has been shown to be inhibited by meropenem, a β-lactam antibiotic targeting PBPs in bacteria [35]. It has been shown in a previous study that MtbPBP4 binds to meropenem [36]. To further analyze the potential for designing inhibitors for MtbPBP4, docking simulations of meropenem and MtbPBP4 were performed using CovDOCK in Glide (Schrödinger, LLC., New York, NY, USA) (Figure 6a,b). The docking results showed that meropenem forms a covalent bond with Ser114 from the first motif STNK, with a docking score of −5.596 kcal/mol. Additionally, meropenem forms hydrogen bonds with Asp218, Asn297, Ser411, and Leu412. These structural insights into the interactions between meropenem and MtbPBP4 can guide the design of effective inhibitors targeting MtbPBP4.

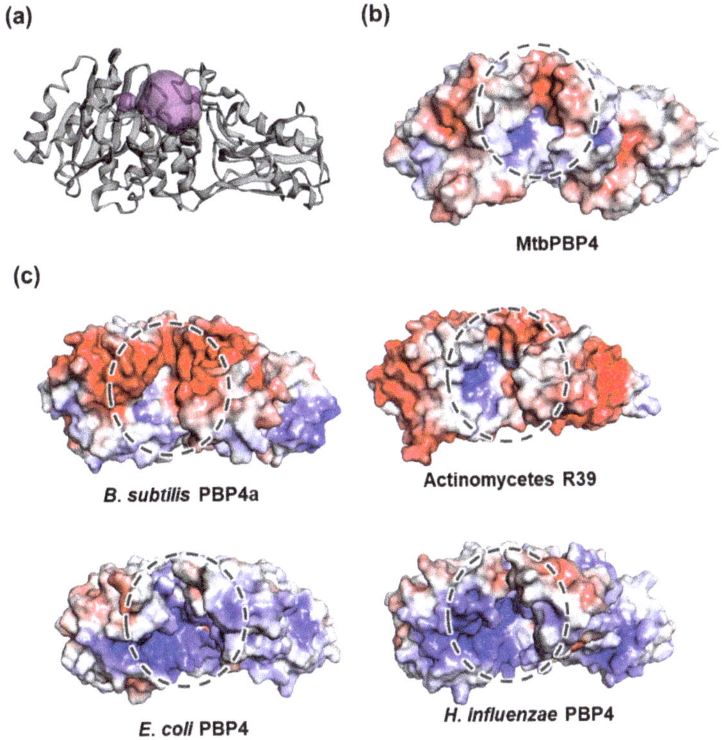

Figure 5. Putative active sites of MtbPBP4 and surface representations. (**a**) The predicted binding pocket of MtbPBP4 is represented in purple spheres. The ESP surface of (**b**) MtbPBP4 and (**c**) PBP4 homologs. Electrostatic surface potential represents different charge distributions on active sites marked as dotted circles. The negative and positive charges are colored in red and blue, respectively.

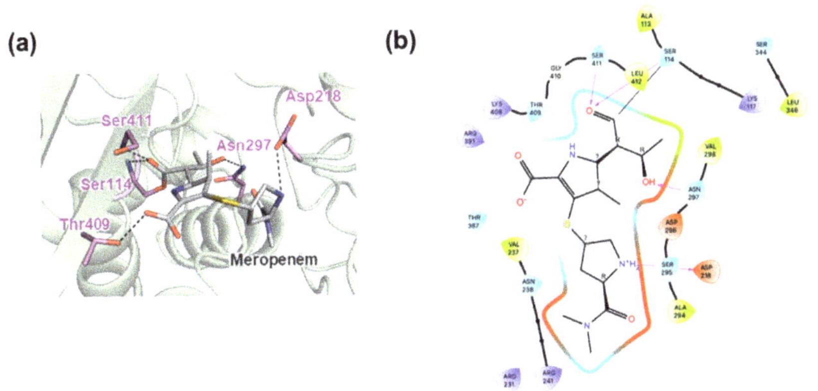

Figure 6. Docking simulation results for MtbPBP4 and a meropenem. (**a**) Covalent binding of a meropenem to MtbPBP4 through Ser114. The residues of MtbPBP4 that make hydrogen bond to a meropenem are represented by light purple sticks, and meropenem is colored in gray. Hydrogen bonds are represented by black dashes. (**b**) Detailed interaction residue information. Hydrogen bonds are represented by pink arrows, and covalent bonds are shown via the black lines. This figure was produced using Maestro of Schrödinger Suite (Schrödinger, LLC).

Our study provides structural and functional insights into MtbPBP4, a critical enzyme involved in peptidoglycan synthesis in *M. tuberculosis*. By understanding the mode of action of β-lactam antibiotics such as meropenem on MtbPBP4, we can use this knowledge to guide the design of targeted inhibitors that specifically disrupt the enzyme's activity. With the identification of the key residues involved in the binding and inhibition, we can focus on developing novel anti-tuberculosis agents that effectively target MtbPBP4, offering a promising vision for the production of new and potent drugs to combat MDR and XDR-TB.

3. Materials and Methods

3.1. Cloning, Expression, and Purification of MtbPBP4

The Rv3627c gene coding MtbPBP4 (NCBI reference sequence: WP_003899610.1) from Ala69 to Thr461 was amplified from the genomic DNA of *M. tuberculosis* H37Rv strain by polymerase chain reaction using two primers: (forward) 5′-CCA GGG AGC AGC CTC G GCC GCT GGC GTG ACC GCG GCG CT-3′ and (reverse) 5′-GCA AAG CAC CGG CCT CGT CA CGT CGT GCA CCC GCA GAA C-3′. The amplified gene was inserted into a pLIC-His vector for ligase-independent cloning (LIC) as previously described [37,38] This resulted in the incorporation of an N-terminal, 6xHis-tag, a TEV cutting site (ENLYFQ), and GAAAS residues for LIC cloning. The MtbPBP4-containing plasmid was transformed into *Escherichia coli* DH5α competent cells for cloning.

To overexpress MtbPBP4, cloned genes were transformed into *E. coli* BL21(DE3) competent cells. The transformed cells were grown in Luria broth with ampicillin until OD500 reached 0.6 at 37 °C, and protein overexpression was induced by 0.5 mM isopropyl-β-D-1-thiogalactopyranoside (IPTG). After further incubation in 15 °C for 20 h, the cells were harvested at 5000 rpm. The cell pellets were suspended in lysis buffer containing 20 mM Tris-HCl, pH 8.0, 500 mM NaCl, and 10% *v*/*v* glycerol and homogenized using ultrasonication. The resulting lysate was centrifuged at 13,000 rpm for 1 h to remove cellular debris. The supernatant was applied to a Ni–nitrilotriacetic acid (Ni-NTA) column for affinity chromatography, and the recombinant proteins were eluted by a stepwise increase in imidazole concentration from 0.1 M to 0.5 M. Subsequently, the protein was further purified using size-exclusion chromatography (SEC) on a Superdex 200 Increase 10/300 GL column (Cytiva) in a buffer containing 20 mM HEPES, pH 7.5, and 300 mM NaCl. The monomeric elution fractions were concentrated using a Vivaspin Turbo15 (Sartorius, Goettingen, Germany) with a 30 K molecular weight cutoff.

3.2. Oligomeric State Determination of MtbPBP4

The oligomeric state of MtbPBP4 was determined through gel filtration chromatography using a Superdex 200 Increase 10/300 GL column (Cytiva). The sample was run at a flow rate of 0.5 mL/min in a buffer containing 20 mM HEPES, pH 7.5, and 300 mM NaCl. To estimate the apparent molecular weight of MtbPBP4, we used a calibration curve based on standard proteins with known molecular weights. This involved comparing the elution volume of the target protein to that of the standard proteins. The gel filtration calibration kit LMW (Cytiva) included ribonuclease A (13.7 kDa), carbonic anhydrase (29 kDa), ovalbumin (43 kDa), conalbumin (75 kDa), and blue dextran 2000. To determine the oligomeric states of the MtbPBP4 with quantitative values, we conducted a SEC-MALS experiment using a fast protein liquid chromatography (FPLC) system (Cytiva) connected to a MiniDAWN TREOS MALS instrument (Wyatt, Santa Barbara, CA, USA). A Superdex 200 Increase 10/300 GL (Cytiva) gel filtration column was pre-equilibrated with a buffer containing 20 mM HEPES, pH 7.5, and 300 mM NaCl. The detector was normalized using 2 mg/mL bovine serum albumin (Thermo Fisher Scientific, Waltham, MA, USA), and 100 μL of the protein solution was injected at a concentration of 3.5 mg/mL. The acquired data were analyzed using ASTRA 8 software (Wyatt). Chromatography was performed on a Superdex 200 Increase 10/300 GL column with a buffer containing 20 mM HEPES, pH 7.5, and 300 mM NaCl at a flow rate of 0.5 mL/min.

3.3. Structure Prediction of MtbPBP4

For the prediction of MtbPBP4, we employed ColabFold, which is a user-friendly and fast implementation of AlphaFold2 [32]. The protein sequence of Rv3627c from *M. tuberculosis* H37Rv (UNIPROT ID O06380) served as the input sequence, and a Multiple Sequence Alignment (MSA) was generated using MMseqs2. The calculations were run using the ColabFold online platform (AlphaFold2.ipynb) with default options. To visualize the predicted structures, we utilized the *PyMOL* program (PyMOL Molecular Graphics System, version 2.5.3; Schrödinger, LLC.).

3.4. MD Simulation

Molecular Dynamics (MD) simulations of MtbPBP4 were conducted using the GROMACS 2023.1 packages [39]. The AMBER99SB-ILDN force field was selected for the simulations. The molecules were placed in a cubic box filled with TIP3P water molecules and neutralized by adding Na^+ and Cl^- ions. Steepest descent minimization was performed to optimize the system. Subsequently, equilibration steps were carried out in both the NVT ensemble (constant number of particles, volume, and temperature) and NPT ensemble (constant number of particles, pressure, and temperature) for 100 ps at 1 atm and 300 K. The V-rescale thermostat and Parrinello-Rahman barostat were used for controlling temperature and pressure in the NVT and NPT ensembles, respectively. After equilibration, MD simulations of the predicted MtbPBP4 were performed for 100 ns at 300 K utilizing the linear constraint solver (LINCS) algorithm for constraining bond lengths and the particle mesh Ewald (PME) method for handling long-range electrostatic interactions. The resulting MD trajectories were analyzed using GROMACS distribution programs, particularly gmx_rms for calculating parameter values such as root mean square deviation (RMSD). The analysis was processed using the XMGrace program (Oregon Graduate Institute of Science and Technology, Hillsboro, OR, USA), and plots were generated using KaleidaGraph 5.0 software (Synergy Software, Reading, PA, USA).

3.5. Covalent Docking

The MtbPBP4 protein was prepared for Glide docking calculations using the Protein Preparation Wizard of Schrödinger Suite (Schrödinger, LLC., New York, NY, USA). Proper protein preparation is crucial for accurate protein–ligand docking simulations. The active site of the protein was identified, and key residues such as Ser114, Ser295, Asn297, and Thr409 were selected for further molecular docking analysis. To prepare the protein, the atoms of the protein were scaled with a van der Waal's radius at a factor of 1 Å, and a partial charge cutoff of 0.25 Å was applied. The ligand structure of meropenem was obtained from the RCSB PDB (http://www.rcsb.org accessed on 6 January 2024) and prepared using LigPrep (Schrödinger, LLC). The stereoisomers' chirality was determined from the 3D structure, and all other parameters were kept as default during the LigPrep preparation. For covalent docking, the CovDock module (Schrödinger, LLC) was used. The amino acid Ser114 was set as the reactive residue, which would have a nucleophilic interaction with the O− and carbonyl group of meropenem acting as the electrophile. The reaction type was predefined as Beta Lactam Addition using the pre-installed options in CovDock [40].

4. Conclusions

Our comparison of MtbPBP4 sequences indicated that this enzyme belongs to the peptidase family S13, specifically D-Ala-D-Ala carboxypeptidase C. From SEC experiments, it was confirmed that MtbPBP4 mainly exists in a monomer state. The predicted three-dimensional structure of MtbPBP4 is similar to that of the known PBP4 homologs, but certain differences were identified, such as the absence of domain III, which is typically inserted between domain II in homologs. MtbPBP4 does contain conserved active site motifs, consistent with other homologs. The docking of β-lactam antibiotics to MtbPBP4 revealed the binding site located in these conserved residues. As a carboxypeptidase, MtbPBP4 is involved in the cleavage of peptidoglycan, which plays a crucial role in mycobacterial morphology and cell division. These structural insights provide information for the development of anti-tuberculosis drugs, as targeting MtbPBP4 may represent a promising approach to disrupting cell wall synthesis and inhibiting mycobacterial growth.

Author Contributions: Conceptualization, D.-H.K. and S.-M.K.; methodology, D.-H.K.; software, D.-H.K.; validation, D.-H.K. and S.-M.K.; formal analysis, D.-H.K.; investigation, D.-H.K. and S.-M.K.; resources, D.-H.K.; data curation, D.-H.K. and S.-M.K.; writing—original draft preparation, D.-H.K.; writing—review and editing, D.-H.K. and S.-M.K.; visualization, D.-H.K. and S.-M.K.; supervision, D.-H.K.; project administration, D.-H.K.; funding acquisition, D.-H.K. All authors have read and agreed to the published version of the manuscript.

Funding: This work was supported by a research grant from Jeju National University in 2021.

Informed Consent Statement: Not applicable.

Data Availability Statement: Data contained within the article.

Conflicts of Interest: The authors declare no conflicts of interest.

References

1. McBryde, E.S.; Meehan, M.T.; Doan, T.N.; Ragonnet, R.; Marais, B.J.; Guernier, V.; Trauer, J.M. The risk of global epidemic replacement with drug-resistant Mycobacterium tuberculosis strains. *Int. J. Infect. Dis.* **2017**, *56*, 14–20. [CrossRef] [PubMed]
2. Briffotaux, J.; Huang, W.; Wang, X.; Gicquel, B. MmpS5/MmpL5 as an efflux pump in Mycobacterium species. *Tuberculosis* **2017**, *107*, 13–19. [CrossRef] [PubMed]
3. Bothra, A.; Arumugam, P.; Panchal, V.; Menon, D.; Srivastava, S.; Shankaran, D.; Nandy, A.; Jaisinghani, N.; Singh, A.; Gokhale, R.S.; et al. Phospholipid homeostasis, membrane tenacity and survival of Mtb in lipid rich conditions is determined by MmpL11 function. *Sci. Rep.* **2018**, *8*, 8317. [CrossRef] [PubMed]
4. Rullas, J.; Dhar, N.; McKinney, J.D.; Garcia-Perez, A.; Lelievre, J.; Diacon, A.H.; Hugonnet, J.E.; Arthur, M.; Angulo-Barturen, I.; Barros-Aguirre, D.; et al. Combinations of beta-Lactam Antibiotics Currently in Clinical Trials Are Efficacious in a DHP-I-Deficient Mouse Model of Tuberculosis Infection. *Antimicrob. Agents Chemother.* **2015**, *59*, 4997–4999. [CrossRef] [PubMed]
5. Cohen, K.A.; El-Hay, T.; Wyres, K.L.; Weissbrod, O.; Munsamy, V.; Yanover, C.; Aharonov, R.; Shaham, O.; Conway, T.C.; Goldschmidt, Y.; et al. Paradoxical Hypersusceptibility of Drug-resistant Mycobacteriumtuberculosis to beta-lactam Antibiotics. *eBioMedicine* **2016**, *9*, 170–179. [CrossRef]
6. Honeyborne, I.; Lipman, M.; Zumla, A.; McHugh, T.D. The changing treatment landscape for MDR/XDR-TB—Can current clinical trials revolutionise and inform a brave new world? *Int. J. Infect. Dis.* **2019**, *80*, S23–S28. [CrossRef]
7. Migliori, G.B.; Tiberi, S.; Zumla, A.; Petersen, E.; Chakaya, J.M.; Wejse, C.; Munoz Torrico, M.; Duarte, R.; Alffenaar, J.W.; Schaaf, H.S.; et al. MDR/XDR-TB management of patients and contacts: Challenges facing the new decade. The 2020 clinical update by the Global Tuberculosis Network. *Int. J. Infect. Dis.* **2020**, *92*, S15–S25. [CrossRef]
8. Dheda, K.; Gumbo, T.; Maartens, G.; Dooley, K.E.; Murray, M.; Furin, J.; Nardell, E.A.; Warren, R.M.; the Lancet Respiratory Medicine drug-resistant tuberculosis Commission group. The Lancet Respiratory Medicine Commission: 2019 update: Epidemiology, pathogenesis, transmission, diagnosis, and management of multidrug-resistant and incurable tuberculosis. *Lancet Respir. Med.* **2019**, *7*, 820–826. [CrossRef]
9. Faridgohar, M. Finding New Ways to Combat Multidrug-Resistant Tuberculosis. *Microb. Drug Resist.* **2020**, *26*, 71–80. [CrossRef]
10. Lu, Z.; Wang, H.; Zhang, A.; Liu, X.; Zhou, W.; Yang, C.; Guddat, L.; Yang, H.; Schofield, C.J.; Rao, Z. Structures of Mycobacterium tuberculosis Penicillin-Binding Protein 3 in Complex with Five beta-Lactam Antibiotics Reveal Mechanism of Inactivation. *Mol. Pharmacol.* **2020**, *97*, 287–294. [CrossRef]
11. Shalaby, M.W.; Dokla, E.M.E.; Serya, R.A.T.; Abouzid, K.A.M. Penicillin binding protein 2a: An overview and a medicinal chemistry perspective. *Eur. J. Med. Chem.* **2020**, *199*, 112312. [CrossRef] [PubMed]
12. Sharifzadeh, S.; Brown, N.W.; Shirley, J.D.; Bruce, K.E.; Winkler, M.E.; Carlson, E.E. Chemical tools for selective activity profiling of bacterial penicillin-binding proteins. *Methods Enzymol.* **2020**, *638*, 27–55. [CrossRef] [PubMed]
13. Sauvage, E.; Kerff, F.; Terrak, M.; Ayala, J.A.; Charlier, P. The penicillin-binding proteins: Structure and role in peptidoglycan biosynthesis. *FEMS Microbiol. Rev.* **2008**, *32*, 234–258. [CrossRef] [PubMed]
14. Bansal, A.; Kar, D.; Murugan, R.A.; Mallick, S.; Dutta, M.; Pandey, S.D.; Chowdhury, C.; Ghosh, A.S. A putative low-molecular-mass penicillin-binding protein (PBP) of Mycobacterium smegmatis exhibits prominent physiological characteristics of DD-carboxypeptidase and beta-lactamase. *Microbiology* **2015**, *161*, 1081–1091. [CrossRef]
15. Chen, X.; Li, Y.; Bai, K.; Gu, M.; Xu, X.; Jiang, N.; Chen, Y.; Li, J.; Luo, L. Class A Penicillin-Binding Protein C Is Responsible for Stress Response by Regulation of Peptidoglycan Assembly in Clavibacter michiganensis. *Microbiol. Spectr.* **2022**, *10*, e0181622. [CrossRef]
16. Murphy, S.G.; Murtha, A.N.; Zhao, Z.; Alvarez, L.; Diebold, P.; Shin, J.H.; VanNieuwenhze, M.S.; Cava, F.; Dorr, T. Class A Penicillin-Binding Protein-Mediated Cell Wall Synthesis Promotes Structural Integrity during Peptidoglycan Endopeptidase Insufficiency in Vibrio cholerae. *mBio* **2021**, *12*, e03596-20. [CrossRef]

17. Hamilton, S.M.; Alexander, J.A.N.; Choo, E.J.; Basuino, L.; da Costa, T.M.; Severin, A.; Chung, M.; Aedo, S.; Strynadka, N.C.J.; Tomasz, A.; et al. High-Level Resistance of Staphylococcus aureus to beta-Lactam Antibiotics Mediated by Penicillin-Binding Protein 4 (PBP4). *Antimicrob. Agents Chemother.* **2017**, *61*, e02727-16. [CrossRef] [PubMed]
18. de Leon, S.R.; Daniels, K.; Clarke, A.J. Production and purification of the penicillin-binding protein 3 from Pseudomonas aeruginosa. *Protein Expr. Purif.* **2010**, *73*, 177–183. [CrossRef]
19. Brambilla, L.; Moran-Barrio, J.; Viale, A.M. Low-molecular-mass penicillin binding protein 6b (DacD) is required for efficient GOB-18 metallo-beta-lactamase biogenesis in Salmonella enterica and Escherichia coli. *Antimicrob. Agents Chemother.* **2014**, *58*, 205–211. [CrossRef]
20. Zhang, W.; Li, S.; Ma, L.; Ding, W.; Xu, Y. Identification of a novel carboxypeptidase encoded by Rv3627c that plays a potential role in mycobacteria morphology and cell division. *Enzyme Microb. Technol.* **2019**, *126*, 32–40. [CrossRef]
21. Chatterjee, S.S.; Chen, L.; Gilbert, A.; da Costa, T.M.; Nair, V.; Datta, S.K.; Kreiswirth, B.N.; Chambers, H.F. PBP4 Mediates beta-Lactam Resistance by Altered Function. *Antimicrob. Agents Chemother.* **2017**, *61*, e00932-17. [CrossRef] [PubMed]
22. Beg, M.A.; Hejazi, I.I.; Thakur, S.C.; Athar, F. Domain-wise differentiation of Mycobacterium tuberculosis H(37) Rv hypothetical proteins: A roadmap to discover bacterial survival potentials. *Biotechnol. Appl. Biochem.* **2022**, *69*, 296–312. [CrossRef] [PubMed]
23. Ealand, C.S.; Asmal, R.; Mashigo, L.; Campbell, L.; Kana, B.D. Characterization of putative DD-carboxypeptidase-encoding genes in Mycobacterium smegmatis. *Sci. Rep.* **2019**, *9*, 5194. [CrossRef] [PubMed]
24. Ogasawara, Y.; Dairi, T. Discovery of an alternative pathway of peptidoglycan biosynthesis: A new target for pathway specific inhibitors. *J. Ind. Microbiol. Biotechnol.* **2021**, *48*, kuab038. [CrossRef] [PubMed]
25. Paysan-Lafosse, T.; Blum, M.; Chuguransky, S.; Grego, T.; Pinto, B.L.; Salazar, G.A.; Bileschi, M.L.; Bork, P.; Bridge, A.; Colwell, L.; et al. InterPro in 2022. *Nucleic Acids Res.* **2023**, *51*, D418–D427. [CrossRef]
26. Kishida, H.; Unzai, S.; Roper, D.I.; Lloyd, A.; Park, S.Y.; Tame, J.R. Crystal structure of penicillin binding protein 4 (dacB) from Escherichia coli, both in the native form and covalently linked to various antibiotics. *Biochemistry* **2006**, *45*, 783–792. [CrossRef] [PubMed]
27. Sauvage, E.; Duez, C.; Herman, R.; Kerff, F.; Petrella, S.; Anderson, J.W.; Adediran, S.A.; Pratt, R.F.; Frere, J.M.; Charlier, P. Crystal structure of the Bacillus subtilis penicillin-binding protein 4a, and its complex with a peptidoglycan mimetic peptide. *J. Mol. Biol.* **2007**, *371*, 528–539. [CrossRef]
28. Sauvage, E.; Herman, R.; Petrella, S.; Duez, C.; Bouillenne, F.; Frere, J.M.; Charlier, P. Crystal structure of the Actinomadura R39 DD-peptidase reveals new domains in penicillin-binding proteins. *J. Biol. Chem.* **2005**, *280*, 31249–31256. [CrossRef]
29. Kawai, F.; Clarke, T.B.; Roper, D.I.; Han, G.J.; Hwang, K.Y.; Unzai, S.; Obayashi, E.; Park, S.Y.; Tame, J.R. Crystal structures of penicillin-binding proteins 4 and 5 from Haemophilus influenzae. *J. Mol. Biol.* **2010**, *396*, 634–645. [CrossRef]
30. Corpet, F. Multiple sequence alignment with hierarchical clustering. *Nucleic Acids Res.* **1988**, *16*, 10881–10890. [CrossRef]
31. Robert, X.; Gouet, P. Deciphering key features in protein structures with the new ENDscript server. *Nucleic Acids Res.* **2014**, *42*, W320–W324. [CrossRef]
32. Mirdita, M.; Schutze, K.; Moriwaki, Y.; Heo, L.; Ovchinnikov, S.; Steinegger, M. ColabFold: Making protein folding accessible to all. *Nat. Methods* **2022**, *19*, 679–682. [CrossRef]
33. Holm, L.; Laiho, A.; Toronen, P.; Salgado, M. DALI shines a light on remote homologs: One hundred discoveries. *Protein Sci.* **2023**, *32*, e4519. [CrossRef] [PubMed]
34. Tian, W.; Chen, C.; Lei, X.; Zhao, J.; Liang, J. CASTp 3.0: Computed atlas of surface topography of proteins. *Nucleic Acids Res.* **2018**, *46*, W363–W367. [CrossRef]
35. Kumar, P.; Arora, K.; Lloyd, J.R.; Lee, I.Y.; Nair, V.; Fischer, E.; Boshoff, H.I.; Barry, C.E., 3rd. Meropenem inhibits D,D-carboxypeptidase activity in Mycobacterium tuberculosis. *Mol. Microbiol.* **2012**, *86*, 367–381. [CrossRef] [PubMed]
36. Levine, S.R.; Beatty, K.E. Investigating beta-Lactam Drug Targets in Mycobacterium tuberculosis Using Chemical Probes. *ACS Infect. Dis.* **2021**, *7*, 461–470. [CrossRef] [PubMed]
37. Cabrita, L.D.; Dai, W.; Bottomley, S.P. A family of E. coli expression vectors for laboratory scale and high throughput soluble protein production. *BMC Biotechnol* **2006**, *6*, 12. [CrossRef] [PubMed]
38. Eschenfeldt, W.H.; Lucy, S.; Millard, C.S.; Joachimiak, A.; Mark, I.D. A family of LIC vectors for high-throughput cloning and purification of proteins. *Methods Mol Biol* **2009**, *498*, 105–115. [CrossRef]
39. Abraham, M.J.; Murtola, T.; Schulz, R.; Páll, S.; Smith, J.C.; Hess, B.; Lindahl, E. GROMACS: High performance molecular simulations through multi-level parallelism from laptops to supercomputers. *SoftwareX* **2015**, *1*, 19–25. [CrossRef]
40. Zhu, K.; Borrelli, K.W.; Greenwood, J.R.; Day, T.; Abel, R.; Farid, R.S.; Harder, E. Docking covalent inhibitors: A parameter free approach to pose prediction and scoring. *J. Chem. Inf. Model.* **2014**, *54*, 1932–1940. [CrossRef]

International Journal of
Molecular Sciences

MDPI

Article

Cortical and Striatal Astrocytes of Neonatal Rats Display Distinct Molecular and Pharmacological Characteristics of Dopamine Uptake

Vesna Sočan [1], Klemen Dolinar [2] and Mojca Kržan [1,*]

1 Institute of Pharmacology and Experimental Toxicology, Faculty of Medicine, University of Ljubljana, 1000 Ljubljana, Slovenia; vesna.socan@mf.uni-lj.si

2 Institute of Pathophysiology, Faculty of Medicine, University of Ljubljana, 1000 Ljubljana, Slovenia; klemen.dolinar@mf.uni-lj.si

* Correspondence: mojca.limpel@mf.uni-lj.si

Abstract: Astrocytes are crucial in the regulation of neurotransmitter homeostasis, and while their involvement in the dopamine (DA) tripartite synapse is acknowledged, it necessitates a more comprehensive investigation. In the present study, experiments were conducted on primary astrocyte cultures from the striatum and cortex of neonatal rats. The pharmacological intricacies of DA uptake, including dependence on time, temperature, and concentration, were investigated using radiolabelled [^{3}H]-DA. The mRNA expression of transporters DAT, NET, PMAT, and OCTs was evaluated by qPCR. Notably, astrocytes from both brain regions exhibited prominent mRNA expression of NET and PMAT, with comparatively lower expression of DAT and OCTs. The inhibition of DA uptake by the DAT inhibitor, GBR12909, and NET inhibitors, desipramine and nortriptyline, impeded DA uptake in striatal astrocytes more than in cortical astrocytes. The mRNA expression of NET and PMAT was significantly upregulated in cortical astrocytes in response to the DA receptor agonist apomorphine, while only the mRNA expression of NET exhibited changes in striatal astrocytes. Haloperidol, a DA receptor antagonist, and L-DOPA, a DA precursor, did not induce significant alterations in transporter mRNA expression. These findings underscore the intricate and region-specific mechanisms governing DA uptake in astrocytes, emphasizing the need for continued exploration to unravel the nuanced dynamics of astrocytic involvement in the DA tripartite synapse.

Keywords: astrocyte; dopamine uptake; NET; PMAT; neonatal rat

1. Introduction

Astrocytes, named after their star-like shape, are one of the most abundant cell types in the central nervous system (CNS) [1]. They are crucial regulators of numerous homeostatic functions as they provide structural and metabolic support to neuronal cells, control ion balance, and maintain the blood–brain barrier [2,3]. Unlike neuronal cells, astrocytes cannot generate an action potential but respond to stimuli through fluctuations in intracellular Ca^{2+} [4]. Astrocytes help maintain neurotransmitter homeostasis in the CNS and modulate synaptic transmission. Together with neuronal cells, they form the tripartite synapse, a concept coined by Perea et. al. describing bidirectional communication between astrocytes and neuronal cells [5]. Astrocytes express various transporter proteins in their plasma membrane for neurotransmitter uptake and are also capable of (glio)transmitter release themselves [4,6–12].

The role of astrocytes in the homeostasis of the neuromodulator dopamine (DA) has been studied to a limited extent. DA is involved in the regulation of several aspects of brain function through multiple DA pathways, each related to a particular brain region [13,14]. DA function in the cortex, particularly the prefrontal cortex, is involved in higher-order cognitive functions, such as attention, working memory, decision-making, and executive control. The mesocortical DA pathway, originating from the ventral tegmental area in the midbrain, projects to the prefrontal cortex, and imbalances in this pathway have been linked to disorders, such as schizophrenia and attention deficit hyperactivity disorder [15]. In contrast, the striatum, a key component of the basal ganglia, is involved in motor control, reward processing, and habit formation. It receives DA input from the nigrostriatal pathway, originating in the substantia nigra. Dysfunctions in the nigrostriatal pathway have been associated with addiction, depression, and other conditions, such as Parkinson's disease (PD) [13]. Astrocytes have been implicated in the development of and in the protection from these diseases [16–18]. Although they have traditionally been viewed as simple, homogenous cells providing support to neurons, we now recognize that astrocytes from different brain regions are heterogeneous [19]. Chai et al. demonstrated that cortical astrocytes and hippocampal astrocytes were similar when comparing RNA sequencing data, but hippocampal and striatal astrocytes were shown to be different populations [20]. Whether astrocytes from distinct brain regions serve different roles in DA homeostasis is yet to be discovered.

DA exerts its effects by binding to metabotropic G protein-coupled DA receptors, which are divided into two subtypes, D1 and D2 receptors [13,21,22]. The DA concentration in the synapse is controlled mainly by its reuptake by the sodium-dependent, low-capacity, high-affinity (uptake 1) dopamine transporter (DAT) [23] and, to a lesser degree, by other transporters. Studies on rodents have reported that the norepinephrine transporter (NET) has an equal affinity for DA and plays a larger role in DA uptake in brain regions where DAT is sparsely distributed, particularly in the cortex [24–26] as well as in the DA-denervated striatum in PD [27]. Sodium-independent transporters with a low affinity for DA, albeit high capacity (uptake 2), such as organic cation transporter 3 (OCT3) [28,29] and plasma membrane monoamine transporter (PMAT) [30,31], have an important role in the removal of excess extracellular DA when the DA concentration is high and has overcome the capacity of DAT [27]. They transport organic cations, zwitterions, and some uncharged compounds and operate as facilitated diffusion systems and/or antiporters [32]. OCTs and PMAT are considered "polyspecific" or "multispecific", as they interact with a wide array of cationic compounds with diverse chemical structures. Nevertheless, there are notable differences in substrate specificity and transport kinetics among these transporters. PMAT has been shown to have a strong kinetic preference for serotonin and DA over other monoamines (i.e., histamine, norepinephrine, and epinephrine), whereas a seemingly opposite preference was observed for OCT3 [33].

Several studies have shown that astrocytes are able to take up DA [21,24,34–38] and the DA precursor, L-DOPA [39–44], and express DA receptors in their membrane [35,45–47]. Astrocytes respond to DA with fluctuations in intracellular calcium [48,49] and may be involved in response to treatment with dopaminergic drugs, such as DA receptor antagonists, haloperidol and clozapine, as well as the DA receptor agonist, apomorphine [50–56]. Whether astrocytes are capable of active uptake of DA as well as their role in the homeostasis of DA is up for debate. The exitance of the primary mechanism responsible for DA reuptake into neuronal cells, DAT, has not been found in rat cortical astrocytes [24]; however, it has been found in striatal rat astrocytes [39] and striatal mouse astrocytes [57]. Inazu et al. have shown that DA is transported into rat cortical astrocytes via NET, rather than DAT, as well as by high-capacity, low-affinity uptake 2 transport mediated by OCT3. Contrarily, Hösli et al. showed astrocytic DA uptake in striatal and cerebellar primary cell cultures of neonatal rats is mediated only by facilitated diffusion, and uptake is independent of Na^+/Cl^- ions [34]. As astrocyte morphology has been shown to be region-dependent, particularly in the human brain [58,59], it may be one of the reasons contributing to the conflicting results of astrocyte DA uptake studies.

The aim of our study was to determine the regional pharmacological and molecular characteristics of DA uptake into neonatal rat astrocytes from two distinct brain regions, the cortex and striatum, to further elucidate the role of astrocytes in the homeostasis of DA. Furthermore, we examined whether the drugs apomorphine, L-DOPA, and haloperidol, used in the treatment of pathologies stemming from aberrations in the dopaminergic system, such as PD and schizophrenia, affect the gene expression of these transporters in astrocytes.

2. Results

2.1. Dependence of [^{3}H]-Dopamine Uptake on Time, Temperature, and Concentration in Neonatal Rat Astrocytes from the Cortex and Striatum

We first examined the time course of [^{3}H]-DA uptake with a 30 nM concentration of DA in cultured cortical and striatal rat astrocytes for the time span of 60 min. As shown in Figure 1a,b, [^{3}H]-DA uptake increased in a time-dependent manner for the initial 20 min and almost reached a plateau at 30 min in both cortical and striatal astrocytes. The accumulation of [^{3}H]-DA in both cortical and striatal astrocytes was significantly greater at 37 °C than the accumulation of [^{3}H]-DA uptake at 4 °C. Based on these findings, we carried out the following [^{3}H]-DA uptake assays for 20 min.

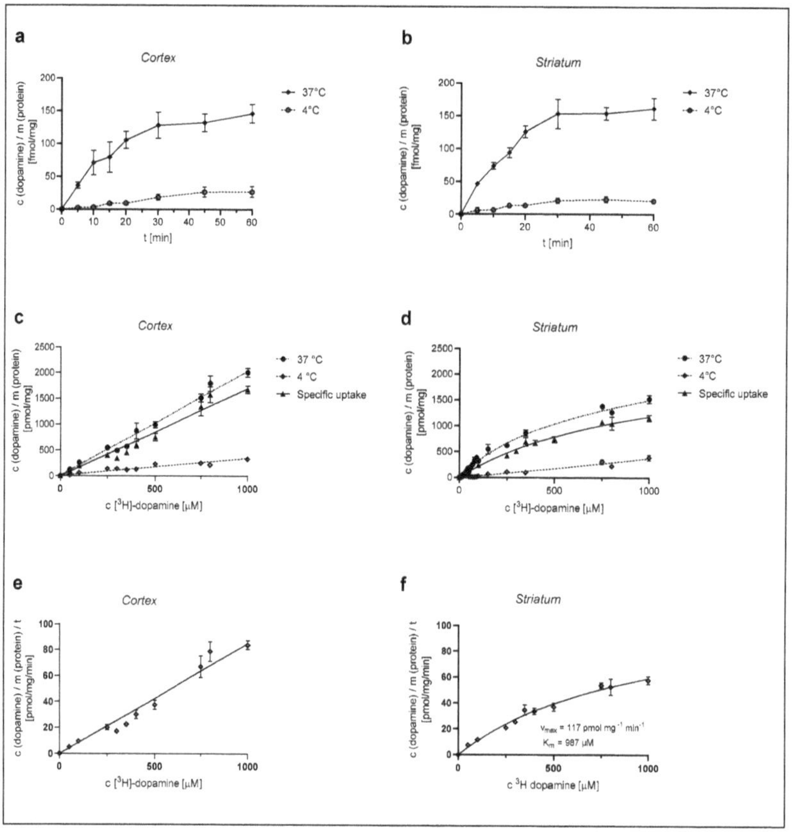

Figure 1. Dependence of [^{3}H]-DA uptake on time, temperature, and concentration in cortical and striatal astrocytes of neonatal rats. Time dependence of total (37 °C) and nonspecific (4 °C) [^{3}H]-DA uptake in cortical (**a**) and striatal (**b**) astrocytes, concentration dependence of [^{3}H]-DA uptake (total, specific, and nonspecific) in cortical (**c**) and striatal (**d**) astrocytes and [^{3}H]-DA uptake velocity in cortical (**e**) and striatal (**f**) astrocytes. Data are presented as mean ± SEM (n = 9) from three separate experiments.

Astrocytes were then exposed to different concentrations of [^{3}H]-DA (0.03–1000 µM) for a time span of 20 min, in line with the time course experiments, and the total (37 °C) and the nonspecific (4 °C) DA uptake were measured from which the specific uptake was calculated (as the difference between total and nonspecific uptake) (Figure 1c,d). DA uptake into cultured cortical astrocytes, presented in Figure 1c, was clearly dependent on the DA concentration. However, it does not appear saturable at even the highest concentration of 1 mM, whereas DA uptake in the striatum (Figure 1d) was saturable at 1 mM concentration. The apparent maximum saturation of the radioligand, B_{max}, of the specific uptake was determined as 2343 ± 251 pmol/mg.

The DA uptake velocity in striatal astrocytes (Figure 1f) was calculated from the specific uptake and the time span of the [^{3}H]-DA incubation (20 min). The kinetic parameters of the uptake velocity were calculated using the Michaelis–Menten equation. The apparent maximal uptake rate, V_{max}, of DA was calculated as 117 ± 13 pmol/mg/min and the apparent Michaelis–Menten constant, K_m, was calculated as 987 ± 176 μM. Cortical astrocytes appear to have a greater uptake capacity for DA in comparison to striatal astrocytes.

2.2. Dependence of [^{3}H]-Dopamine Uptake in Cortical and Striatal Astrocytes of Neonatal Rats on the Presence of Ouabain amd Sodium Ions

A 1 mM concentration of the Na^+/K^+-ATPase pump inhibitor ouabain reduced 30 nM [^{3}H]-DA uptake to 68 ± 2% in cortical and 66 ± 2% in striatal astrocytes in comparison to the control (Figure 2a). In contrast, 30 nM [^{3}H]-DA uptake was reduced to 57 ± 8% of the control in cortical astrocytes and to 41 ± 6% of the control in striatal astrocytes in the absence of Na^+ (Figure 2b). The DA uptake was significantly reduced by both ouabain and the absence of Na^+. However, there was no significant difference in the DA uptake between the studied brain regions (Student *t*-test, $p > 0.05$).

2.3. qPCR Analysis of Transporter mRNA Expression in Neonatal Rat Astrocytes from the Cortex and Striatum

We examined the mRNA expression of high-affinity and low-affinity transporters involved in DA uptake in neonatal rat astrocytes from the cortex and striatum as well as neonatal rat cerebral cortical and striatal tissue. qPCR was performed after the extraction of total RNA from three-weeks-old, confluent cultures and tissue samples for the comparison of transporter mRNA expression between tissue (additionally used as a positive control) and astrocyte cell culture samples.

The mRNA expression profile of the transporters in neonatal rat tissue samples (Figure 3, Tissue) was in line with our expectations. OCT1 and OCT2 mRNA expression was lower than OCT3 mRNA expression in the two studied brain regions. PMAT tissue expression was prominent in both tissue samples. Active uptake transporters, DAT and NET, were expressed in both tissue samples. Conversely, astrocyte cultures from both brain regions displayed a distinct transporter mRNA expression profile, with OCT1 and OCT2 being more prominently expressed than OCT3. Among the studied transporters, PMAT mRNA expression appeared to be the greatest in both brain tissue samples, as well in cortical and striatal astrocytes. The striatal astrocyte mRNA expression of PMAT was statistically significantly greater than the cortical PMAT mRNA expression ($p < 0.05$) (Figure 3, Astrocytes). The mRNA expression of uptake 1 transporter, NET, exceeded the expression of DAT in astrocyte cell cultures from both brain regions, the cortex and striatum. Interestingly, DAT mRNA expression was lower in striatal than in cortical neonatal rat tissue, whereas striatal astrocytes displayed low, but present, mRNA expression of DAT. This was contrary to cortical astrocytes, where DAT mRNA expression was at the limit of detection of our qPCR method (Figure 3, Astrocytes). The differences in transporter expression in astrocytes from each brain region were determined separately by the Brown–Forsythe ANOVA test with post-hoc Dunnett's T3 multiple comparisons test (Cortex: $F = 17.54$ (5.00, 35.28), $p < 0.0001$, Striatum $F = 23.21$ (5.00, 17.05), $p < 0.0001$). Striatal PMAT mRNA expression was significantly greater than all other transporters ($p < 0.05$), whereas NET mRNA expression was significantly greater than DAT, OCT2, and OCT3 ($p < 0.05$). Cortical astrocytes displayed no significant difference between mRNA expression of PMAT and NET; however, the mRNA expression of both transporters was significantly greater than the expression of OCT1, OCT2, OCT3, and DAT.

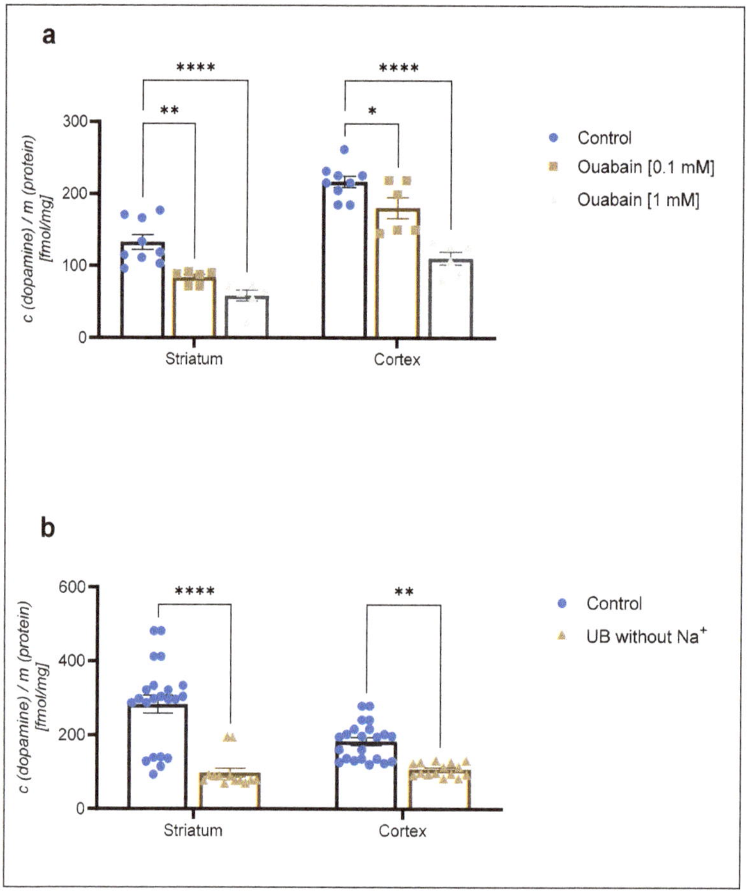

Figure 2. Inhibition of (**a**) 30 nM concentration of [^{3}H]-DA uptake by 0.1 mM and 1 mM concentration of ouabain and (**b**) 30 nM concentration of [^{3}H]-DA uptake in uptake buffer (UB) without Na^+ in cortical and striatal astrocytes from neonatal rats. Results are presented as percent of the control; each bar expressed as mean ± SEM from three or more separate experiments ((**a**): $n = 9$, (**b**): $n = 22$). Statistical significance of reduction of DA uptake in comparison to the control was determined by One-Way ANOVA: * $p < 0.05$, ** $p < 0.01$, **** $p < 0.0001$. Difference between cortical and striatal DA uptake was nonsignificant (Student t-test, $p > 0.05$).

2.4. Inhibition of [^{3}H]-Dopamine Uptake by Antidepressants, Desipramine, Nortriptyline, Amitriptyline, DAT Inhibitor GBR12909, Corticosterone, and Decynium 22

We evaluated the effect of DAT inhibitor GBR12909, OCT and PMAT inhibitors, D22 and corticosterone, and NET selective inhibitors, tricyclic antidepressants, amitriptyline, nortriptyline, and desipramine on DA uptake into striatal and cortical astrocytes. Inhibition curves are presented in Figure 4. The apparent affinity for the transporters involved in DA uptake in cortical and striatal neonatal rat astrocytes was examined by determining their IC_{50} values for the inhibition of transporter-mediated [^{3}H]-DA uptake.

[^{3}H]-DA uptake was more prominently inhibited in striatal than in cortical astrocytes by the selective DAT inhibitor, GBR12909, and NET inhibitor, desipramine, whereas inhibition by nortriptyline and amitriptyline was slight or nonsignificant in astrocytes from both brain regions. Corticosterone had no significant effect on [^{3}H]-DA uptake in both striatal and cortical astrocytes, whereas the PMAT inhibitor, D22, inhibited [^{3}H]-DA uptake similarly in both striatal and cortical astrocytes.

2.5. Changes in mRNA Expression of Plasma Membrane Monoamine Transporter and Norepinephrine Transporter in Cultured Neonatal Rat Astrocytes after 24 h Treatment with Apomorphine, Haloperidol, and L-DOPA

Cortical and striatal astrocyte cultures of neonatal rats were treated with apomorphine, haloperidol, and L-DOPA for 24 h. First, the cell viability after exposure to these three compounds was examined by MTS, Figure 5e, f. The astrocytes' viability was not affected in comparison to the control even with the highest concentration used, 150 µM. We proceeded to perform qPCR analysis of astrocyte cell cultures after a 24 h treatment with the chosen concentration of 100 µM (Figure 5a–d).

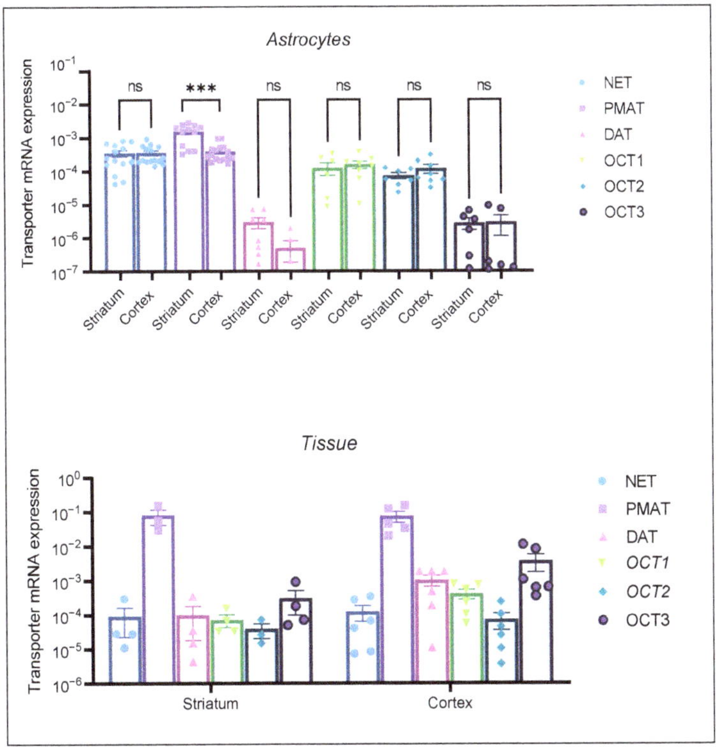

Figure 3. mRNA expression of transporters NET, PMAT, DAT, OCT1, OCT2, and OCT3 in striatal and cortical astrocytes cultured from neonatal rats (Astrocytes) as well as in striatal and cortical tissue samples of neonatal rats (Tissue). mRNA expression is normalized to the mRNA expression of the endogenous control, β-actin. Data are presented as mean ± SEM (astrocytes: $n \geq 6$, from at least two separate experiments, tissue: $n \geq 4$ from 4–6 animal tissue samples). Statistical analysis of comparative mRNA expression of transporters between striatal and cortical astrocytes (Astrocytes), was performed by multiple unpaired Mann–Whitney U tests with multiple testing correction two-stage step-up (Benjamin, Kriger, and Yekutieli) to control the false discovery rate (FDR) *** $p = 0.001$, ns—nonsignificant.

Apomorphine induced the greatest fold change in mRNA expression in both striatal and cortical astrocytes of neonatal rats (Table 1). Apomorphine upregulated mRNA expression of NET mRNA in both cortical and striatal astrocytes but had an effect only on PMAT mRNA expression in cortical, not in striatal, astrocytes. Haloperidol and L-DOPA induced downregulation of PMAT, but the result was statistically nonsignificant (Table 1).

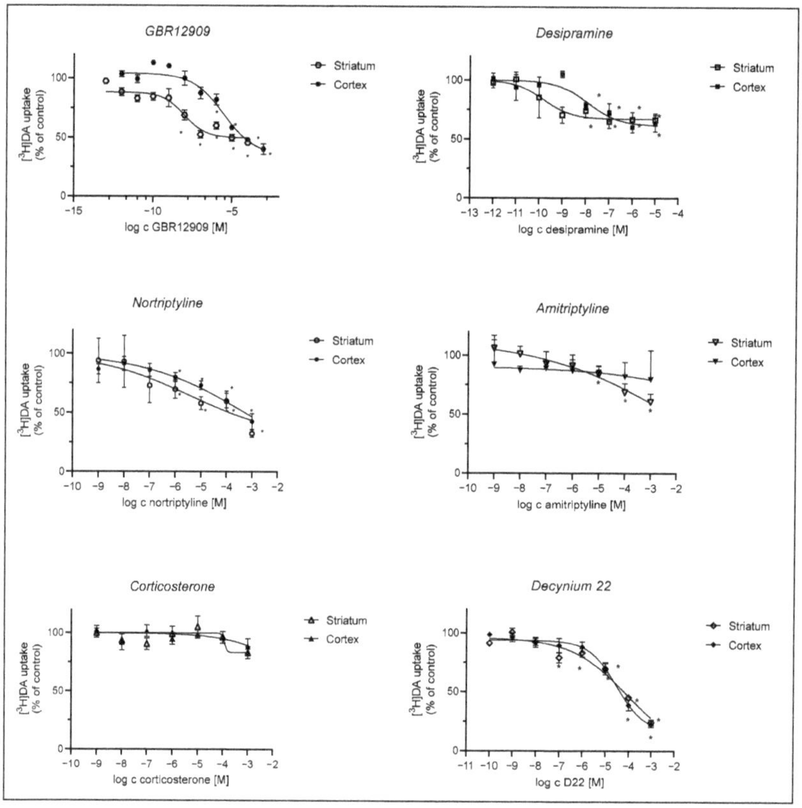

Figure 4. Inhibition of [^{3}H]-DA uptake by GBR12909, desipramine, nortriptyline, amitriptyline, corticosterone, and D22. Astrocytes were preincubated with various compounds for 20 min, which was followed by a 20 min incubation with 30 nM [^{3}H]-DA at 37 °C. Results are presented as percent (mean ± SEM) of the total [^{3}H]-DA uptake of the control from three to four separate experiments carried out in triplicates ($n \geq 9$). IC_{50} and pIC_{50} (calculated as the negative log of the corresponding IC_{50} values) of compounds inhibiting the total [^{3}H]-DA uptake into cultured adult rat astrocytes were calculated from the corresponding inhibition curves. GBR12909: cortex: IC_{50} = 2.97 ± 2.46 μM, pIC_{50} = 5.5, striatum: IC_{50} = 0.00956 ± 0.00571 μM, pIC_{50} = 8.0, desipramine: cortex IC_{50} = 0.0125 ± 0.0137 μM, pIC_{50} = 7.9, striatum: IC_{50} = 0.000157 ± 0.000223 μM, pIC_{50} = 9.8; nortriptyline: cortex ns, striatum: IC_{50} = 3.5 ± 32.1 μM, pIC_{50} = 5.5, amitriptyline: ns, corticosterone: cortex and striatum: ns, D22: cortex—IC_{50} = 30 ± 17 μM, pIC_{50} = 4.5, striatum—IC_{50} = 81 ± 235 μM, pIC_{50} = 4.1. Statistical analysis of the difference in DA uptake in comparison to the control was carried out using an unpaired *t*-test with Welch control, * $p < 0.05$.

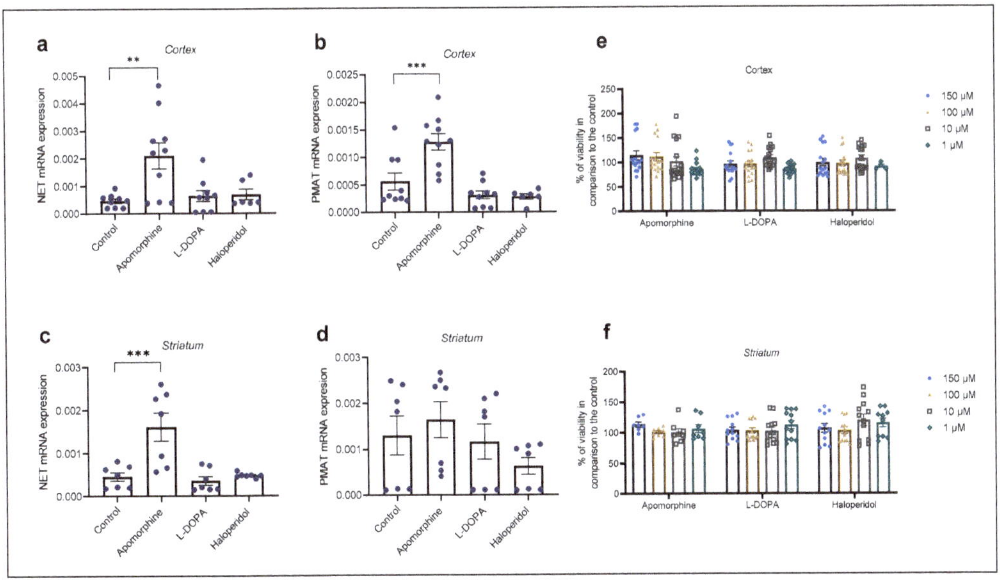

Figure 5. Changes in mRNA expression of transporters NET (**a**,**c**) and PMAT (**b**,**d**) after 24 h treatment with apomorphine, L-DOPA, and haloperidol. Data are presented relative to the expression of endogenous control, β-actin, as mean ± SEM of at least two separate experiments ((**a**,**b**): $n = 9$; (**c**,**d**): $n = 7$). Statistical analysis was performed using One-Way ANOVA with post-hoc Dunnett's test (Cortex: NET: $F(3,30) = 6.701$, $p = 0.0014$, PMAT: $F(3,30) = 13.88$, $p < 0.0001$; striatum: NET: $F(3,24) = 10.98$, $p < 0.0001$, PMAT: $F(3,24) = 1.392$, $p = 0.2695$); ** $p = 0.0013$, *** $p < 0.001$. Cell viability of cortical (**e**) and striatal (**f**) astrocytes after 24 h treatment with various concentrations (1, 10, 100, 150 μM) of apomorphine, L-DOPA, and haloperidol presented as percent of the control, mean ± SEM ((**e**): $n = 16$, 4 separate experiments, (**f**): $n = 12$, 3 separate experiments). Differences in cell viability between treatments and control were determined as nonsignificant by One-Way ANOVA with post-hoc Dunnett's test, $p > 0.05$.

Table 1. Fold changes in mRNA expression of transporters PMAT and NET after 24 h treatment with apomorphine, haloperidol, and L-DOPA. Significant ($p < 0.05$) fold changes in mRNA expression by apomorphine are emphasized by bold text.

Transporter	**PMAT**				**NET**			
Brain Region	**Cortex**		**Striatum**		**Cortex**		**Striatum**	
	Fold Change	***p* Value**	**Fold Change**	***p* Value**	**Fold Change**	***p* Value**	**Fold Change**	***p* Value**
Control	1 ± 0.4		1 ± 0.5		1 ± 0.2		1 ± 0.3	
Apomorphine	**2.3 ± 0.7**	**0.0008**	1.3 ± 0.5	0.8	**4.5 ± 1.3**	**0.02**	**3.5 ± 1.0**	**0.03**
Haloperidol	0.6 ± 0.2	0.7	0.9 ± 0.4	1.0	1.4 ± 0.5	0.8	0.8 ± 0.3	0.9
L-DOPA	0.5 ± 0.2	0.4	0.5 ± 0.2	0.4	1.5 ± 0.5	0.6	1.0 ± 0.2	1.0

3. Discussion

In the present study, we examined the pharmacological and molecular characteristics of DA uptake in cortical and striatal astrocytes of neonatal rats to characterize the involvement of astrocytes from these two brain regions in the dopaminergic tripartite synapse. Additionally, we examined whether mRNA expression of the observed DA uptake transporters in astrocytes is sensitive to treatment with dopaminergic drugs and whether this is brain region-dependent.

DA uptake in cortical astrocytes of neonatal rats has been studied previously, but there is little research concerning the characterization of striatal DA uptake, particularly compared to the DA uptake in the cortical brain region. Research from Inazu et al. [37,38] and Pelton et al. [60] on cortical astrocytes of neonatal rats suggest that DA uptake is time, temperature-, concentration- and Na^+-dependent as well as inhibited by ouabain, a Na^+/K^+-ATPase inhibitor; however, Inazu et al. have not found the DA uptake to be saturable at a 1 mM concentration of radiolabelled DA [38]. Similarly, Hansson et al. [61,62] and Hösli et al. [34] reported that astrocytes do not show saturable kinetics for DA, which they found to be Na^+-independent, leading them to question the existence of a high-affinity carrier system in astrocytes. Our results suggest, similar to the results of Inazu et al. and Pelton et al., that both striatal and cortical astrocyte DA uptake is time-, temperature- and concentration-dependent. In the presence of nM concentrations of DA, uptake is significantly reduced by the absence of sodium ions, both in the uptake medium as well as by the inhibition of the Na^+/K^+-ATPase by 1 mM concentration of ouabain in both brain regions, indicating a presence of an active carrier system in astrocytes from both brain regions. The capacity of active carrier systems, such as DAT and NET, for DA transport is exceeded in the presence of higher concentrations of DA when the uptake 2 transporters, such as the OCTs and PMAT, become vital, due to their high-capacity, albeit low-affinity, for DA. Astrocytes from both brain regions display the ability to take up DA in the millimolar range, which indicates the prominent involvement of high-capacity transporters in both brain regions. Interestingly, striatal-specific DA uptake appears to be saturable, albeit at high concentrations of DA in the millimolar range, whereas cortical astrocytes-specific DA uptake appears to be nonsaturable. Based on these results, we presume astrocytes from both brain regions possess uptake 1 and uptake 2 carrier systems. Striatal astrocyte DA uptake is more dependent on uptake 1 transporters, whereas cortical astrocyte DA uptake is mediated primarily by facilitated diffusion. Nonetheless, astrocytes from both brain regions are capable of continuous DA uptake in the presence of higher concentrations (mM) of DA. As synaptic concentrations of DA have been observed to reach the millimolar range [63], the present study highlights the importance of astrocytes in the removal of excess DA from the synapse as well as DA homeostasis, serving as a sort of neuronal back up system.

To further identify which particular DA carrier system is involved in astrocyte DA uptake, we performed qPCR analysis. The presence of the main neuronal active DA uptake carrier system, DAT, in astrocytes has been reported by Asanuma et al. [39] in primary striatal astrocyte cultures of neonatal rats and in primary striatal cultures of neonatal BALB/c mice by Karakaya et al. [57]. The mRNA expression of DAT in astrocyte cultures from both brain regions in the present study appears to be low. Interestingly, striatal astrocytes exhibited greater mRNA expression of DAT compared to cortical astrocytes, although the expression of DAT appeared to be more prominent in cortical, rather than striatal neonatal rat tissue. GBR12909 inhibited striatal DA uptake but had less of an effect on cortical DA uptake. The correlation between levels of RNA and protein products of specific genes may vary [64]. As low mRNA levels of DAT may not reflect its protein expression and role in astrocyte DA uptake, strong inhibition of nanomolar concentrations of DA uptake in striatal astrocytes by DAT specific inhibitor, GBR12909, indicate its presence in our striatal astrocyte cultures. Nonetheless, our qPCR data show a more prominent mRNA expression of NET in astrocyte cultures from both brain regions, similar to the results of Takeda et al., who reported only NET, and not DAT, mRNA in cortical astrocytes of neonatal rats [24]. Its significance in both striatal and cortical astrocyte DA uptake is supported by the significant inhibition of DA uptake by desipramine, a NET selective inhibitor (pK_i 8.1–8.7 [25,65]). Interestingly, nortriptyline, a NET selective, but less potent, inhibitor (pK_i 8.2 [66,67]) had little effect on cortical compared to striatal DA uptake. The least potent NET inhibitor of the tricyclic antidepressants, amitriptyline, (pKi 6.5 [68]) produced only a slight inhibition of DA uptake in striatal, not cortical, astrocytes. Although we may presume astrocyte DA uptake is mediated to some extent by NET in both brain regions, its role may be of greater importance in striatal than in cortical astrocytes of neonatal rats.

Inhibition of the nonspecific uptake carriers, such as the OCTs and PMAT, was less potent. Specifically, the sensitivity of [^{3}H]-DA uptake to the inhibition by corticosterone, an OCT inhibitor, was very low in astrocytes from both brain regions. Together with our qPCR data, this suggests a low involvement of the OCTs. PMAT is relatively insensitive to corticosterone (Ki = 450 μM [69]) but more potently inhibited by D22 than the OCTs [70] (pKi 7.0 [33,71]). D22-induced inhibition of DA uptake at higher concentrations (around 10 μM), indicates that PMAT could be involved in astrocyte DA uptake in both striatal and cortical astrocytes, albeit to a limited extent. Additionally, it is worth noting the effect of desipramine on [^{3}H]-DA uptake may be attributed to reported nonspecific inhibitory effects of high concentrations of desipramine on other transporters, such as PMAT, which has been shown to be inhibited by a large array of antidepressants, albeit with affinities in the 5–200 μM range [71,72].

The overlapping expression and selectivity of monoamine neurotransmitter transporters as well as transporter inhibitors pose a difficult challenge in determining the contributions of each individual transporters in neurotransmitter uptake studies [73]. Nevertheless, results of the present study indicate neonatal rat striatal and cortical astrocytes are capable of DA uptake, mediated by similar uptake carrier systems, in the millimolar range, which primarily relies on NET and PMAT in cortical astrocytes, and on DAT, NET and PMAT in striatal astrocytes.

Astrocytes participate in DA homeostasis and aid in DA removal from the synapse. Studies have shown astrocytes respond to various dopaminergic drugs [21]; however, research on whether astrocytes may serve as a potential therapeutic target of currently available treatment options is lacking. We used three different compounds: apomorphine, haloperidol, and L-DOPA for the treatment of astrocyte cell cultures in the present study. L-DOPA, a prodrug of DA administered to patients with PD, is metabolized to DA and supplements the low endogenous levels of DA. L-DOPA can interact with D1 or D2 receptors independent of its conversion to endogenous dopamine [74]. Haloperidol competitively blocks post-synaptic D2 in the brain, eliminating DA neurotransmission and leading to the relief of delusions and hallucinations that are commonly associated with psychosis [75]. Apomorphine is the oldest dopaminergic drug available for PD and remains the only drug with efficacy comparable to that of L-DOPA [76]. Like L-DOPA and DA, apomorphine acts as a potent, direct, and broad-spectrum DA agonist, activating all DA receptor subtypes, serotonin receptors, and α-adrenergic receptors [77]. Several studies have investigated the effects of apomorphine [56,78,79], haloperidol [50,52,54,80,81], and L-DOPA [39,82] on astrocytes. Apomorphine has been observed to enhance the biosynthesis of multiple trophic growth factors and is known for its ability to promote neuronal survival [56]. Haloperidol [54] and L-DOPA have been observed to induce a proinflammatory response. Particularly, L-DOPA has been investigated from the aspect of L-DOPA-induced dyskinesia associated with glial activation [83,84]. DA receptor susceptibility to dopaminergic drugs [75,85,86] as well as pathological conditions, such as hypoxia [87], have been explored by multiple studies. The role of dopaminergic drugs in astrocyte DA homeostasis and transport is, however, underexplored. Therefore, we have investigated whether they may affect the role of astrocytes in DA uptake, more specifically, the mRNA expression of transporters, NET and PMAT, involved in astrocyte DA uptake. In the present study, astrocyte mRNA expression of transporter, NET, was upregulated by treatment with apomorphine in astrocytes from both brain regions, whereas PMAT mRNA expression was upregulated only in cortical, but not in striatal astrocytes. Haloperidol and L-DOPA had no significant effect on mRNA expression of either transporter. Our findings indicate apomorphine may induce brain region-specific changes in astrocyte DA transport.

In the present study, we confirmed that astrocytes cultivated from neonatal rat cortex and striatum differ in the molecular and pharmacological characteristics of DA transport. The transportation of DA into astrocytes from both brain regions depends on time, temperature, and exposure to sodium ions and ouabain; but, only DA uptake in striatal astrocytes is saturable and more sensitive to inhibition by desipramine and GBR12909. Astrocytes from both brain regions express DAT, NET, PMAT, and OCTs, but only apomorphine treatment affects their expression. Apomorphine increased the expression of NET and PMAT in cortical astrocytes but only NET in striatal astrocytes.

Although in vitro studies are a vital step in the study of new, unexplored concepts and areas, they inevitably pose some limitations. The prolonged lack of dopaminergic stimulation under nonphysiological conditions may affect the expression of transporters involved in DA uptake and may induce some changes in the characteristics of the observed astrocyte DA uptake. Nevertheless, the findings of the present study indicate that astrocytes from different brain regions have distinct characteristics in regard to DA homeostasis and are sensitive to treatment with dopaminergic drugs, such as apomorphine, which may open our horizons to potential new therapeutic targets in the treatment of neurodegenerative diseases, such as PD, that have yet to be managed successfully.

4. Materials and Methods

4.1. Materials

All tissue culture reagents, except fetal bovine serum, which was from Cambrex IEP GmbH (Wiesbaden, Germany), were obtained from Gibco Invitrogen (Paisley, Scotland, UK). [^{3}H]-DA (2220 GBq/mmol) was purchased from Perkin Elmer (Waltham, MA, USA). The E.Z.N.A.® HP Total RNA Kit was from Omega Bio-tek (Norcross, GA, USA), and the High Capacity cDNA Reverse Transcription Kit, TaqMan Gene Expression Assays, and TaqMan® Gene Expression Master Mix were from Applied Biosystems (Carlsbad, CA, USA). GBR12909 was from Tocris (Bristol, UK), and decynium 22 (D22), corticosterone, L-DOPA, haloperidol, and apomorphine HCl were obtained from Sigma Aldrich (St. Louis, MO, USA). Nortriptyline HCl, desipramine HCl, and amitriptyline HCl were purchased from Sandoz (Cham, Switzerland). The CellTiter 96® AQueous One Solution Cell Proliferation Assay (MTS) was obtained from Promega (Madison, WI, USA).

4.2. Animals and Primary Cell Culture Preparation

Astrocyte cell cultures were obtained from neonatal (3-day-old) rats, species *Rattus norvegicus*, strain Wistar, in accordance with the Administration of the Republic of Slovenia for Food Safety, Veterinary and Plant Protection issue U34401-23/2022/6. This study was approved by the National Veterinary Administration (approval numbers U34401-20/2017/2, approval date 20 June 2017 and U34401-23/2022/6, approval date 23 December 2022). Collectively, 30 animals were sacrificed using decapitation, and all necessary measures were taken to reduce discomfort and suffering of the animals according to the 3R principle. Astrocyte cultures were prepared as described [88,89]. Briefly, primary cultures derived from the striatum or cerebral cortex were grown in high-glucose Dulbecco's Modified Eagle Medium (DMEM), containing 10% FBS, 1 mM pyruvate, 2 mM glutamine, and 25 µg/mL streptomycin at 37 °C in humidified 95% air/5% CO_2. To reduce the number of contaminating microglial cells, confluent cultures were exposed to overnight shaking at 225 RPM. Medium containing detached cells was removed the next morning, and fresh growth medium was added. The whole procedure was repeated three times. Upon reaching confluence, the cells were seeded in 12-well plates and grown for an additional 3 weeks before being used for DA uptake or qPCR experiments. For cell viability experiments, cells were plated in 96-well plates and grown until reaching confluence (3–4 days).

4.3. Dopamine Uptake Experiments

4.3.1. Dependence of [^{3}H]-Dopamine Uptake on Temperature, Time, and Concentration

Monolayer cultures in 12-well plates were preincubated for 30 min in the uptake buffer (25 mM HEPES, 125 mM NaCl, 4.8 mM KCl, 1.2 mM KH_2PO_4, 1.2 mM $MgSO_4$, 1.4 mM $CaCl_2$, and 5.6 mM glucose, pH 7.4) at 37 °C (total uptake) or at 4 °C (nonspecific uptake). To determine the time-dependence of DA uptake, cultured astrocytes were incubated with

30 nM [^{3}H]-DA for different time intervals (0, 5, 10, 15, 20, 30, 45, 60 min). The concentration dependence of DA uptake was determined by exposing cultured astrocytes to different DA concentrations (up to 1 mM), as indicated in the Results, for 20 min. Specific DA uptake was calculated as the difference between the total and nonspecific [^{3}H]-DA uptake. The apparent uptake velocity, V_{max}, was determined using the Michaelis–Menten equation. Experiments were terminated by placing the plates on ice. [^{3}H]-DA was removed quickly, and the plates were washed twice with ice-cold uptake buffer without Ca^{2+}. The cells were subsequently lysed in 300 µL of 0.5 M NaOH. An aliquot (250 µL) of each sample was transferred to a scintillation vial to measure the radioactivity. The amount of transported DA was normalized to the total protein content, which was determined in the remaining aliquots (50 µL) of each sample using the Bradford method using the Bio-Rad Protein Assay (Hercules, CA, USA) measured with the Biotek Synergy HT Microplate Reader (Agilent, Santa Clara, CA, USA).

4.3.2. Inhibition of [^{3}H]-Dopamine Uptake

To determine the sensitivity of DA transport to different uptake inhibitors (corticosterone, D22, GBR12909, and antidepressants, nortriptyline, amitriptyline, and desipramine) or Na^+/K^+-ATPase inhibitor ouabain, cultured astrocytes were first preincubated in uptake buffer in the presence or absence of the inhibitor for 20 min at 37 °C. Astrocytes were then exposed to 30 nM [^{3}H]-DA for 20 min. To assess the dependency of DA uptake on the presence of Na^+, cultured astrocytes were incubated in the normal (Na^+-containing) or Na^+-free uptake buffer (125 mM NaCl in the uptake buffer was substituted with equimolar choline chloride ($(CH_3)_3N(Cl)CH_2CH_2OH$)) with 30 nM [^{3}H]-DA for 20 min. Samples were harvested and processed for measurements of radioactivity and protein content as described above.

4.4. *Quantitative Polymerase Chain Reaction (qPCR)*

The total RNA from neonatal rat cortical and striatal tissue and cultured neonatal rat cortical and striatal astrocytes was extracted using EZNA HP Total RNA kit. Cell cultures were collected from 12-well plates, each well representing one sample, whereas striatal and cortical brain tissue samples were collected together from multiple animals. The striatal tissue was used whole. Due to its size, the cortical tissue was dissected into smaller sections, and the frontal part of the cortex was used for qPCR analysis. Brain tissue samples used for qPCR analysis were flash frozen in liquid nitrogen after excision and stored at −70 °C. Treated and nontreated astrocyte cell cultures plated onto 12-well plates were placed on ice and washed with sterile PBS buffer three times, which was then quickly and completely removed from each well. The plates were then stored at −70 °C until the total RNA extraction was performed. The quantity and quality of the extracted RNA was evaluated using absorbance as measured by the Biotek Synergy HT Microplate Reader on a Take3 microvolume plate. cDNA was synthesized using 1 µg of total RNA from each sample. qPCR analysis of cell culture samples from 12-well plates was performed in triplicates or quadruplicates, each well representing one sample. Experiments were repeated at least twice, and the analysis of tissue samples was performed using small sections (up to 30 mg) of previously flash frozen tissue from 3–5 animals. qPCR was performed using TaqMan Gene Expression Assays (PMAT (SLC29A4) Rn01453824_m1, NET (SLC6A2) Rn00580207_m1, DAT (SLC6A3) Rn00562224_m1, OCT1 (SLC22A1) (Rn00562250_m1), OCT2 (SLC22A2) (Rn00580893_m1), OCT3 (SLC22A3) Rn00570264_m1, and ß-actin (Rn00667869_m1)) according to the manufacturer's instructions in the QuantStudio™ 3 System (Thermo Fisher Scientific, Carlsbad, CA, USA). Expression of target genes was normalized to the expression of β-actin according to the equation [target/reference] = $[EFF_{reference}\hat{}Cq_{reference}]/[EFF_{target}\hat{}Cq_{target}]$, where Cq is the quantification cycle and EFF is the amplification efficiency (expressed as a value between 1 and 2). EFF was determined with the LinRegPCR software (Version 2014.7) [90].

4.5. Cell Culture Treatment and Cell Viability

Cell viability was assessed using confluent astrocyte cell cultures plated in 96-well plates and treated with various concentrations of apomorphine, L-DOPA, and haloperidol (from 1–150 μM) for 24 h. The cell viability was measured using the CellTiter 96™ AQueous One Solution Cell Proliferation Assay (MTS) (Promega) according to the manufacturer's instructions. For the qPCR analysis, confluent three-weeks-old neonatal rat cortical and striatal astrocyte cell cultures plated in 12-well plates were treated with 100 μM concentration of the three compounds, haloperidol, apomorphine, and L-DOPA, for 24 h, based on cell viability experiments.

4.6. Data Analysis

The uptake experiments and qPCR experiments on cell cultures were routinely carried out in triplicates or quadruplicates and each experiment was repeated at least twice. Triplicate and quadruplicate of each experiment consisted of the number of wells (3 or 4) used from the cell cultures plated onto 12-well plates. The total number of samples (n) represented the total number of wells used, pooled from multiple experiments. Similarly, in cell viability experiments, cell culture in each well plated onto 96-well plate represented one sample. All data are presented as arithmetic means ± SEM. GraphPad Prism 9.5 was used for processing, presentation, and statistical analysis of the data. The kinetic parameters (K_m and V_{max}) and IC_{50} values were calculated using a nonlinear regression method using the GraphPad Prism software 9.5.0. The normality of the data distribution was investigated by the Shapiro–Wilk test. The comparison of data among groups was carried out using ANOVA (One-Way ANOVA or Brown–Forsythe ANOVA) with Dunnett's or Dunnett's T3 multiple comparisons post-hoc test. If only two groups of data were compared, the Mann–Whitney U test, an unpaired Student's *t*-test or Welch's *t*-test was used, depending on the normality of the data distribution and the variance. All *t*-tests were two-tailed. p values < 0.05 were considered statistically significant.

5. Conclusions

In conclusion, our study investigated the molecular and pharmacological features of DA uptake in cortical and striatal astrocytes from neonatal rats, shedding light on their distinct roles in the dopaminergic tripartite synapse. Notably, we identified time-, temperature-, and concentration-dependent DA uptake mechanisms in both regions, with striatal astrocytes exhibiting saturable kinetics. Our mRNA expression analysis revealed low DAT levels but prominent NET and PMAT expression in astrocytes from both studied brain regions. Treatment with dopaminergic drugs influenced NET and PMAT expression differentially in cortical and striatal astrocytes.

Author Contributions: Conceptualization, V.S.; methodology, V.S. and M.K.; validation, V.S.; formal analysis, V.S. and K.D.; investigation, V.S., M.K. and K.D.; resources, V.S., M.K. and K.D.; writing—original draft preparation, V.S.; writing—review and editing, V.S., M.K. and K.D.; visualization, V.S.; supervision, M.K.; project administration, M.K.; funding acquisition, M.K. All authors have read and agreed to the published version of the manuscript.

Funding: The research was supported by Slovenian research agency grants: P3-0067, P3-0043 and J7-3153 and V.S.'s young researchers grant.

Institutional Review Board Statement: The study was conducted in accordance with the Administration of the Republic of Slovenia for Food Safety, Veterinary and Plant Protection issue U34401-23/2022/6. This study was approved by the National Veterinary Administration (approval numbers U34401-20/2017/2, approval date 20 June 2017 and U34401-23/2022/6, approval date 23 December 2022).

Informed Consent Statement: Not applicable.

Data Availability Statement: All data are contained within the manuscript. The data presented in this study are available on request from the corresponding author.

Acknowledgments: The authors are grateful for technical assistance of Jožica Košir.

Conflicts of Interest: The authors declare no conflicts of interest. The funders had no role in the design of the study, in the collection, analysis, or interpretation of data, in the writing of the manuscript, or in the decision to publish the results.

References

1. von Bartheld, C.S.; Bahney, J.; Herculano-Houzel, S. The search for true numbers of neurons and glial cells in the human brain: A review of 150 years of cell counting. *J. Comp. Neurol.* **2016**, *524*, 3865–3895. [CrossRef]
2. Verkhratsky, A.; Nedergaard, M. Physiology of Astroglia. *Physiol. Rev.* **2018**, *98*, 239–389. [CrossRef]
3. Verkhratsky, A.; Semyanov, A.; Zorec, R. Physiology of Astroglial Excitability. *Function* **2020**, *1*, zqaa016. [CrossRef]
4. Goenaga, J.; Araque, A.; Kofuji, P.; Herrera Moro Chao, D. Calcium signaling in astrocytes and gliotransmitter release. *Front. Synaptic Neurosci.* **2023**, *15*, 1138577. [CrossRef]
5. Araque, A.; Parpura, V.; Sanzgiri, R.P.; Haydon, P.G. Tripartite synapses: Glia, the unacknowledged partner. *Trends Neurosci.* **1999**, *22*, 208–215. [CrossRef]
6. Santello, M.; Calì, C.; Bezzi, P. Gliotransmission and the tripartite synapse. *Adv. Exp. Med. Biol.* **2012**, *970*, 307–331. [CrossRef] [PubMed]
7. Perea, G.; Araque, A. Communication between astrocytes and neurons: A complex language. *J. Physiol. Paris* **2002**, *96*, 199–207. [CrossRef]
8. Perea, G.; Araque, A. GLIA modulates synaptic transmission. *Brain Res. Rev.* **2010**, *63*, 93–102. [CrossRef]
9. Perea, G.; Araque, A. Glial calcium signaling and neuron-glia communication. *Cell Calcium* **2005**, *38*, 375–382. [CrossRef]
10. Araque, A.; Perea, G. Glial modulation of synaptic transmission in culture. *Glia* **2004**, *47*, 241–248. [CrossRef] [PubMed]
11. Perea, G.; Araque, A. Synaptic regulation of the astrocyte calcium signal. *J. Neural Transm.* **2005**, *112*, 127–135. [CrossRef] [PubMed]
12. Perea, G.; Navarrete, M.; Araque, A. Tripartite synapses: Astrocytes process and control synaptic information. *Trends Neurosci.* **2009**, *32*, 421–431. [CrossRef] [PubMed]
13. Klein, M.O.; Battagello, D.S.; Cardoso, A.R.; Hauser, D.N.; Bittencourt, J.C.; Correa, R.G. Dopamine: Functions, Signaling, and Association with Neurological Diseases. *Cell Mol. Neurobiol.* **2019**, *39*, 31–59. [CrossRef] [PubMed]
14. Speranza, L.; di Porzio, U.; Viggiano, D.; de Donato, A.; Volpicelli, F. Dopamine: The Neuromodulator of Long-Term Synaptic Plasticity, Reward and Movement Control. *Cells* **2021**, *10*, 735. [CrossRef]
15. Ott, T.; Nieder, A. Dopamine and Cognitive Control in Prefrontal Cortex. *Trends Cogn. Sci.* **2019**, *23*, 213–234. [CrossRef]
16. Turk, A.Z.; Lotfi Marchoubeh, M.; Fritsch, I.; Maguire, G.A.; SheikhBahaei, S. Dopamine, vocalization, and astrocytes. *Brain Lang.* **2021**, *219*, 104970. [CrossRef]
17. Linnerbauer, M.; Rothhammer, V. Protective Functions of Reactive Astrocytes Following Central Nervous System Insult. *Front. Immunol.* **2020**, *11*, 573256. [CrossRef]
18. Miyazaki, I.; Asanuma, M. Neuron-Astrocyte Interactions in Parkinson's Disease. *Cells* **2020**, *9*, 2623. [CrossRef]
19. Saba, J.; López Couselo, F.; Turati, J.; Carniglia, L.; Durand, D.; de Laurentiis, A.; Lasaga, M.; Caruso, C. Astrocytes from cortex and striatum show differential responses to mitochondrial toxin and BDNF: Implications for protection of striatal neurons expressing mutant huntingtin. *J. Neuroinflammation* **2020**, *17*, 290. [CrossRef]
20. Chai, H.; Diaz-Castro, B.; Shigetomi, E.; Monte, E.; Octeau, J.C.; Yu, X.; Cohn, W.; Rajendran, P.S.; Vondriska, T.M.; Whitelegge, J.P.; et al. Neural Circuit-Specialized Astrocytes: Transcriptomic, Proteomic, Morphological, and Functional Evidence. *Neuron* **2017**, *95*, 531–549.e9. [CrossRef]
21. Jennings, A.; Rusakov, D.A.; Rusakov, D.A. Do astrocytes respond to dopamine? *Opera Med. Physiol. Opera Medica Physiol.* **2016**, *2*, 34–43.
22. Beaulieu, J.M.; Espinoza, S.; Gainetdinov, R.R. Dopamine receptors—IUPHAR Review 13. *Br. J. Pharmacol.* **2015**, *172*, 1–23. [CrossRef]
23. Maiya, R.; Mayfield, R.D. Dopamine Transporter Network and Pathways. In *International Review of Neurobiology*; Academic Press: Cambridge, MA, USA, 2004; Volume 61, pp. 79–96.
24. Takeda, H.; Inazu, M.; Matsumiya, T. Astroglial dopamine transport is mediated by norepinephrine transporter. *Naunyn-Schmiedeberg's Arch. Pharmacol.* **2002**, *366*, 620–623. [CrossRef]
25. Inazu, M.; Takeda, H.; Matsumiya, T. Functional expression of the norepinephrine transporter in cultured rat astrocytes Norepinephrine transporter in astrocytes. *J. Neurochem.* **2003**, *84*, 136–144. [CrossRef]
26. Morón, J.A.; Brockington, A.; Wise, R.A.; Rocha, B.A.; Hope, B.T. Dopamine uptake through the norepinephrine transporter in brain regions with low levels of the dopamine transporter: Evidence from knock-out mouse lines. *J. Neurosci.* **2002**, *22*, 389–395. [CrossRef]
27. Nishijima, H.; Tomiyama, M. What mechanisms are responsible for the reuptake of levodopa-derived dopamine in parkinsonian striatum? *Front. Neurosci.* **2016**, *10*, 575. [CrossRef]
28. Gasser, P.J. Organic Cation Transporters in Brain Catecholamine Homeostasis. *Handb. Exp. Pharmacol.* **2021**, *266*, 187–197. [CrossRef]

29. Gasser, P.J. Roles for the uptake(2) transporter OCT3 in regulation of dopaminergic neurotransmission and behavior. *Neurochem. Int.* **2019**, *123*, 46–49. [CrossRef]
30. Itagaki, S.; Ganapathy, V.; Ho, H.T.B.; Zhou, M.; Babu, E.; Wang, J. Electrophysiological characterization of the polyspecific organic cation transporter plasma membrane monoamine transporter. *Drug Metab. Dispos. Biol. Fate Chem.* **2012**, *40*, 1138–1143. [CrossRef]
31. Dahlin, A.; Xia, L.; Kong, W.; Hevner, R.; Wang, J. Expression and immunolocalization of the plasma membrane monoamine transporter in the brain. *Neuroscience* **2007**, *146*, 1193–1211. [CrossRef]
32. Koepsell, H.; Schmitt, B.M.; Gorboulev, V. Organic cation transporters. *Rev. Physiol. Biochem. Pharmacol.* **2003**, *150*, 36–90. [CrossRef]
33. Wang, J. The plasma membrane monoamine transporter (PMAT): Structure, function, and role in organic cation disposition. *Clin. Pharmacol. Ther.* **2016**, *100*, 489–499. [CrossRef]
34. Hösli, E.; Hösli, L. Autoradiographic studies on the uptake of 3H-dopamine by neurons and astrocytes in explant and primary cultures of rat CNS: Effects of uptake inhibitors. *Int. J. Dev. Neurosci.* **1997**, *15*, 45–53. [CrossRef]
35. Hösli, E.; Hösli, L. Binding sites for [3H]dopamine and dopamine-antagonists on cultured astrocytes of rat striatum and spinal cord: An autoradiographic study. *Neurosci. Lett.* **1986**, *65*, 177–182. [CrossRef]
36. Inazu, M.; Takeda, H.; Matsumiya, T. Expression and functional characterization of the extraneuronal monoamine transporter in normal human astrocytes. *J. Neurochem.* **2003**, *84*, 43–52. [CrossRef]
37. Inazu, M.; Kubota, N.; Takeda, H.; Zhang, J.; Kiuchi, Y.; Oguchi, K.; Matsumiya, T. Pharmacological characterization of dopamine transport in cultured rat astrocytes. *Life Sci.* **1999**, *64*, 2239–2245. [CrossRef]
38. Inazu, M.; Takeda, H.; Ikoshi, H.; Uchida, Y.; Kubota, N.; Kiuchi, Y.; Oguchi, K.; Matsumiya, T. Regulation of dopamine uptake by basic fibroblast growth factor and epidermal growth factor in cultured rat astrocytes. *Neurosci. Res.* **1999**, *34*, 235–244. [CrossRef]
39. Asanuma, M.; Miyazaki, I.; Murakami, S.; Diaz-Corrales, F.J.; Ogawa, N. Striatal astrocytes act as a reservoir for L-DOPA. *PLoS ONE* **2014**, *9*, e106362. [CrossRef]
40. Li, X.M.; Juorio, A.V.; Paterson, I.A.; Walz, W.; Zhu, M.Y.; Boulton, A.A. Gene expression of aromatic L-amino acid decarboxylase in cultured rat glial cells. *J. Neurochem.* **1992**, *59*, 1172–1175. [CrossRef]
41. Inyushin, M.Y.; Huertas, A.; Kucheryavykh, Y.V.; Kucheryavykh, L.Y.; Tsydzik, V.; Sanabria, P.; Eaton, M.J.; Skatchkov, S.N.; Rojas, L.V.; Wessinger, W.D. L-DOPA Uptake in Astrocytic Endfeet Enwrapping Blood Vessels in Rat Brain. *Park. Dis.* **2012**, *2012*, 321406. [CrossRef]
42. Tsai, M.J.; Lee, E.H. Characterization of L-DOPA transport in cultured rat and mouse astrocytes. *J. Neurosci. Res.* **1996**, *43*, 490–495. [CrossRef]
43. Juorio, A.V.; Li, X.M.; Walz, W.; Paterson, I.A. Decarboxylation of L-dopa by cultured mouse astrocytes. *Brain Res.* **1993**, *626*, 306–309. [CrossRef]
44. Sampaio-Maia, B.; Serrão, M.P.; Soares-da-Silva, P. Regulatory pathways and uptake of L-DOPA by capillary cerebral endothelial cells, astrocytes, and neuronal cells. *Am. J. Physiol. Cell Physiol.* **2001**, *280*, C333–C342. [CrossRef]
45. Miyazaki, I.; Asanuma, M.; Diaz-Corrales, F.J.; Miyoshi, K.; Ogawa, N. Direct evidence for expression of dopamine receptors in astrocytes from basal ganglia. *Brain Res.* **2004**, *1029*, 120–123. [CrossRef]
46. Verharen, J.P.H.; de Jong, J.W.; Lammel, S. Dopaminergic Control over the Tripartite Synapse. *Neuron* **2020**, *105*, 954–956. [CrossRef]
47. Hösli, L.; Hösli, E. Receptors for dopamine and serotonin on astrocytes of cultured rat central nervous system. *J. Physiol.* **1987**, *82*, 191–195.
48. Fischer, T.; Scheffler, P.; Lohr, C. Dopamine-induced calcium signaling in olfactory bulb astrocytes. *Sci. Rep.* **2020**, *10*, 631. [CrossRef]
49. Liu, J.; Wang, F.; Huang, C.; Long, L.H.; Wu, W.N.; Cai, F.; Wang, J.H.; Ma, L.Q.; Chen, J.G. Activation of phosphatidylinositol-linked novel D1 dopamine receptor contributes to the calcium mobilization in cultured rat prefrontal cortical astrocytes. *Cell Mol. Neurobiol.* **2009**, *29*, 317–328. [CrossRef]
50. Hsu, S.S.; Liang, W.Z. Ca^{2+} signaling as a mechanism of haloperidol-induced cytotoxicity in human astrocytes and assessing the protective role of a Ca^{2+} chelator. *Naunyn Schmiedebergs Arch. Pharmacol.* **2020**, *393*, 2117–2127. [CrossRef] [PubMed]
51. Shao, Z.; Dyck, L.E.; Wang, H.; Li, X.-M. Antipsychotic drugs cause glial cell line-derived neurotrophic factor secretion from C6 glioma cells. *J. Psychiatry Neurosci. JPN* **2006**, *31*, 32–37.
52. Tanahashi, S.; Yamamura, S.; Nakagawa, M.; Motomura, E.; Okada, M. Clozapine, but not haloperidol, enhances glial D-serine and L-glutamate release in rat frontal cortex and primary cultured astrocytes. *Br. J. Pharmacol.* **2012**, *165*, 1543–1555. [CrossRef]
53. Steiner, J.; Schroeter, M.L.; Schiltz, K.; Bernstein, H.G.; Müller, U.J.; Richter-Landsberg, C.; Müller, W.E.; Walter, M.; Gos, T.; Bogerts, B.; et al. Haloperidol and clozapine decrease S100B release from glial cells. *Neuroscience* **2010**, *167*, 1025–1031. [CrossRef]
54. Quincozes-Santos, A.; Bobermin, L.D.; Tonial, R.P.; Bambini-Junior, V.; Riesgo, R.; Gottfried, C. Effects of atypical (risperidone) and typical (haloperidol) antipsychotic agents on astroglial functions. *Eur. Arch. Psychiatry Clin. Neurosci.* **2010**, *260*, 475–481. [CrossRef]
55. Hösli, L.; Hösli, E.; Baggi, M.; Bassetti, C.; Uhr, M. Action of dopamine and serotonin on the membrane potential of cultured astrocytes. *Exp. Brain Res.* **1987**, *65*, 482–485. [CrossRef]

56. Li, A.; Guo, H.; Luo, X.; Sheng, J.; Yang, S.; Yin, Y.; Zhou, J.; Zhou, J. Apomorphine-induced activation of dopamine receptors modulates FGF-2 expression in astrocytic cultures and promotes survival of dopaminergic neurons. *Faseb J.* **2006**, *20*, 1263–1265. [CrossRef]
57. Karakaya, S.; Kipp, M.; Beyer, C. Oestrogen regulates the expression and function of dopamine transporters in astrocytes of the nigrostriatal system. *J. Neuroendocrinol.* **2007**, *19*, 682–690. [CrossRef]
58. Zhou, B.; Zuo, Y.X.; Jiang, R.T. Astrocyte morphology: Diversity, plasticity, and role in neurological diseases. *CNS Neurosci. Ther.* **2019**, *25*, 665–673. [CrossRef]
59. Khakh, B.S.; Deneen, B. The Emerging Nature of Astrocyte Diversity. *Annu. Rev. Neurosci.* **2019**, *42*, 187–207. [CrossRef]
60. Pelton, E.W., 2nd; Kimelberg, H.K.; Shipherd, S.V.; Bourke, R.S. Dopamine and norepinephrine uptake and metabolism by astroglial cells in culture. *Life Sci.* **1981**, *28*, 1655–1663. [CrossRef] [PubMed]
61. Hansson, E.; Eriksson, P.; Nilsson, M. Amino acid and monoamine transport in primary astroglial cultures from defined brain regions. *Neurochem. Res.* **1985**, *10*, 1335–1341. [CrossRef] [PubMed]
62. Hansson, E. Transport of monoamine and amino acid neurotransmitters by primary astroglial cultures. *Neurochem. Res.* **1985**, *10*, 667–675. [CrossRef]
63. Daws, L.C. Organic Cation Transporters in Psychiatric Disorders. *Handb. Exp. Pharmacol.* **2021**, *266*, 215–239. [CrossRef]
64. Buccitelli, C.; Selbach, M. mRNAs, proteins and the emerging principles of gene expression control. *Nat. Rev. Genet.* **2020**, *21*, 630–644. [CrossRef]
65. O'Neill, D.J.; Adedoyin, A.; Alfinito, P.D.; Bray, J.A.; Cosmi, S.; Deecher, D.C.; Fensome, A.; Harrison, J.; Leventhal, L.; Mann, C.; et al. Discovery of Novel Selective Norepinephrine Reuptake Inhibitors: 4-[3-Aryl-2,2-dioxido-2,1,3-benzothiadiazol-1(3H)-yl]-1-(methylamino)butan-2-ols (WYE-103231). *J. Med. Chem.* **2010**, *53*, 4511–4521. [CrossRef] [PubMed]
66. Glennon, R.A.; Lee, M.; Rangisetty, J.B.; Dukat, M.; Roth, B.L.; Savage, J.E.; McBride, A.; Rauser, L.; Hufeisen, S.; Lee, D.K. 2-Substituted tryptamines: Agents with selectivity for 5-HT(6) serotonin receptors. *J. Med. Chem.* **2000**, *43*, 1011–1018. [CrossRef]
67. Paczkowski, F.A.; Bryan-Lluka, L.J.; Pörzgen, P.; Brüss, M.; Bönisch, H. Comparison of the pharmacological properties of cloned rat, human, and bovine norepinephrine transporters. *J. Pharmacol. Exp. Ther.* **1999**, *290*, 761–767.
68. Arunotayanun, W.; Dalley, J.W.; Huang, X.P.; Setola, V.; Treble, R.; Iversen, L.; Roth, B.L.; Gibbons, S. An analysis of the synthetic tryptamines AMT and 5-MeO-DALT: Emerging 'Novel Psychoactive Drugs'. *Bioorg Med. Chem. Lett.* **2013**, *23*, 3411–3415. [CrossRef] [PubMed]
69. Zhou, M.; Xia, L.; Engel, K.; Wang, J. Molecular determinants of substrate selectivity of a novel organic cation transporter (PMAT) in the SLC29 family. *J. Biol. Chem.* **2007**, *282*, 3188–3195. [CrossRef]
70. Fraser-Spears, R.; Krause-Heuer, A.M.; Basiouny, M.; Mayer, F.P.; Manishimwe, R.; Wyatt, N.A.; Dobrowolski, J.C.; Roberts, M.P.; Greguric, I.; Kumar, N.; et al. Comparative analysis of novel decynium-22 analogs to inhibit transport by the low-affinity, high-capacity monoamine transporters, organic cation transporters 2 and 3, and plasma membrane monoamine transporter. *Eur. J. Pharmacol.* **2019**, *842*, 351–364. [CrossRef]
71. Engel, K.; Wang, J. Interaction of Organic Cations with a Newly Identified Plasma Membrane Monoamine Transporter. *Mol. Pharmacol.* **2005**, *68*, 1397. [CrossRef]
72. Haenisch, B.; Bönisch, H. Interaction of the human plasma membrane monoamine transporter (hPMAT) with antidepressants and antipsychotics. *Naunyn-Schmiedeberg's Arch. Pharmacol.* **2010**, *381*, 33–39. [CrossRef]
73. Koepsell, H. General Overview of Organic Cation Transporters in Brain. *Handb. Exp. Pharmacol.* **2021**, 1–39. [CrossRef]
74. Viaro, R.; Longo, F.; Vincenzi, F.; Varani, K.; Morari, M. l-DOPA promotes striatal dopamine release through D1 receptors and reversal of dopamine transporter. *Brain Res.* **2021**, *1768*, 147583. [CrossRef]
75. Lidow, M.S.; Goldman-Rakic, P.S. A common action of clozapine, haloperidol, and remoxipride on D1- and D2-dopaminergic receptors in the primate cerebral cortex. *Proc. Natl. Acad. Sci. USA* **1994**, *91*, 4353–4356. [CrossRef]
76. Carbone, F.; Djamshidian, A.; Seppi, K.; Poewe, W. Apomorphine for Parkinson's Disease: Efficacy and Safety of Current and New Formulations. *CNS Drugs* **2019**, *33*, 905–918. [CrossRef] [PubMed]
77. Jenner, P.; Katzenschlager, R. Apomorphine—Pharmacological properties and clinical trials in Parkinson's disease. *Park. Relat. Disord.* **2016**, *33*, S13–S21. [CrossRef]
78. Yang, F.; Liu, Y.; Tu, J.; Wan, J.; Zhang, J.; Wu, B.; Chen, S.; Zhou, J.; Mu, Y.; Wang, L. Activated astrocytes enhance the dopaminergic differentiation of stem cells and promote brain repair through bFGF. *Nat. Commun.* **2014**, *5*, 5627. [CrossRef]
79. Hoenicka, J.; Quiñones-Lombraña, A.; España-Serrano, L.; Alvira-Botero, X.; Kremer, L.; Pérez-González, R.; Rodríguez-Jiménez, R.; Jiménez-Arriero, M.Á.; Ponce, G.; Palomo, T. The ANKK1 Gene Associated with Addictions Is Expressed in Astroglial Cells and Upregulated by Apomorphine. *Biol. Psychiatry* **2010**, *67*, 3–11. [CrossRef] [PubMed]
80. Reuss, B.; Unsicker, K. Atypical neuroleptic drugs downregulate dopamine sensitivity in rat cortical and striatal astrocytes. *Mol. Cell Neurosci.* **2001**, *18*, 197–209. [CrossRef]
81. Nardin, P.; Tramontina, A.C.; Quincozes-Santos, A.; Tortorelli, L.S.; Lunardi, P.; Klein, P.R.; Wartchow, K.M.; Bobermin, L.D.; Gottfried, C.; Elisabetsky, E.; et al. In vitro S100B secretion is reduced by apomorphine: Effects of antipsychotics and antioxidants. *Progress. Neuro-Psychopharmacol. Biol. Psychiatry* **2011**, *35*, 1291–1296. [CrossRef]
82. Asanuma, M.; Miyazaki, I. 3-O-Methyldopa inhibits astrocyte-mediated dopaminergic neuroprotective effects of L-DOPA. *BMC Neurosci.* **2016**, *17*, 52. [CrossRef]

83. Del-Bel, E.; Bortolanza, M.; Dos-Santos-Pereira, M.; Bariotto, K.; Raisman-Vozari, R. l-DOPA-induced dyskinesia in Parkinson's disease: Are neuroinflammation and astrocytes key elements? *Synapse* **2016**, *70*, 479–500. [CrossRef]
84. Carta, A.R.; Mulas, G.; Bortolanza, M.; Duarte, T.; Pillai, E.; Fisone, G.; Vozari, R.R.; Del-Bel, E. l-DOPA-induced dyskinesia and neuroinflammation: Do microglia and astrocytes play a role? *Eur. J. Neurosci.* **2017**, *45*, 73–91. [CrossRef]
85. Damask, S.P.; Bovenkerk, K.A.; de la Pena, G.; Hoversten, K.M.; Peters, D.B.; Valentine, A.M.; Meador-Woodruff, J.H. Differential effects of clozapine and haloperidol on dopamine receptor mRNA expression in rat striatum and cortex. *Mol. Brain Res.* **1996**, *41*, 241–249. [CrossRef]
86. D'Souza, U.; McGuffin, P.; Buckland, P.R. Antipsychotic regulation of dopamine D1, D2 and D3 receptor mRNA. *Neuropharmacology* **1997**, *36*, 1689–1696. [CrossRef]
87. Nikolić, B.; Trnski Levak, S.; Kosic, K.; Drlje, M.; Banovac, I.; Hranilovic, D.; Jovanov Milosevic, N. Lasting mesothalamic dopamine imbalance and altered exploratory behavior in rat after mild neonatal hypoxic event. *Front. Integr. Neurosci.* **2023**, *17*, 1304338. [CrossRef]
88. Krzan, M.; Schwartz, J.P. Histamine transport in neonatal and adult astrocytes. *Inflamm. Res.* **2006**, *55* (Suppl. 1), S36–S37. [CrossRef]
89. Schwartz, J.P.; Wilson, D.J. Preparation and characterization of type 1 astrocytes cultured from adult rat cortex, cerebellum, and striatum. *Glia* **1992**, *5*, 75–80. [CrossRef]
90. Ruijter, J.M.; Ramakers, C.; Hoogaars, W.M.; Karlen, Y.; Bakker, O.; van den Hoff, M.J.; Moorman, A.F. Amplification efficiency: Linking baseline and bias in the analysis of quantitative PCR data. *Nucleic Acids Res.* **2009**, *37*, e45. [CrossRef]

International Journal of
Molecular Sciences

MDPI

Article

Evaluation of the Structure–Function Relationship of SGNH Lipase from *Streptomyces rimosus* by Site-Directed Mutagenesis and Computational Approach †

Želimira Filić [1,‡], Ana Bielen [2,‡], Ela Šarić [1], Mirsada Ćehić [1,§], Ivo Crnolatac [3], Sanja Tomić [3,*], Dušica Vujaklija [1,*] and Marija Abramić [3]

1 Division of Physical Chemistry, Institute Ruđer Bošković, 10000 Zagreb, Croatia; zelimira.filic@irb.hr (Ž.F.); ela.saric@irb.hr (E.Š.); mirsada@vss.hr (M.Ć.)
2 Faculty of Food Technology and Biotechnology, University of Zagreb, 10000 Zagreb, Croatia; ana.bielen@pbf.unizg.hr
3 Division of Organic Chemistry and Biochemistry, Institute Ruđer Bošković, 10000 Zagreb, Croatia; ivo.crnolatac@irb.hr (I.C.); marija.abramic@irb.hr (M.A.)
* Correspondence: sanja.tomic@irb.hr (S.T.); dusica.vujaklija@irb.hr (D.V.)
† Dedicated to the memory of Dr. Jasenka Pigac.
‡ These authors contributed equally to this work.
§ Current address: Department of Environmental Protection, University of Applied Sciences in Security and Safety, VSS, 10000 Zagreb, Croatia.

Abstract: *Streptomyces rimosus* extracellular lipase (SrL) is a multifunctional hydrolase belonging to the SGNH family. Here site-directed mutagenesis (SDM) was used for the first time to investigate the functional significance of the conserved amino acid residues Ser10, Gly54, Asn82, Asn213, and His216 in the active site of SrL. The hydrolytic activity of SrL variants was determined using *para*-nitrophenyl (*p*NP) esters with C4, C8, and C16 fatty acid chains. Mutation of Ser10, Asn82, or His216, but not Gly54, to Ala abolished lipase activity for all substrates. In contrast, the Asn213Ala variant showed increased enzymatic activity for C8 and C16 *p*NP esters. Molecular dynamics (MD) simulations showed that the interactions between the long alkyl chain substrate (C16) and Ser10 and Asn82 were strongest in Asn213Ala SrL. In addition to Asn82, Gly54, and Ser10, several new constituents of the substrate binding site were recognized (Lys28, Ser53, Thr89, and Glu212), as well as strong electrostatic interactions between Lys28 and Glu212. In addition to the H bonds Ser10–His216 and His216–Ser214, Tyr11 interacted strongly with Ser10 and His216 in all complexes with an active enzyme form. A previously unknown strong H bond between the catalytically important Asn82 and Gly54 was uncovered, which stabilizes the substrate in an orientation suitable for the enzyme reaction.

Keywords: SGNH/GDSL-hydrolase; site-directed mutagenesis; catalytic residues; molecular docking; molecular dynamics; protein–substrate interactions; substrate binding site; catalytic efficiency; *Streptomyces rimosus*

1. Introduction

Streptomycetes are Gram-positive mycelial bacteria that possess remarkable capacity for the synthesis of bioactive compounds and a wide range of extracellular hydrolytic enzymes that they use to gain nutrients by degrading a complex organic material in their natural habitat [1,2]. Secreted enzymes that support the complex life cycle of *Streptomyces* spp. have been used in various industries and agriculture [3–5]. Specifically, applications of lipases from the genus *Streptomyces* are related to detergent, food, cosmetics, and pharmaceutical industries [5]. Recently, great interest has been attracted by the potential application of *Streptomyces* lipases in biofuel production due to their transesterification activity [3,6].

Extracellular lipase from *Streptomyces rimosus* (SrL) is among the best characterized streptomycetes' lipolytic enzymes. This enzyme was first purified from the culture filtrate of a mutant of an industrial streptomycete, *S. rimosus* R6-554 W, and biochemically determined as a monomeric, basic protein that hydrolyzes triolein and *p*NP esters, preferring those of medium-size (C8–C12) acyl chain length [7]. The highest lipase activity was at 50–60 °C and in alkaline conditions (pH 9–10). Significant thermostability and pH stability was also the property of the isolated lipase from *S. rimosus*. Additional biochemical study showed that SrL possessed pronounced lipolytic activity toward various triacylglycerols and oils of vegetable and animal origin [8]. Glycerol esters of fatty acids with medium chain length (C8–C12) were hydrolyzed most efficiently by SrL where primary and secondary ester bonds were cleaved. An enzyme preference for the glycerol esters of unsaturated fatty acids over those of C16 and C18 saturated fatty acids was observed. SrL exerted also Tween-hydrolyzing activity with a rate comparable to that for hydrolysis of triacylglycerols and oils [8]. Furthermore, significant thioesterase and phospholipase activity of SrL was revealed [9]. Stability in organic solvent mixtures containing 50% ethanol, 1,4-dioxane, acetonitrile, or acetone was shown for SrL, which also catalyzed transesterification in *n*-hexane [8].

Amino acid sequencing of purified native SrL protein allowed the design of the primers and the cloning of the corresponding gene [10]. Sequence analysis revealed protein of 268 amino acid residues, including 34 amino acids of the signal peptide. In silico analysis of the deduced amino acid sequence established that SrL belongs to the GDS(L) subfamily of lipolytic enzymes of the family II [11], which have five conserved blocks (I–V) and four invariant, catalytically important residues: Ser, Gly, Asn, and His. The first consensus motif in block I, GDS(L), contains the putative active-site Ser and is located close to the N-terminus, different from the GxSxG motif found in many lipases situated near the center of the protein sequence. The recognition of four conserved residues (Ser, Gly, Asn, and His) in the GDS(L) subfamily led to a new designation of these enzymes as the SGNH-hydrolase superfamily [12].

Heterologous expression of SrL was obtained in *S. lividans* TK23 where maximal lipase activity was detected in culture filtrates of the late stationary phase [13]. The crystal structure of SrL was determined at the resolution of 1.75 Å [14,15]. The mature SrL protein comprises 234 amino acid residues and a three-layered $\alpha\beta\alpha$-sandwich fold typical of the SGNH-hydrolase superfamily formed by five parallel β-strands surrounded by 11 α-helices and by two 3_{10} helices. The 3D structure of SrL is stabilized by the three disulfide bonds and bears the closest similarity to that of the phospholipase A_1 from *S. albidoflavus* (SaPLA1; 64.4% amino acid sequence identity) and the esterase from *S. scabies* (SsEst; 22% sequence identity). Indeed, the crystal structure of SrL was resolved by molecular replacement using a homology model based on the crystal structure of SaPLA1 [16]. In the SrL active site, analogous interactions to the ones in SsEst were observed, indicating that Ser10 and His216 were catalytic residues. Namely, earlier reports on the crystal structures of the free enzyme and enzyme-inhibitor complexes revealed that the *S. scabies* esterase active site contains a dyad of Ser14 and His283, which closely resemble the two components of typical Ser–His–Asp(Glu) triads of serine hydrolases, but lacks the carboxylic acid component [17]. These studies confirmed the nucleophilic role of Ser14 and identified the oxyanion hole in SsEst. Similar to SsEst, analysis of the crystal structure of SrL did not show any charged residue (i.e., a third residue of the typical lipase catalytic triad) in the vicinity of His216. SaPLA1 also possesses a catalytic dyad (Ser11–His218). An additional Ser (at position 216) is suggested to stabilize the imidazole moiety of catalytic His218 in SaPLA1.

While experimental data supported the crucial roles of Ser and His in the *S. scabies* esterase active site, those for SrL are missing. Since until now no crystal structure of SrL in complex with ligand (inhibitor or substrate) is resolved, experimental evidence or confirmation is missing regarding the functional importance of conserved amino acid residues indicated by structural analogy or by in silico studies [15,18]. To gain insight into the structure–function relationship of extracellular lipase from *S. rimosus*, this research used an experimental approach based on the site-directed mutagenesis and determination of the hydrolytic activity of purified enzyme variants for three substrates of different fatty acid chain lengths. In addition, MD simulations were used to investigate substrate binding for the wild-type (WT) and mutated variants of SrL.

2. Results

2.1. Selection of Mutations Predicted to Affect Enzyme Functionality

Sequence analysis placed SrL into the SGNH-hydrolase superfamily (Pfam: CL0264) and GDSL-2 family (Pfam: PF13472) composed primarily of bacterial sequences [19]. The crystal structure determination revealed its similarity to the 3D structures of phospholipase A1 from *Streptomyces albidoflavus* and esterase from *Streptomyces scabies*. Several biochemically well-characterized and structurally resolved enzymes belonging to the GDSL-2 family were selected (Figure 1), and a multiple-sequence alignment was performed. Figure 1 shows the sequence conservation within the four blocks and the presence of catalytically important residues. To question the functional importance of conserved residues Ser10, Gly54, Asn82, and His216, belonging to blocks I, II, III, and V (Figure 1), lipase variants in which those residues were substituted by alanine were generated. Since SGNH hydrolases from genus *Streptomyces* (three out of four) did not have a highly conserved Asp in block V, additional lipase variants Asn213Ala and Asn213Asp were prepared to investigate whether the Asn213, a residue near the assumed catalytic His 216, influenced SrL activity.

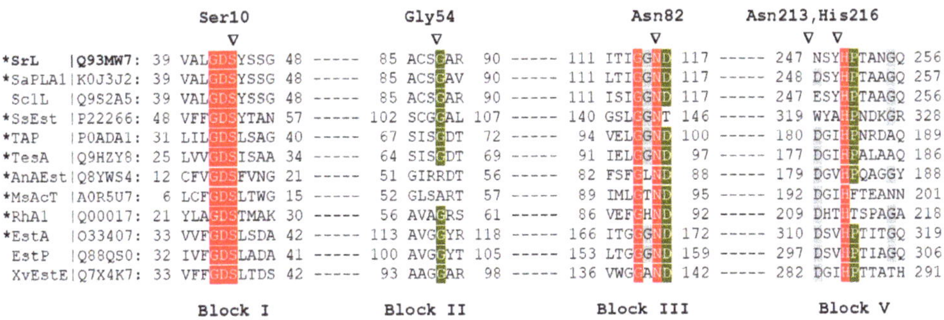

Figure 1. Multiple sequence alignment of selected protein sequences that belong to the GDS(L) family. Enzyme names and UniProtKB accession numbers are shown (SrL, extracellular lipase from *S. rimosus*; SaPLA1, phospholipase from *S. albidoflavus*; Sc1L, lipase from *Streptomyces coelicolor*; SsEst, esterase from *S. scabies*; TAP, thioesterase/protease/phospholipase from *Escherichia coli*; TesA, lysophospholipase from *Pseudomonas aeruginosa*; AnAEst, lipase from *Nostoc* sp.; MsAcT, arylesterase from *Mycolicibacterium smegmatis*; RhA1, rhamnogalacturonan acetylesterase from *Aspergillus aculeatus*; EstA, esterase from *Pseudomonas aeruginosa*; EstP, esterase from *P. putida*; XvEstE, esterase from *Xanthomonas vesicatoria*). The asterisk indicates an enzyme with a determined 3D structure. The conservation of marked amino acids in the alignment is color-coded (red, 100%; green, 80%; gray, 60%). Amino acids of SrL that were changed in this study are marked with triangles. The numbers next to the aa shown above the blocks indicate their position in mature SrL [10]. The alignment was constructed in MAFFT [20,21] using the G-INS-1 strategy and MAFFT homologs to improve the aligning of distantly related sequences.

2.2. The Impact of Targeted Amino Acid Mutations on Enzyme Functionality

Recombinant SrL and its mutants were produced in *S. lividans* TK23 and purified to homogeneity as described in Section 4. First, it was analyzed whether the introduced mutations led to conformational changes in the SrL variants. Structural characteristics of purified enzyme variants together with WT lipase were evaluated by circular dichroism (CD). As shown in Figure 2, almost identical CD spectra were obtained for all proteins, suggesting that the introduced mutations did not lead to changes in the secondary structures of the proteins.

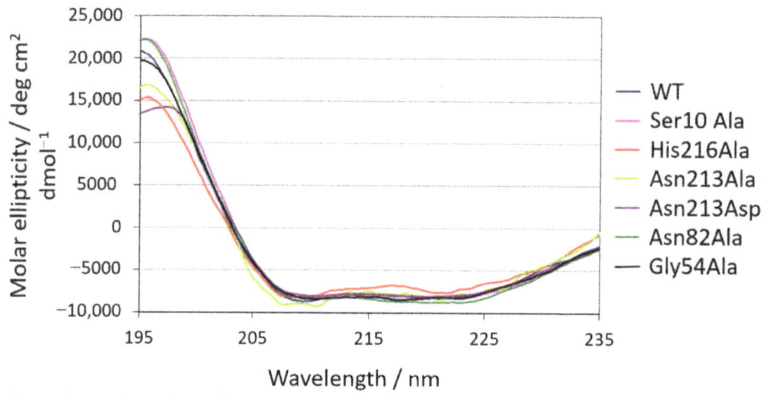

Figure 2. Circular dichroism spectra of SrL and its variants.

Next, the specific activity of the purified enzymes was determined for three *p*NP esters with different fatty acid chain lengths: *p*-nitrophenyl butyrate (*p*NPB; C), *p*-nitrophenyl caprylate (*p*NPC; C8), and *p*-nitrophenyl palmitate (*p*NPP; C16). As shown in Figure 3, hydrolytic activity of Ser10Ala, Asn82Ala, and His216Ala was abolished for all examined substrates. The replacement of Gly54 with alanine residue decreased lipase activity with *p*NPC by about 40%, while the hydrolysis of *p*NPP and *p*NPB was almost unchanged (~10% decrease), compared with WT SrL. The substitution of Asn213 with Ala produced an enzyme variant with significantly higher activity for *p*NPP (~55%) and *p*NPC (~29%), while the activity toward *p*NPB remained unchanged in comparison with WT. The activity of mutant Asn213Asp was moderately enhanced toward *p*NPB (~29%) and *p*NPC (~34%) but decreased by ~35% with long acyl chain length substrate *p*NPP, compared with WT (Figure 3).

Out of the three tested substrates, *p*NPC was the preferred one in all active enzyme variants: the specific activities of WT, Asn213Ala, and Gly54Ala were 797, 1026, and 467 U/mg protein, respectively. Activity toward *p*NPP was somewhat lower for WT and Asn213Ala (31% and 17%, respectively), and that with *p*NPB was two orders of magnitude lower than the activity toward *p*NPC.

2.3. Biophysical Characterization

To evaluate the thermal stability of active lipases, differential scanning calorimetry (DSC) was used as the most direct method to obtain the melting temperature and enthalpy of the thermal unfolding process. As organic solvents are broadly utilized in biocatalysis, the thermal stability of WT enzyme and its active variants in buffer with the addition of 20% DMSO (*v*/*v*) was evaluated. The results are presented in Table 1, while baseline subtracted thermograms are shown in Figure S1. According to the obtained melting temperatures (Table 1), the WT is the most stable enzyme form (T_m 66.0 °C and 61.3 °C, without and with DMSO, respectively), followed by Asn213Asp, Gly54Ala, and Asn213Ala (T_m 62.1 °C and 54.0 °C, respectively). In terms of molar enthalpies of the unfolding process, there were no considerable differences between different active SrL variants. The same is true for the measurements in the buffer with 20% DMSO. Although the molar enthalpies are considerably lower (>80 kJ/mol), when the organic solvent is present, the different lipases show similar unfolding enthalpy values.

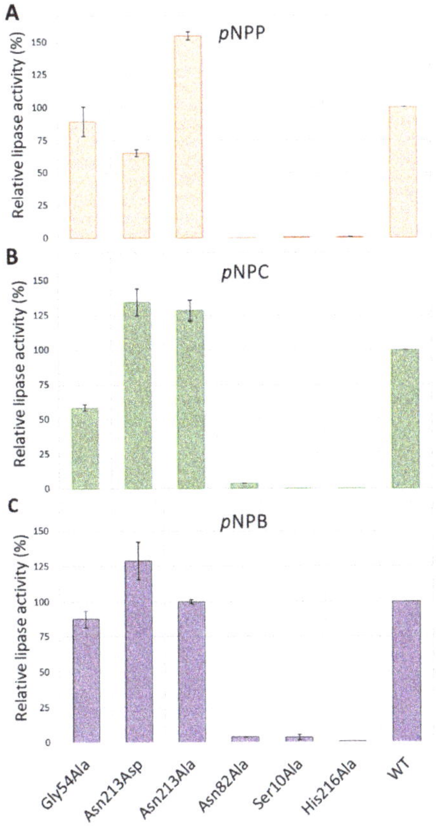

Figure 3. Comparison of the enzyme activities toward *p*NP esters with different acyl chain lengths (**A–C**). The activities are expressed relative to the specific activity of the wild type (100% = 797.40 U/mg protein for *p*NPC, 549.62 U/mg protein for *p*NPP, and 2.70 U/mg protein for *p*NPB) and represent the average of three measurements. Numerical values of standard deviations (SD) are given in Table S1.

Table 1. Melting temperatures (T_m) and molar enthalpy ($\Delta_r H$) of the thermal unfolding process for WT and enzyme variants: Asn213Ala, Asn213Asp, and Gly54Ala. All proteins were measured in phosphate buffer and buffer/DMSO mixture.

Buffer (10 mM Phosphate, 200 mM NaCl, pH 8)				
SrL	***c*** **(mg/mL)**	M_w **(kDa)**	T_m **(°C)**	$\Delta_r H$ **(kJ/mol)**
WT	1	25	66.0	534.8
Asn213Ala	1	25	62.1	534.8
Asn213Asp	1	25	63.9	537.6
Gly54Ala	1	25	63.0	535.1
Buffer/DMSO (10 mM Phosphate, 200 mM NaCl, pH 8, 20% DMSO, ***v/v*****)**				
WT	1	25	61.3	456.7
Asn213Ala	1	25	54.0	450.5
Asn213Asp	1	25	60.0	452.1
Gly54Ala	1	25	58.5	449.9

2.4. Molecular Simulations

In total, more than 5 μs of MD simulations were performed to investigate whether mutations lead to deformation of the protein conformation and to determine the binding mode of the long fatty acid chain substrate (C16) to SrL. Since the experimental 3D structure of an SrL–ligand complex has not yet been determined, the adaptive steered MD simulations (ASMD) were combined with MD simulations to determine the substrate binding to the active site of SrL.WT SrL and its variants Gly54Ala, Asn82Ala, and Asn213Ala were simulated both as free proteins and in complex with the long-chain substrate *p*NPP. The complexes were constructed using the atomic coordinates of ligand-free SrL and ASMD simulations as described in Section 4.

2.5. MD Simulations of Free Enzyme Forms

Figure 4 shows an overlay of the WT with the SrL variants, indicating that the mutations did not cause protein deformation. These results were in good agreement with the circular dichroism result (Figure 2). Also, during the 300 ns of MD simulations, the structure of WT SrL and its variants remained stable, and the RMSD values, which described the deviation of the protein structures sampled during the simulations from the equilibrated structure, were mostly ~1.0 Å (Figure S2). According to the MM/GBSA free energy calculations of the ligand-free lipase variants (Table S2), the WT and Gly54Ala variants were the most stable, and the Asn213Ala variant appeared to be the least stable variant during the 300 ns of MD simulations of the ligand-free proteins at 300 K. The Ser10–His216 distance was mostly in the range of 2–4 Å for all variants, although it was generally shorter for WT SrL and its Asn213Ala variant than for the Gly54Ala and Asn82Ala variants (Figure S3).

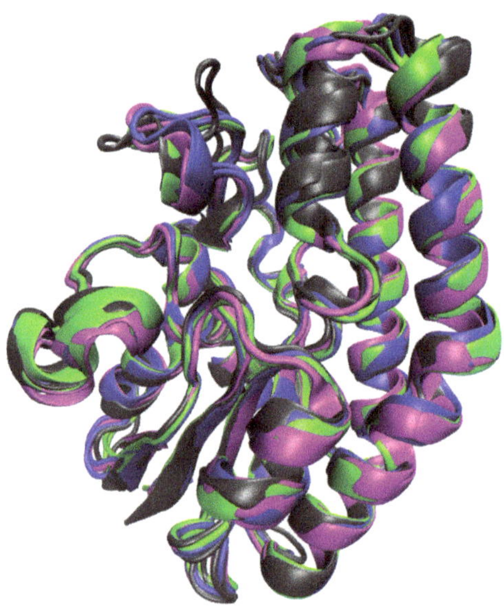

Figure 4. Overlay of the structures of WT SrL (colored magenta) and its mutants Asn213Ala (colored blue), Gly54Ala (colored gray), and Asn82Ala (colored green) obtained after 300 ns of MD simulations at 300 K in water (OPC water molecules) solution.

2.6. MD Simulations of the Lipase–Substrate Complex

The web tool CAVER 3.0 (CAVER Web—tool for the analysis of tunnels and channels in protein structures (https://caver.cz/, accessed on 17 August 2021) [22] was used to determine the catalytic cavity in WT SrL. In addition, ASMD simulations were used to determine the possible orientations of *p*NPP in the cavity (see Section 4 for details of the procedure). The ASMD simulations were carried out for two different initial structures, i.e., for two different positions of the *p*NPP. In one, *p*NPP was placed in tunnel 3 (cyan) with the tail of the alkyl chain solvated (ASMD-T3 set), and in the other, the alkyl chain of *p*NPP was placed in tunnel 1 (green) with the phenyl ring and NO_2 group solvated (ASMD-T1 set) (Figure S4). The SrL–*p*NPP complexes obtained as a result of the ASMD simulations were visually examined, and it was found that the orientation of *p*NPP in the complex obtained starting from the structure accommodated in tunnel 1 was more suitable for hydrolysis than the *p*NPP orientation in the complexes obtained starting from the alkyl chain occupying tunnel 3. In both cases, the most suitable conformation of the ligand was in the complexes obtained with a puling step of 0.5 Å. These structures were further simulated for 100 ns at 290 K. During the simulation of the structure from the ASMD-T3 set, *p*NPP moved away from Ser10, and most of the alkyl chain remained solvated (Figure S5). On the other hand, *p*NPP from the ASMD-T1 set remained close to the initial position and was nicely accommodated in the SrL catalytic pocket (see Figure 5) during the 100 ns of unconstrained MD simulations. MM/GBSA binding free energies also revealed higher stability of the complex from the ASMD-T1 set than from the ASMD-T3 set (-65.6 ± 4.2 kcal/mol vs. -43.6 ± 4.3 kcal/mol).

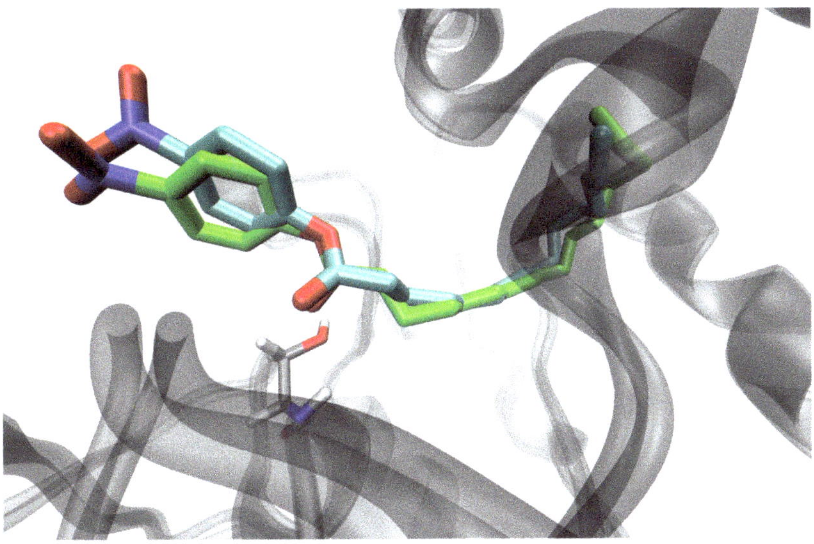

Figure 5. Orientation of the ligand in the SrL–*p*NPP complex, obtained by ASMD simulation (cyan), for the *p*NPP with alkyl chain accommodated in tunnel 1 and after additional 100 ns of unconstrained MD simulation of the solvated complex at 290 K (green). The position of Ser10 in the later structure is shown (light gray sticks).

Before further simulations, all structures were optimized and slowly heated (0 to 290 K), and the density of the system was equilibrated as described in Section 4. Three independent 200 ns long MD simulations were performed for each complex. With the exception of the Asn82Ala–*p*NPP complex, all other complexes were stable during each of the 200 ns of MD simulations. In the complex with N82A, *p*NPP migrated from the catalytic cleft into solvent after 77 ns in one simulation and after 55 ns in the other, so further simulations of this complex were not run. Based on the results, the stability of the complexes during the MD simulations, and the binding free energies calculated using the MM/GBSA [23] approach, as well as the distance between the C atom of the cleavable bond and Oγ of Ser10 (Table S3), the initial structures were selected for the following two independent, 300 ns MD simulations (for the WT complex and the complexes with the variants Gly54Ala and Asn213Ala). These complexes remained stable throughout the simulation period with RMSD values mostly below 1.5 Å (two independent 300 ns long MD simulations performed for each complex) (Figure S6), and *p*NPP remained close to its initial position (Figure S7). Interestingly, helices α6 and α11 approached each other (Figure S8), and the strong electrostatic interaction between Lys28 and Glu212 has been established in all variants (Figure 6) with geometry that often (>60% of the simulation time) fulfilled the conditions for hydrogen bonding. Moreover, these two residues interacted with *p*NPP.

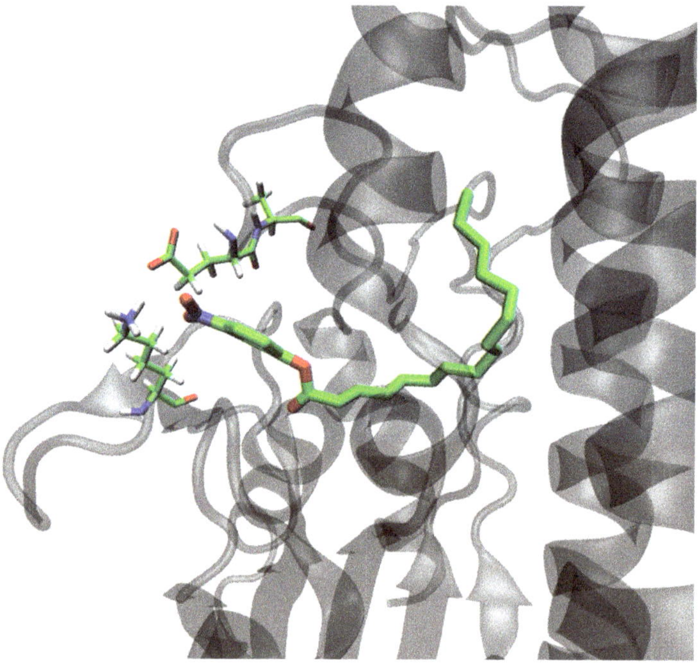

Figure 6. Interaction between Lys28 and Glu212 in the Asn213Ala–*p*NPP complex obtained after 300 ns of MD simulation in water (OPC water molecules) at 300 K.

To evaluate the substrate binding affinities, the MM/GBSA, MM/PBSA [24], and LIE energies (Tables 2, S4 and S5) were calculated. Although the calculations predicted similar binding affinities for all variants, the LIE method [25] revealed a slightly higher affinity of *p*NPP for the WT protein and MM/GBSA for the Asn213Ala mutant. In this stable binding mode, the hydrophilic ester moiety (O=C–O–Phe–NO_2) of *p*NPP interacted with Ser10, Lys28, Ser53, Gly54, Asn82 (side chain), and His216, and occasionally with Glu212, while the hydrophobic, alkyl chain interacted strongly with Tyr11, Gly81, Asn82 (backbone), Ala84, Gly54, Phe86, Ala108 Ala167, Tyr141, Phe144, and Ile163. Significant stabilization was also achieved by Thr89 in the complexes with the WT protein and in the complexes with the mutants Gly54Ala and Asn213Ala. The alkyl chain was additionally stabilized by Met90 in mutant variants (Figures S9 and S10 and Table 3). The strongest interaction between *p*NPP and protein was achieved via the *p*NPP(O2)–Ser10(OH) hydrogen bond (Table 3 and Figure S11). In addition to Ser10, *p*NPP interacted strongly with Ser53, Gly54, Asn82, and Tyr141, forming either H bonds or strong electrostatic interactions with them during the simulation of all complexes (Table 3). With Phe86, Tyr141, and Phe144, the substrate had CH–π interactions during most of the simulation time (Figure 7), and with the other amino acids mentioned above, the substrate underwent electrostatic and van der Waals interactions that occasionally became hydrogen bonds.

Table 2. LIE, MM/GBSA, and MM/PBSA free energy approximations for the binding of *p*NPP to WT SrL and its single-point mutants Gly54Ala and Asn213Ala. The energies (kcal/mol) and standard deviations were calculated for the sets of conformers sampled throughout the 300 ns and during the last 100 ns of the MD simulations in water (OPC water molecules) at 300 K. MM/GBSA and MM/PBSA energies were calculated for ionic strengths of 0 (100–300 ns) and 0.1 M at smaller intervals and for the entire 300 ns, respectively. Values obtained by averaging values energies calculated on structures sampled during two independent simulations of the same system are shown in bold.

SrL, Variants and Run	Method				
	LIE	MM/GBSA (1–300 ns)	MM/PBSA (1–300 ns)	MM/GBSA (200–300 ns)	MM/PBSA (200–300 ns)
WT-1	−11.9	−67.5 ± 4.2	−26.7 ± 3.7	−68.0 ± 3.3	−27.2 ± 4.1
WT-2	−13.7	−67.6 ± 4.2	−29.0 ± 4.1	−66.2 ± 3.6	−28.3 ± 4.2
WT-average	**−12.8**	**−67.6**	**−27.9**		
Asn213Ala-1	−11.8	−69.5 ± 3.3	−28.2 ± 3.3	−68.9 ± 3.4	−28.2 ± 3.7
Asn213Ala-2	−12.0	−68.9 ± 3.5	−27.2 ± 3.4	−69.5 ± 3.3	−27.1 ± 3.5
Asn213Ala-average	**−11.9**	**−69.2**	**−27.7**		
Gly54Ala-1	−11.3	−67.8 ± 3.2	−27.7 ± 3.0	−68.5 ± 3.4	−27.5 ± 3.0
Gly54Ala-2	−11.6	−67.1 ± 3.6	−26.9 ± 3.3	−68.5 ± 3.4	−26.4 ± 3.1
Gly54Ala-average	**−11.5**	**−67.4**	**−27.3**		

Table 3. Intermolecular, *p*NPP—SrL hydrogen bonds during MD simulations of the complexes with WT SrL and its mutants Asn213Ala and Gly54Ala. Indicated is the percentage of simulation time during which a particular hydrogen bond exists. In cases where an amino acid residue interacts with the substrate through several different H bonds, the percentage of its population may exceed 100. The results are for the simulations are performed at 300 K solvated in OPC water.

	HB Population					
AA/Variant	**WT-1**	**WT-2**	**Asn213Ala-1**	**Asn213Ala-2**	**Gly54Ala-1**	**Gly54Ala-2**
Ser10	179	196	184	202	153	196
Tyr11	27	27	8	10	6	11
Lys28	18	71	-	-	-	-
Ser53	67	-	86	65	52	55
Gly54	30	52	31	26	-	-
Asn82	55	19	88	88	68	90
Thr89	27	56	-	-	-	-
Met90	-	-	42	39	17	24
Tyr141	44	20	38	25	25	24

Throughout the simulation, Ser10 and His216 were properly oriented and formed strong hydrogen bonds in all complexes. Interestingly, Tyr11, which was involved in stabilizing the substrate, also interacted strongly with Ser10 and His216 (Figure S12).

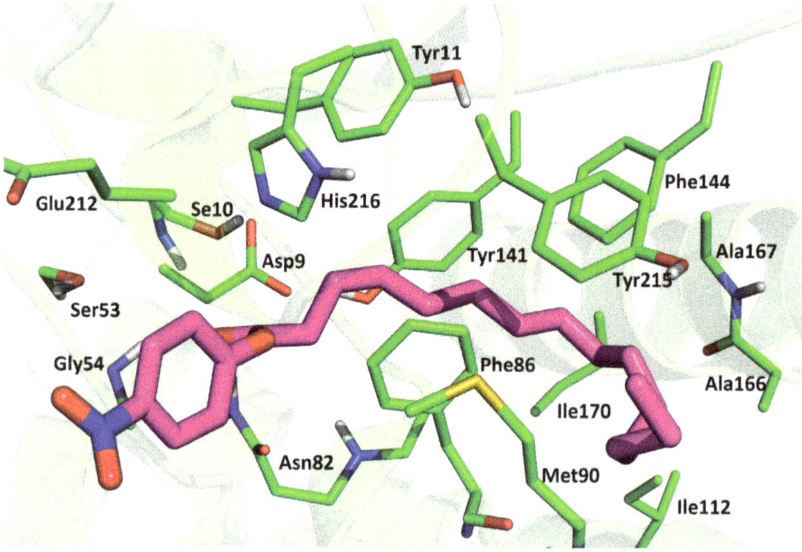

Figure 7. Position of *p*NPP in the Asn213Ala-*p*NPP complex after 300 ns of MD simulation in water (OPC-water molecules) at 300 K. The amino acid residues that have close contacts (distance less than 4 Å) with *p*NPP are indicated.

3. Discussion

The SGNH superfamily encompasses a large group of phylogenetically broadly distributed enzymes in all domains of life. The GDSL-like lipases/acylhydrolases are the most abundant in this superfamily. Most members of this group belong to bacteria, followed by plants and fungi [19]. Although some of them exhibit multifunctional properties, many members have not been functionally investigated, such as in *Arabidopsis*, where more than 100 GDSL lipases are recognized [12,26,27].

The native extracellular SGNH/GDSL lipase of *S. rimosus* was previously purified and biochemically characterized, the corresponding gene was cloned, and a high level of heterologous expression was obtained [7,8,10,13]. In the present study, site-directed mutagenesis was applied for the first time to investigate the function of amino acid residues suggested to be important for its hydrolytic activity. The assumption of catalytic relevance of Ser10, Gly54, Asn82, and His216 resulted from the primary structure analysis, which placed SrL in the SGNH hydrolase superfamily, and from the comparison of the crystal structure of ligand-free SrL [15] with similar 3D structures of two SGNH-hydrolyses (esterase from *S. scabies* and phospholipase from *S. albidoflavus*) [16,17].

Our results provide experimental evidence for the crucial importance of Ser10, Asn82, and His216 for the hydrolytic activity of SrL (Figure 3) as their substitution with Ala yielded inactive enzyme variants. On the other hand, point mutations (Gly54 to Ala and Asn213 to Ala) did not change the substrate selectivity of SrL. All active mutants preferred the C8 acyl chain length in *p*NP esters.

The activity of the His-tagged WT SrL enzyme was in very good agreement with previously reported specific activity of native SrL for *p*NPP (659 U/mg) and *p*NPC (850 U/mg) [7]. Replacement of Gly54, a conserved constituent of the block II in the SGNH family enzymes, with alanine caused only a slight to moderate decrease (depending on the *p*NP substrate) in the activity of the variant Gly54Ala compared with the WT enzyme. The GDSL protein family has been reported to contain a significant proportion of sequences with distinct variations in motifs that sometimes lack catalytically important residues. Such mutations can generate pseudoenzymes or enzymes with new properties [28,29]. Natural variations of conserved Gly to Ala and Arg have been reported for two members of the SGNH superfamily (Figure 1), acyltransferase from *Mycobacterium smegmatis* [30] recently reclassified as *Mycolicibacterium smegmatis* [31] and a serine hydrolase from cyanobacterium *Anabaena* sp. PCC 7120, respectively. By site-directed mutagenesis it was shown that Arg54 in *Anabaena* hydrolase has a role in substrate binding and catalytic activity [32].

In lipases with the catalytic triad Ser–Asp–His, the catalytic Asp is located at the position of Asn213 of WT SrL (block V); therefore, a protein mutated at this position was generated and analyzed. The SrL variant Asn213Asp showed slightly higher activity with *p*NPB and *p*NPC but lower activity with *p*NPP compared with the WT enzyme (Figure 3). On the other hand, the replacement of Asn213 with Ala improved the hydrolytic activity toward the medium- and long-chain substrates *p*NPC and *p*NPP. Although Asn213 did not interact directly with *p*NPP, it interacted with Lys28 and, together with Ser214, with the catalytic His216 (hydrogen bonds and electrostatically) in the MD simulations. Asn213 and Lys28 were located at turns (unstructured regions of SrL) that faced each other and separated the catalytic part of the substrate binding pocket from the surrounding solvent. The interactions between Lys28 and Asn213 and Glu212 consolidated and shaped the catalytic site of SrL and ensured proper alignment of the substrate through the hydrogen bonding network (in conjunction with strong electrostatic interactions) with the amino acid residues of the binding site.

MD simulations showed that the interaction between *p*NPP and the catalytic residue Ser10, as well as with Asn82 from the oxyanion hole, was stronger in the complex of *p*NPP with Asn213Ala than in the complex with the wild-type protein.

Molecular dynamics simulations revealed the key interactions between lipase and the long-chain *p*NP ester substrate, including H bonds and electrostatic interactions with Asn82 and H bonds with Gly54 and Ser10. The observed stabilization of the substrate carbonyl oxygen with Asn82 and Gly54 suggested that they acted as oxyanion hole residues, stabilizing the negatively charged tetrahedral intermediates during the hydrolytic reaction. Moreover, Asn82 (NH_2 group from the side chain) formed a strong H bond with the carbonyl oxygen of Gly54 and stabilized it in the orientation suitable for the enzymatic reaction. In addition, stable H-bonding between His216 and Ser214 was observed during the MD simulations. Overall, our computational results emphasized the crucial importance of Ser10, Asn82, and His216 for the catalytic activity of SrL, which was consistent with the experimental analysis of SrL variants. The obtained results supported the hypothesis of a variation in the active site of *S. rimosus* lipase [15], consisting of Ser10 and His216 but lacking the Asp (or Glu) residue that ensured the correct orientation of the imidazole ring of His in typical serine hydrolases. Such variation has been reported for the two SGNH hydrolases structurally closest to SrL, *S. scabies* esterase and *S. albidoflavus* phospholipase A1 [16,17]. In addition, the molecular modeling results of our study showed previously unreported interactions occurring within the active site of SrL lipase in the presence of substrate, thus providing a good basis for future research on this interesting class of SGNH enzymes.

4. Materials and Methods

4.1. Bacterial Strains and Cultivation Conditions

The bacterial strains and culture media used in this study are listed in Table S6. *E. coli* was cultivated at 37 °C in LB medium [33]. *Streptomyces lividans* TK23 was grown at 30 °C in liquid CRM medium [34] for plasmid isolation or inoculum preparation, solid MS medium for sporulation, solid R5 medium for protoplast regeneration, and liquid GR_2d medium for lipase production, as described previously [35,36]. When appropriate, antibiotics were added to the media in the following final concentrations: 100 mg/mL ampicillin, 25 mg/mL thiostreptone, and 10 mg/mL kanamycin (Sigma-Aldrich, St. Louis, MO, USA). Only to MS medium thiostreptone was added at a concentration of 50 mg/mL.

4.2. Cloning and Site-Directed Mutagenesis

The wild-type *srl* gene was amplified by PCR using forward primer srlEco, reverse primer srlHind (Table S7), and plasmid pDJ5 as a template ([10], Table S8) according to the procedure detailed in Supplementary Materials. The amplified 840 bp DNA fragment was ligated into the pGEM-T vector (Promega, Madison, WI, USA), which was used to transform *E. coli* XL1 cells. The recombinant plasmid, SrLpGEM, was purified from selected transformants with a QIAprep Spin Miniprep Kit (Qiagen, Hilden, Germany), and the sequence of the lipase gene was verified by sequencing. The plasmid, SrLpGEM, was used either to clone the *srl* gene into the bifunctional vector pANT849pWB19N [35] or as a template to generate targeted mutations in the lipase gene using the site-directed mutagenesis method [37]. Modifications introduced to this protocol are described in detail in Supplementary Materials. The obtained SrLpGEM constructs and pANT849pWB19N [38] were used for the transformation of methylation-deficient *E. coli* GM119. Non-methylated plasmids were digested with *Eco*RI and *Hin*dIII (Fermentas, Waltham, MA, USA), and DNA fragments (5300 bp pANT849 and 840 bp *srl* gene or its mutated variants) were gel purified (QIAquick Gel Extraction Kit, Qiagen, Hilden, Germany) and ligated using T4 DNA ligase (Fermentas, Waltham, MA, USA). The ligation mixture was used to transform *Streptomyces lividans* TK23 protoplasts, as described [35,36]. The recombinant plasmids, SrLpANT, carrying the *srl* gene or its variants were purified from selected transformants by an alkaline lysis procedure [36], and the presence of the *srl* insert was confirmed by PCR and sequencing.

4.3. Biosynthesis and Purification of SrL Lipase Variants

The purification of His-tagged SrL lipase variants was performed according to the protocol developed previously [35], which includes (i) growth of *S. lividans*-SrLpANT in GR_2d liquid medium for ~6 days or until the extracellular lipase in the culture supernatant showed maximum activity toward *p*NPP (Sigma-Aldrich, St. Louis, MO, USA), (ii) precipitation of extracellular proteins with 80% ammonium sulfate and subsequent dialysis, and (iii) purification of His-tagged lipases by Ni-NTA Agarose gravity flow chromatography (Qiagen, Hilden, Germany). The only change compared with the described method was the increase in imidazole (30 mM) in the washing step, which allowed the removal of non-specifically bound proteins, thus avoiding gel filtration in the purification protocol. The purified proteins were analyzed by SDS-PAGE, and their concentration was determined using NanoDrop 2000 (Thermo Fisher Scientific, Fermentas, Waltham, MA, USA).

4.4. Circular Dichroism (CD) Spectrometry

Circular dichroism (CD) spectra of purified lipase variants were recorded in a range of 235 to 195 nm by using a J-715 Spectropolarimeter (JASCO, Tokyo, Japan) at room temperature. The concentration of the samples was ~0.5 mg/mL (in 10 mM phosphate buffer pH 6.8, 0.2 M NaCl).

4.5. Enzyme Activities

Lipase activity was measured spectrophotometrically at room temperature (22 °C) by monitoring the release of *p*NP from three different substrates: *p*NPP, *p*NPC, and *p*NPB. Enzymatic reaction with buffered *p*NPP or *p*NPC emulsion was performed as described previously [35]. The reaction mixture was created by mixing a substrate stock solution in dioxane and buffer (50 mM Tris-HCl pH 8, 5 mM sodium deoxycholate) to a final substrate concentration of 0.33 mM *p*NPP or 0.36 mM *p*NPC and 2.5% dioxane (v/v). The hydrolysis of *p*NP esters was initiated by adding 10 µL of the enzyme (0.01 mg/mL) into 1 mL of the reaction mixture. The catalytic activity of lipase toward *p*NPP and *p*NPC was measured at 410 nm by UV/Visible Varian Cary 100 Bio spectrophotometer. The spontaneous hydrolysis of each of the substrates was also examined. A similar protocol was used for *p*NPB, except that the hydrolysis was carried out at pH 7.2. The substrate concentration in the reaction mixture was 1 mM, and the dioxane concentration was 1% (v/v). Enzyme reactions were started by adding 10 μL of 1 mg/mL enzyme into a 1 mL reaction mixture. One unit (1U) of lipase activity was defined as the amount of enzyme needed to liberate 1 μmol of *p*NP per minute at the assay conditions.

4.6. Enzyme Stability Measurements

Thermal stability measurements of SrL and its active variants (Asn213Ala, Asn213Asp, and Gly54Ala) were completed in a Nano-DSC (TA Instruments, Waters Corporation, New Castle, DE, USA) differential scanning microcalorimeter. Samples were prepared by diluting the purified protein to a concentration of 1 mg/mL in buffer (10 mM phosphate buffer, 200 mM NaCl, pH 8) and buffer/DMSO mixtures (20% DMSO, *v*/*v*), followed by degassing for 15 min on a degassing station (TA Instruments, Waters Corporation, New Castle, DE, USA). The working volume of the calorimeter was 300 μL, and the thermal cycle was performed with a 60 s equilibration time in the temperature range of 10–90 °C at a scan rate of 1 °C/min. The buffer–buffer and buffer/20% DMSO–buffer/20% DMSO thermal scans were subtracted from the protein measurements. The results were assessed with Nano Analyze software, 3.11.0 (TA Instruments, Waters Corporation, New Castle, DE, USA), and thermal unfolding characteristics (T_m—melting temperature, ΔH—enthalpy) were evaluated.

4.7. Selection of the Crystallographically Determined Structure of SrL for Molecular Modeling and Parameterization of the System

The crystallographically determined structure of SrL, PDB_id 5MAL, was used as the initial structure for our molecular modeling study. The asymmetric unit comprised two monomeric SrL molecules, A and B, which did not reveal significant conformational differences: the secondary structural elements overlapped well (RMSD of 0.54 Å). For the molecular modeling study, molecule A was utilized. The structure showed two possible orientations of the amino acid residues Ser129, Gln135, and Glu212. Orientation A was chosen for Ser129 and Gln135 because in this orientation they had more stable interactions with the rest of the protein, whereas for Glu212 both orientations were studied, orientation A with Cα-Cβ-Cγ-Cσ equal to −96.5° and orientation B with Cα-Cβ-Cγ-Cσ equal to −63.7°.

The SrL structure had six Cys residues all connected by disulfide bonds. In the model, all Arg and Lys residues were positively charged (+1e), while Glu and Asp residues were negatively charged (1e), as expected under physiological (experimental) conditions. According to their ability to form hydrogen bonds with surrounding residues, His residues 42, 195, and 216 were protonated at Nδ, and His181 and His205 were protonated at Nε. Parameterization was performed using the FF19SB force field [39] for protein and the compatible GAFF2 force field [40] for ligand. The AM1-BCC atomic charges [41] were assigned to the ligand. The mutated proteins, Gly54Ala, Asn82Ala, and Asn213Ala, were prepared using tleap, a basic preparation program for Amber simulations available within the AMBER20 package (http://ambermd.org, accessed on 20 August 2021). Proteins and protein–substrate complexes were neutralized with Na^+ ions and solvated in an octahedral box filled with OPC water molecules [42] recommended for use with the ff19SB force field. The minimum distance between the solvated complex and the edge of the box was 11 Å. The solvated systems were minimized, followed by heating, density equalization, and productive MD simulations.

4.8. MD Simulations Details

All simulations were carried out using the AMBER20 suite of programs [43,44]. The minimization of the prepared systems (ligand-free protein variants and their complexes with *p*NPP) was performed in three cycles. In the first cycle of optimization (1500 steps), water molecules were relaxed, while the rest of the system was harmonically restrained with a force constant of 32 kcal mol^{-1} Å^{-2}; in the second cycle (3500 steps), the protein backbone atoms were restrained with a force constant of 12 kcal mol^{-1} Å^{-2}; and in the third cycle (2500 steps), the system was completely relaxed. The energy-optimized system was heated in three steps (from 0 to 100 K, from 100 K to 200 K, and from 200 K to 300 K) (each interval 10 ps) using the NVT ensemble and the time step of 0.5 fs, followed by 3 ns of density equilibration (NPT ensemble and the time step of 1 fs). The two independent MD simulations of 300 ns duration were performed for the ligand-free protein variants, and for their complexes with *p*NPP, obtained by Docking, three independent MD simulations (each 200 ns long) were performed (see below). For each complex of the variant the most stable structure obtained during these 200 ns long MD simulations was selected, and its two replicas were simulated for an additional 300 ns. During productive MD simulations, the time step was 2 fs, and the SHAKE algorithm was used to constrain covalent bonds with hydrogen atoms. The pressure was maintained at 1 atm with a pressure relaxation time of 1 ps using the Berendsen barostat [45], while the system temperature was kept constant at 300 K with a collision frequency of 5 ps^{-1}, using the Langevin thermostat [46]. Simulations were performed using periodic boundary conditions (PBCs) with a cutoff of 11 Å, while the particle mesh Ewald (PME) method was used to calculate long-range electrostatic interactions.

4.9. Elucidation of the Binding of pNPP to SrL Using Adaptive Steered MD Simulations (ASMD)

For binding of the substrate (*p*NPP) to the active site of lipase, we performed ASMD simulations. In ASMD, an external force is applied to the selected reaction coordinate to allow the system to move in the desired direction during simulations of MD [47,48]. The average nonequilibrium work exerted on the system during ASMD is referred to as the potential of mean force (PMF) and, according to the Jarzynski equation, reflects the relative free energy of binding (ΔG) [49]. ASMD simulations are useful for investigating possible pathways for ligand binding and release to macromolecules but can also be used to tune ligand binding into the active site of an enzyme if the active site is known.

Experimental study on SrL combined with previous computational studies on SrL showed that Ser10 and His216 make the catalytic dyad responsible for ester hydrolysis in SrL and that acyl chains most probably bind in the hydrophobic tunnel and in the hydrophobic "gorge" [15,18]. The starting point for our ASMD simulations was a fully solvated system consisting of the optimized and equilibrated structure of the ligand-free protein variant with the ligand (*p*NPP) in the extended conformation manually placed in the hydrophobic pocket with the C atom of the scissile C-O(R) bond about 6 Å away from Ser10. Three series of ASMD simulations were performed; in one of them, the velocity of the substrate pooling into the proposed binding funnel was 2.1 Å/ns, and in two ASMD simulations it was 0.5 Å/ns (during the simulations, C16 was pooled toward Oγ(Ser10)). Each of the ASMD simulations was performed in seven steps.

The obtained complexes were minimized, heated, and density equilibrated, and three independent 200 ns long productive MD simulations were performed for each complex (a detailed description of the procedure is given in Supplementary Materials). The ASMD simulations as the following productive MD simulations were performed using the TIP3P water model. The structure for which the lowest binding free energy was determined was used as a starting point for 300 ns long MD simulation with OPC water molecules. For this purpose, the system was reparameterized using the OPC water box, and after minimization and equilibration (the procedure equivalent to that used for the ligand-free enzyme), a 300 ns long MD productive simulation was performed.

4.10. MM/PBSA and MM/GBSA Calculations

The substrate binding MM/PBSA and MM/GBSA energies were obtained using the MM/PBSA py script as implemented in the AMBER20 program [43]. MM/GBSA calculations were performed using "igb = 5" [23]. MM/GBSA energies were calculated on 20 ns long intervals sampled throughout the trajectories sampled during three independent 200 ns long productive MD simulations. For the final trajectory obtained during 300 ns of productive MD simulations, both MM/GBSA and the MM/PBSA [24] energies were calculated for the structures sampled during the intervals 80–100 ns, 180–200 ns, 280–300 ns,

100–300 ns, and 1–300 ns using the single-trajectory approach. To test the influence of the frequency of sampling on the results, we performed a few calculations using both 100 and 1000 structures. Since the difference was negligible, all other calculations were performed using 100 structures and, in the case of the 1–300 ns interval, 150 structures. The MM/PBSA calculations were performed using a salt concentration of 0.1 M. The dielectric constants were 80.0 for the solvent (water) and 2.0 for the solute. The polar component of the enthalpy of solvation was calculated by the Poisson–Boltzmann method, and the nonpolar component was determined by $\Delta H_{nonpol} = \gamma SASA + \beta$, where the solvent accessible surface area (SASA) was calculated using the MolSurf program [50].

4.11. Data Analysis

To trace the conformational space spanned by the SrL structures during the MD simulations and to determine the amino acid residues that most strongly stabilized the binding of *p*NPP, a series of analyses were performed. Detailed analyses of the geometry, the population of intermolecular hydrogen bonds, and LIE calculations (linear interaction energy approach introduced by Åqvist at al. [25]) were performed using the CPPTRAJ module of the AmberTools program package [43]. LIE calculations were performed using cutoff values 9 Å and 11 Å, for Van der Waals and electrostatic energy calculations.

5. Conclusions

The site-directed mutagenesis, performed for the first time on conserved residues in the active site of the SGNH lipase from *S. rimosus*, confirmed the essential importance of Ser10, Asn82, and His216, while Gly54 could be substituted by Ala without a significant decrease in hydrolytic activity. The replacement of Asn213 with Ala increased SrL activity, suggesting that SDM can improve the function of this enzyme. MD simulations identified the key interactions between the lipase and the long-chain substrate (*p*NPP). In addition, a previously unrecognized H bond Asn82–Gly54 and a strong electrostatic interaction between Lys28 and Glu212 were uncovered in the enzyme–substrate complexes, the significance of which should be investigated for catalysis.

Supplementary Materials: The supporting information can be downloaded at https://www.mdpi.com/article/10.3390/ijms25010595/s1.

Author Contributions: Conceptualization, M.A., S.T. and D.V.; formal analysis, M.A., S.T., I.C. and D.V.; investigation, Ž.F., A.B., E.Š., M.Ć., I.C. and S.T.; writing the first version, M.A.; writing—review and editing, M.A., S.T., A.B. and D.V. with input of all the authors; visualization, S.T., I.C., Ž.F., A.B., E.Š., M.Ć. and D.V.; supervision, M.A. and D.V.; funding acquisition, D.V. All authors have read and agreed to the published version of the manuscript.

Funding: This work is supported by the Croatian Government and the European Union through the European Regional Development Fund—the Competitiveness and Cohesion Operational Programme: Project KK.01.1.1.01.0002 granted to The Scientific Centre of Excellence for Bioprospecting–BioProCro.

Institutional Review Board Statement: Not applicable.

Informed Consent Statement: Not applicable.

Data Availability Statement: Data is contained within the article and supplementary material.

Conflicts of Interest: The authors declare no conflicts of interest.

References

1. Chater, K.F.; Biró, S.; Lee, K.J.; Palmer, T.; Schrempf, H. The Complex Extracellular Biology of Streptomyces: Review Article. *FEMS Microbiol. Rev.* **2010**, *34*, 171–198. [CrossRef]
2. Morosoli, R.; Shareck, F.; Kluepfel, D. Protein Secretion in Streptomycetes. *FEMS Microbiol. Lett.* **1997**, *146*, 167–174. [CrossRef]
3. Spasic, J.; Mandic, M.; Djokic, L.; Nikodinovic-Runic, J. *Streptomyces* spp. in the Biocatalysis Toolbox. *Appl. Microbiol. Biotechnol.* **2018**, *102*, 3513–3536. [CrossRef]
4. Olanrewaju, O.S.; Babalola, O.O. Bacterial Consortium for Improved Maize (*Zea mays* L.) Production. *Microorganisms* **2019**, *7*, 519. [CrossRef]

5. Kumar, M.; Kumar, P.; Das, P.; Solanki, R.; Kapur, M.K. Potential Applications of Extracellular Enzymes from *Streptomyces* spp. in Various Industries. *Arch. Microbiol.* **2020**, *202*, 1597–1615. [CrossRef]
6. Cho, S.S.; Park, D.J.; Simkhada, J.R.; Hong, J.H.; Sohng, J.K.; Lee, O.H.; Yoo, J.C. A Neutral Lipase Applicable in Biodiesel Production from a Newly Isolated *Streptomyces* sp. CS326. *Bioprocess Biosyst. Eng.* **2011**, *35*, 227–234. [CrossRef]
7. Abramić, M.; Leščić, I.; Korica, T.; Vitale, L.; Saenger, W.; Pigac, J. Purification and Properties of Extracellular Lipase from *Streptomyces rimosus*. *Enzym. Microb. Technol.* **1999**, *25*, 522–529. [CrossRef]
8. Leščić, I.; Vukelić, B.; Majerić-Elenkov, M.; Saenger, W.; Abramić, M. Substrate Specificity and Effects of Water-Miscible Solvents on the Activity and Stability of Extracellular Lipase from *Streptomyces rimosus*. *Enzym. Microb. Technol.* **2001**, *29*, 548–553. [CrossRef]
9. Leščić Ašler, I.; Ivić, N.; Kovačić, F.; Schell, S.; Knorr, J.; Krauss, U.; Wilhelm, S.; Kojić-Prodić, B.; Jaeger, K.E. Probing Enzyme Promiscuity of SGNH Hydrolases. *ChemBioChem* **2010**, *11*, 2158–2167. [CrossRef] [PubMed]
10. Vujaklija, D.; Schröder, W.; Abramić, M.; Zou, P.; Leščić, I.; Franke, P.; Pigac, J. A Novel Streptomycete Lipase: Cloning, Sequencing and High-Level Expression of the *Streptomyces rimosus* GDS(L)-Lipase Gene. *Arch. Microbiol.* **2002**, *178*, 124–130. [CrossRef]
11. Arpigny, J.L.; Jaeger, K.E. Bacterial Lipolytic Enzymes: Classification and Properties. *Biochem. J.* **1999**, *343*, 177–183. [CrossRef]
12. Akoh, C.C.; Lee, G.C.; Liaw, Y.C.; Huang, T.H.; Shaw, J.F. GDSL Family of Serine Esterases/Lipases. *Prog. Lipid Res.* **2004**, *43*, 534–552. [CrossRef]
13. Vujaklija, D.; Abramić, M.; Leščić, I.; Maršić, T.; Pigac, J. *Streptomyces rimosus* GDS(L) Lipase: Production, Heterologous Overexpression and Structure-Stability Relationship. *Food Technol. Biotechnol.* **2003**, *41*, 89–93.
14. Leščić Ašler, I.; Pigac, J.; Vujaklija, D.; Luić, M.; Štefanić, Z. Crystallization and Preliminary X-Ray Diffraction Studies of a Complex of Extracellular Lipase from *Streptomyces rimosus* with the Inhibitor 3,4-Dichloroisocoumarin. *Acta Crystallogr. Sect. F Struct. Biol. Cryst. Commun.* **2011**, *67*, 1378–1381. [CrossRef]
15. Leščić Ašler, I.; Štefanić, Z.; Maršavelski, A.; Vianello, R.; Kojić-Prodić, B. Catalytic Dyad in the SGNH Hydrolase Superfamily: In-Depth Insight into Structural Parameters Tuning the Catalytic Process of Extracellular Lipase from *Streptomyces rimosus*. *ACS Chem. Biol.* **2017**, *12*, 1928–1936. [CrossRef]
16. Murayama, K.; Kano, K.; Matsumoto, Y.; Sugimori, D. Crystal Structure of Phospholipase A1 from *Streptomyces albidoflavus* NA297. *J. Struct. Biol.* **2013**, *182*, 192–196. [CrossRef]
17. Wei, Y.; Schottel, J.L.; Derewenda, U.; Swenson, L.; Patkar, S.; Derewenda, Z.S. A Novel Variant of the Catalytic Triad in the *Streptomyces scabies* Esterase. *Nat. Struct. Biol.* **1995**, *2*, 218–223. [CrossRef]
18. Maršavelski, A.; Sabljić, I.; Sugimori, D.; Kojić-Prodić, B. The Substrate Selectivity of the Two Homologous SGNH Hydrolases from *Streptomyces* Bacteria: Molecular Dynamics and Experimental Study. *Int. J. Biol. Macromol.* **2020**, *158*, 222–230. [CrossRef]
19. Anderson, A.C.; Stangherlin, S.; Pimentel, K.N.; Weadge, J.T.; Clarke, A.J. The SGNH Hydrolase Family: A Template for Carbohydrate Diversity. *Glycobiology* **2022**, *32*, 826–848. [CrossRef]
20. Katoh, K.; Rozewicki, J.; Yamada, K.D. MAFFT Online Service: Multiple Sequence Alignment, Interactive Sequence Choice and Visualization. *Brief. Bioinform.* **2018**, *20*, 1160–1166. [CrossRef]
21. Kuraku, S.; Zmasek, C.M.; Nishimura, O.; Katoh, K. ALeaves Facilitates On-Demand Exploration of Metazoan Gene Family Trees on MAFFT Sequence Alignment Server with Enhanced Interactivity. *Nucleic Acids Res.* **2013**, *41*, W22–W28. [CrossRef] [PubMed]
22. Chovancova, E.; Pavelka, A.; Benes, P.; Strnad, O.; Brezovsky, J.; Kozlikova, B.; Gora, A.; Sustr, V.; Klvana, M.; Medek, P.; et al. CAVER 3.0: A Tool for the Analysis of Transport Pathways in Dynamic Protein Structures. *PLoS Comput. Biol.* **2012**, *8*, e1002708. [CrossRef] [PubMed]
23. Anandakrishnan, R.; Drozdetski, A.; Walker, R.C.; Onufriev, A.V. Speed of Conformational Change: Comparing Explicit and Implicit Solvent Molecular Dynamics Simulations. *Biophys. J.* **2015**, *108*, 1153–1164. [CrossRef] [PubMed]
24. Swanson, J.M.J.; Henchman, R.H.; McCammon, J.A. Revisiting Free Energy Calculations: A Theoretical Connection to MM/PBSA and Direct Calculation of the Association Free Energy. *Biophys. J.* **2004**, *86*, 67–74. [CrossRef] [PubMed]
25. Åqvist, J.; Medina, C.; Samuelsson, J.E. A New Method for Predicting Binding Affinity in Computer-Aided Drug Design. *Protein Eng. Des. Sel.* **1994**, *7*, 385–391. [CrossRef] [PubMed]
26. Xiao, C.; Guo, H.; Tang, J.; Li, J.; Yao, X.; Hu, H. Expression Pattern and Functional Analyses of *Arabidopsis* Guard Cell-Enriched GDSL Lipases. *Front. Plant Sci.* **2021**, *12*, 748543. [CrossRef]
27. Lai, C.P.; Huang, L.M.; Chen, L.F.O.; Chan, M.T.; Shaw, J.F. Genome-Wide Analysis of GDSL-Type Esterases/Lipases in *Arabidopsis*. *Plant Mol. Biol.* **2017**, *95*, 181–197. [CrossRef]
28. Eyers, P.A.; Murphy, J.M. The Evolving World of Pseudoenzymes: Proteins, Prejudice and Zombies. *BMC Biol.* **2016**, *14*, 98. [CrossRef]
29. Biđin, S.; Vujaklija, I.; Paradžik, T.; Bielen, A.; Vujaklija, D. Leitmotif: Protein Motif Scanning 2.0. *Bioinformatics* **2020**, *36*, 3566–3567. [CrossRef]
30. Mathews, I.; Soltis, M.; Saldajeno, M.; Ganshaw, G.; Sala, R.; Weyler, W.; Cervin, M.A.; Whited, G.; Bott, R. Structure of a Novel Enzyme That Catalyzes Acyl Transfer to Alcohols in Aqueous Conditions. *Biochemistry* **2007**, *46*, 8969–8979. [CrossRef]
31. Dovbnya, D.V.; Bragin, E.Y.; Ivashina, T.V.; Donova, M.V. Draft Genome Sequence of *Mycolicibacterium smegmatis* VKM Ac-1171 Contains Full Set of Sterol Catabolic Genes. *Microbiol. Resour. Announc.* **2022**, *11*, e0077222. [CrossRef] [PubMed]
32. Bakshy, K.; Gummadi, S.N.; Manoj, N. Biochemical Characterization of Alr1529, a Novel SGNH Hydrolase Variant from *Anabaena* sp. PCC 7120. *Biochim. Biophys. Acta (BBA)-Proteins Proteomics* **2009**, *1794*, 324–334. [CrossRef] [PubMed]

33. Sambrook, J.; Fritsch, E.F.; Maniatis, T. *Molecular Cloning: A Labortaroy Manual*; Cold Spring Harbor Laboratory Press: New York, NY, USA, 1989; Volume 68, ISBN 0879695773.
34. Pigac, J.; Schrempf, H. A Simple and Rapid Method of Transformation of *Streptomyces rimosus* R6 and Other Streptomycetes by Electroporation. *Appl. Environ. Microbiol.* **1995**, *61*, 352–356. [CrossRef] [PubMed]
35. Bielen, A.; Ćetković, H.; Long, P.F.; Schwab, H.; Abramić, M.; Vujaklija, D. The SGNH-Hydrolase of *Streptomyces coelicolor* Has (Aryl)Esterase and a True Lipase Activity. *Biochimie* **2009**, *91*, 390–400. [CrossRef] [PubMed]
36. Kieser, T.; Bibb, M.J.; Buttner, M.J.; Chater, K.F.; Hopwood, D.A. *Practical Streptomyces Genetics*; The John Innes Foundation: Norwich, England, 2000.
37. Xiao, Y.-H.; Yin, M.-H.; Hou, L.; Luo, M.; Pei, Y. Asymmetric Overlap Extension PCR Method Bypassing Intermediate Purification and the Amplification of Wild-Type Template in Site-Directed Mutagenesis. *Biotechnol. Lett.* **2007**, *29*, 925–930. [CrossRef] [PubMed]
38. DeSanti, C.L.; Strohl, W.R. Characterization of the *Streptomyces* sp. Strain C5 Snp Locus and Development of Snp-Derived Expression Vectors. *Appl. Environ. Microbiol.* **2003**, *69*, 1647–1654. [CrossRef] [PubMed]
39. Tian, C.; Kasavajhala, K.; Belfon, K.A.A.; Raguette, L.; Huang, H.; Migues, A.N.; Bickel, J.; Wang, Y.; Pincay, J.; Wu, Q.; et al. Ff19SB: Amino-Acid-Specific Protein Backbone Parameters Trained against Quantum Mechanics Energy Surfaces in Solution. *J. Chem. Theory Comput.* **2020**, *16*, 528–552. [CrossRef] [PubMed]
40. He, X.; Man, V.H.; Yang, W.; Lee, T.S.; Wang, J. A Fast and High-Quality Charge Model for the next Generation General AMBER Force Field. *J. Chem. Phys.* **2020**, *153*, 114502. [CrossRef]
41. Jakalian, A.; Jack, D.B.; Bayly, C.I. Fast, Efficient Generation of High-Quality Atomic Charges. AM1-BCC Model: II. Parameterization and Validation. *J. Comput. Chem.* **2002**, *23*, 1623–1641. [CrossRef]
42. Izadi, S.; Anandakrishnan, R.; Onufriev, A.V. Building Water Models: A Different Approach. *J. Phys. Chem. Lett.* **2014**, *5*, 3863–3871. [CrossRef]
43. Case, D.A.; Belfon, K.; Ben-Shalom, I.Y.; Brozell, S.R.; Cerutti, D.S.; Cheatham, T.E., III; Cruzeiro, V.W.D.; Darden, T.A.; Duke, R.E.; Giambasu, G.; et al. *AMBER 2020*; University of California: San Francisco, CA, USA, 2020.
44. Salomon-Ferrer, R.; Case, D.A.; Walker, R.C. An Overview of the Amber Biomolecular Simulation Package. *Wiley Interdiscip. Rev. Comput. Mol. Sci.* **2013**, *3*, 198–210. [CrossRef]
45. Berendsen, H.J.C.; Postma, J.P.M.; Van Gunsteren, W.F.; Dinola, A.; Haak, J.R. Molecular Dynamics with Coupling to an External Bath. *J. Chem. Phys.* **1984**, *81*, 3684–3690. [CrossRef]
46. Loncharich, R.J.; Brooks, B.R.; Pastor, R.W. Langevin Dynamics of Peptides: The Frictional Dependence of Isomerization Rates of N-acetylalanyl-N′-methylamide. *Biopolymers* **1992**, *32*, 523–535. [CrossRef] [PubMed]
47. Ozer, G.; Valecv, E.F.; Quirt, S.; Hernandez, R. Adaptive Steered Molecular Dynamics of the Long-Distance Unfolding of Neuropeptide Y. *J. Chem. Theory Comput.* **2010**, *6*, 3026–3038. [CrossRef] [PubMed]
48. Ozer, G.; Keyes, T.; Quirk, S.; Hernandez, R. Multiple Branched Adaptive Steered Molecular Dynamics. *J. Chem. Phys.* **2014**, *141*, 064101. [CrossRef]
49. Jarzynski, C. Nonequilibrium Equality for Free Energy Differences. *Phys. Rev. Lett.* **1997**, *78*, 2690–2693. [CrossRef]
50. Connolly, M.L. Analytical Molecular Surface Calculation. *J. Appl. Crystallogr.* **1983**, *16*, 548–558. [CrossRef]

International Journal of
Molecular Sciences

MDPI

Article

Discovery and Visualization of the Hidden Relationships among N-Glycosylation, Disulfide Bonds, and Membrane Topology

Manthan Desai [1,2], Amritpal Singh [2], David Pham [2], Syed Rafid Chowdhury [2] and Bingyun Sun [1,3,*]

[1] Department of Molecular Biology and Biochemistry, Simon Fraser University, Burnaby, BC V5A 1S6, Canada; manthan_desai@sfu.ca
[2] Department of Computing Science, Simon Fraser University, Burnaby, BC V5A 1S6, Canada; aamritpa@sfu.ca (A.S.); ddkpham@gmail.com (D.P.); rafid_haider@sfu.ca (S.R.C.)
[3] Department of Chemistry, Simon Fraser University, Burnaby, BC V5A 1S6, Canada
* Correspondence: bingyun_sun@sfu.ca

Abstract: Membrane proteins (MPs) are functionally important but structurally complex. In particular, MPs often carry three structural features, i.e., transmembrane domains (TMs), disulfide bonds (SSs), and N-glycosylation (N-GLYCO). All three features have been intensively studied; however, how the three features potentially correlate has been less addressed in the literature. With the growing accuracy from computational prediction, we used publicly available information on SSs and N-GLYCO and analyzed the potential relationships among post-translational modifications (PTMs) and the predicted membrane topology in the human proteome. Our results suggested a very close relationship between SSs and N-GLYCO that behaved similarly, whereas a complementary relation between the TMs and the two PTMs was also revealed, in which the high SS and/or N-GLYCO presence is often accompanied by a low TM occurrence in a protein. Furthermore, the occurrence of SSs and N-GLYCO in a protein heavily relies on the protein length; however, TMs seem not to possess such length dependence. Finally, SSs exhibits larger potential dynamics than N-GLYCO, which is confined by the presence of sequons. The special classes of proteins possessing extreme or unique patterns of the three structural features are comprehensively identified, and their structural features and potential dynamics help to identify their susceptibility to different physiological and pathophysiological insults, which could help drug development and protein engineering.

Keywords: post-translational modifications; transmembrane domains; N-glycosylation; disulfide bonds; membrane protein topology; endoplasmic reticulum; protein structure and function

1. Introduction

Membrane proteins (MPs) are a vital component of the cellular proteome. This class of proteins constitutes close to 30% of the proteome and functions in many critical cellular processes, such as solute and ion transport, cell–cell interactions, adhesion, and signal transduction [1]. Therefore, MPs account for more than 60% of drug targets [2] and are frequently used as diagnostic markers for various diseases.

The topology and folding of MPs are key to their functions due to a strong correlation between them [3]. The arrangement of structural elements such as alpha helices, beta strands, and loops across the lipid bilayer heavily influences protein functions such as ion transport and signal transduction. Protein topology and folding can dictate the location of the protein active sites, thereby modulating its accessibility to ligands, ions, and other proteins [4]. Topology can also influence conformational changes that typically occur in MP functional cycles [5]. Most MPs perform their functions ranging from small movements within the transmembrane (TM) regions to large interdomain movements involving multiple other proteins. Therefore, understanding the link between topology and conformations can be key to interpreting protein functions.

The unique properties of membrane topology, folding, expression, and cellular localization render MPs technically challenging to study. The low solubility, low expression, and high PTMs have prevented highly informative X-ray and NMR structural analyses [6,7] of these proteins. The dynamic nature and complex topogenic signals, such as protein–substrate and protein–lipid interactions as well as cell signaling-regulated PTMs, introduce ample variability in topology formation. In addition, translocon dynamics further complicate topology development, because many accessory factors of translocon only transiently or sub-stoichiometrically associate with the core machinery [8]. As a result, polypeptides with identical sequences can span the membrane differently in a concept known as topological heterogeneity [9].

To overcome the challenges in the experimental determination of MP topology and structure, computational algorithms such as machine learning (ML) have been developed to predict these features. Formally, ML is defined as the automated obtaining of patterns and correlations from data using complex but efficient algorithms [10]. In the scope of transmembrane topology, the latest ML models excel in predicting transmembrane protein segments by analyzing sequence, hydrophobicity patterns, and structural features. A major benefit of ML models in this field is their throughput, which allows for the structural interpretation of the entire MP class.

ML models have experienced several generations of evolution, from the early ones developed by Jack Kyte and Rusell Doolittle in 1981 [11] to the dense alignment surface (DAS) method that appeared in 1997 [12] and the support vector machine (SVM) model in the 2000s [13]. The previously best-performing prediction used the hidden Markov model (HMM) developed by Anders Krogh et al., in 2001 [14]. Neural networks were applied to MP structural prediction in the 1990s [15]. With recent developments in both hardware and computing software, neural networks have been revived and have advanced numerous data-driven research fields. Krogh et al., recently applied deep learning in a neural network to optimize TMHMM and resolved Deep TMHMM, which achieved the highest accuracy in the prediction of TM protein topology so far [16] and recognizes signal peptides and beta-barrels in MPs. It is foreseeable that the future structural challenges in MPs will lie in the discovery of their dynamics (i.e., conformational changes) when interacting with their binding partners and/or upon environmental insults.

In addition to TMs, growing evidence has pinpointed that PTMs are critical in maintaining membrane protein topology and structure. For example, via proteolytic cleavage or the addition of structural- or charge-modifying groups to one or more residues, the topology of MPs can be altered [17]. The most common PTMs in MPs are disulfide bridges (SSs) and N-linked glycosylation (N-GLYCO) bonds. The former forms between thiol groups in two cysteine residues intramolecularly or intermolecularly, and the latter typically forms on sequons [17] even though atypical N-GLYCO is frequently observed [18]. Both PTMs have compartment specificity. Their location can be used to determine the topology of MPs [19].

With the large amount of structural information available at the human proteome level, we performed a series of statistical analyses with a particular focus on SSs and N-GLYCO acquired from the UniProt database in conjunction with membrane topology predicted by Deep TMHMM. The reason to choose these three structural features for analysis is because all of them are chiefly formed in the ER. Membrane insertion of TMs mainly depends on the lateral exit through translocon in the ER membrane [8], which also complexes with the N-GLYCO machinery—oligosaccharyl transferase (OST) [20]. An important paralog to the catalytic enzyme STT3A in OST is STT3B, which has oxidoreductase activity that can form mixed disulfide bonds with target proteins through its components MAGT1 or TUSC3 [20,21]. In addition, a number of ER-resident glycoenzymes and glycochaperons also form complexes with thiol-disulfide oxidoreductases or carry oxidoreductive activity themselves [22,23]. We therefore hypothesize that a close relationship may exist among these structural features.

The analyses allowed us to discover several hidden relationships among SSs, N-glycans, and TMs, including the correlation between their occurrence with protein length and their relative location in a protein. Functional enrichment analyses further revealed important protein classes that possess these structural features, such as the GPCRs, ion channels, adhesion molecules, enzymes, and receptors functioning in signaling and immune responses. In addition, using the compartmental specificity in SSs and N-GLYCO, we identified a set of proteins with conflicting topology from topology prediction. These conflicts allowed us to discover the potential dynamics of PTMs. To allow other researchers to visualize these hidden correlations, we developed a display portal for the proteins showing extreme TMs, PTMs, and conflicts. We further provided supplemental lists on the proteins possessing special TMs and PTMs. The obtained information helps to elucidate how the complex topology and structures form and evolve from the primary sequence of proteins. Such global information can support a better functional understanding of MPs and help a potential prediction of the susceptibility of selective MP classes to particular environmental and physiological insults. The implications for structural flexibility and constraints in MPs can also facilitate drug design and protein engineering [24–26].

2. Results and Discussion

2.1. Length and Count Distribution

The examination of the length distribution in various features shows some interesting properties. The length distribution of the human proteome segregated by the TMs predicted by Deep TMHMM is shown in Figure 1A,B. Clearly, the proteins with TMs possess different length distributions than those without TMs. The non-TM proteins peak at a shorter length of 102–202 amino acids than that of TM proteins, whose length peaks at 228–328 amino acids. Otherwise, the average protein lengths in the non-TM proteome and TM proteome are very similar, with TM proteins having 558.42 (N = 4947) and non-TM proteins having 557.83 (N = 15,464) amino acids.

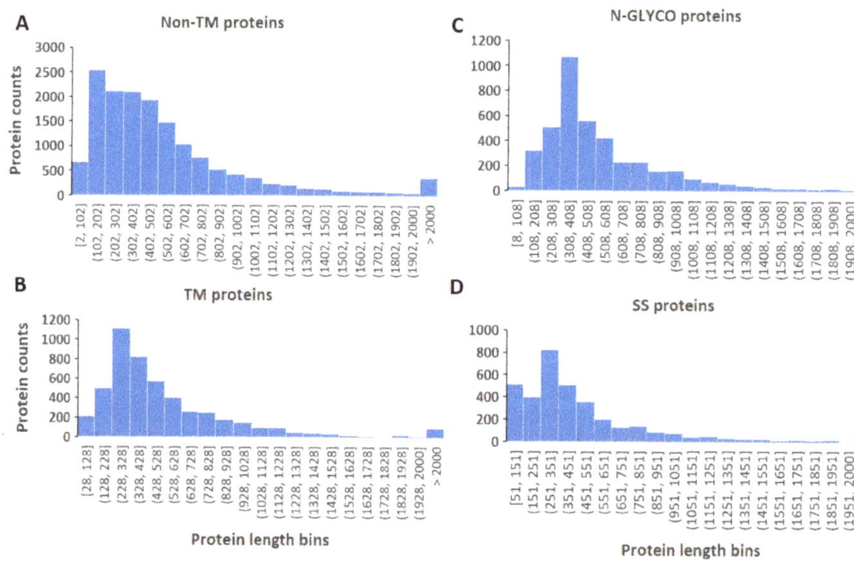

Figure 1. The length distribution of proteins in non-TM (A), TM (B), N-GLYCO (C), and SS (D) proteomes.

The length distribution in the N-GLYCO and SS proteomes was also analyzed and is shown in Figure 1C,D, respectively. Differences were also observed between these two PTMs, with N-GLYCO preferring longer proteins over SS. The respective peak lengths for the N-GLYCO proteome and SS proteome are 308–408 (N = 4214) and 251–351 (N = 3588) amino acids. The overall mean length of the two proteomes is also similar, with the N-GLYCO proteome of 611.93 amino acids and the SS proteome of 559.69 amino acids. The mean lengths in the two PTM proteomes are close to those of TM and non-TM proteomes; however, the peak distributions among these four proteomes are different. Therefore, the length distribution is a more sensitive analysis in disclosing the hidden differences among these large proteomes.

After length analysis, we examined the frequency of TMs, SSs, and N-glycans in each protein of their corresponding proteomes. Figure 2A summarizes the results. Among all three structural features, the single occurrence had the dominant population in all proteomes. In contrast, some proteins possess an extremely high occurrence of a single feature. The highest number of TMs in a single protein is 38, carried by Piezo-type mechanosensitive ion channel component 1 (or membrane protein induced by beta-amyloid treatment). The highest SS bond in a protein is 159, carried by Prolow-density lipoprotein receptor-related protein 1 (LRP-1); for N-GLYCO, it is 102, carried by Mucin-16 (or ovarian cancer-related tumor marker CA125, ovarian carcinoma antigen CA125). Some of these extreme cases, such as Mucin-16, would skew our later analyses and have been left out as to be mentioned below.

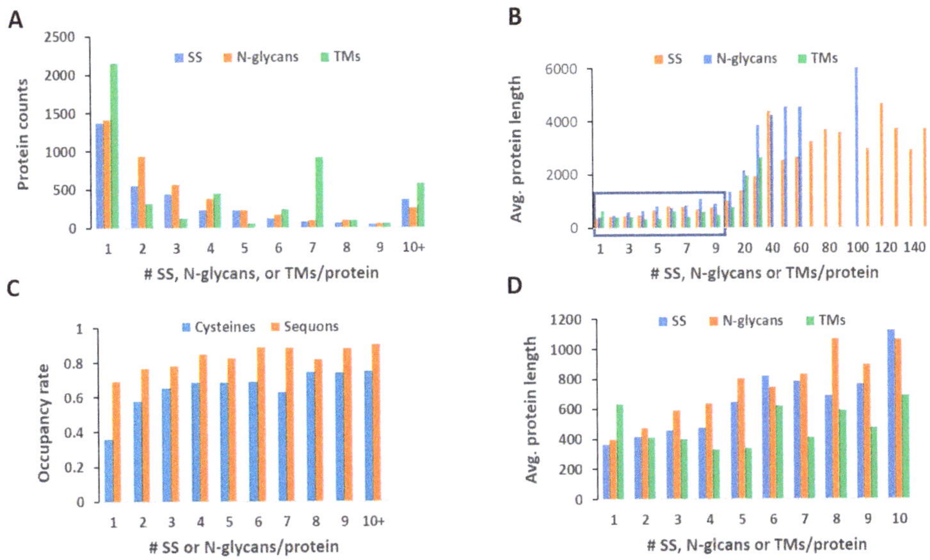

Figure 2. Distribution of protein counts, average protein length, and occupancy rates as a function of feature occurrence in one protein. (**A**) Protein counts vs. numbers of SSs, N-glycans, or TMs/protein; (**B**) Average protein length vs. numbers of SS, N-glycans, or TMs/protein; (**C**) Occupancy rates of cysteines and sequons as a function of the numbers of SS and N-glycans/protein; (**D**) A zoom-in view of the boxed area in (**B**).

Overall, the occurrence of SSs and N-glycans showed a clear decrease in protein counts when the multiplication of these features increased in a protein but not in TMs. For TMs, the protein populations carrying 1 and 7 TMs show large spikes; otherwise, the distribution of other TM counts in proteins seems random. The large population of MPs carrying seven TMs clearly indicates the specialty of this particular class of MPs. A further functional enrichment analysis indicated that the majority of the proteins are G protein-coupled receptors (GPCRs), specifically olfactory receptors, which have the largest population of GPCRs and are poorly studied due to their extremely low abundance. GPCRs are known to mediate most cellular responses to hormones and neurotransmitters [27]. The malfunction of GPCRs can cause serious disorders such as heart failure, stroke, and cancer; hence, they are extensively targeted via drugs. As of 2017, 134 GPCRs are targets for an estimated 700 approved drugs in the US or the EU [28].

As discovered above, proteins carrying these features tend to peak at a relatively longer length than otherwise, and we wanted to evaluate whether the length also correlates with the frequency of occurrence of these features. Figure 2B shows the average protein length as a function of feature multiplication/protein in each respective proteome. In the cases of SSs and N-GLYCO, the protein length shows a positive trend for the modification frequency up to 30 counts/protein. When above this cutoff, the protein length seems to stagger except for Mucin-16 (only one incidence in the N-GLYCO proteome contributed to the 110–100 bin). An investigation of the high-multiplication proteins (counts > 30) shows that the features are typically packed in dense repeats, such as EGF and CUB repeats, which have minimized the demand in protein length. Therefore, the proteins with high multiplication features can be cataloged into special classes that also possess unique functions to be analyzed later in the Section 3.5.

When zooming in to the low multiplication region with counts fewer than 10, as shown in Figure 2D, TMs do not follow the length correlation exhibited by SSs and N-glycans. The shortest mean length was found in those carrying four and five TMs, rather than single-TM carriers. This unique behavior might originate from the working principle of membrane insertion chiefly via the translocon that functions co-translationally with primarily linear amino acid chains that emerge from ribosomes [8]. Therefore, a high number of TMs can be stacked densely within a small span connected only by short loops [8]. It is also recognized that helical hairpins can form in closely spaced TMs after ribosomes [29] and insert into the membrane together instead of sequentially. Both computational simulations and experimental evaluation show that TM oligomerization in the membrane is sequence-dependent and energetically favored [30]. Therefore, in many cases no additional enzymes are necessary in the insertion processes. Furthermore, a novel theory also emerged to view TM oligomerization from the local concentration perspective [31], in which membrane tethering by TMs reduced the 3D free localization space and confined the TMs in 2D to increase local contact for oligomerization. Helix–helix TM interactions are known to occur in both low-TM and high-TM proteins, such as ion channels. Sodium ion channels with 24 TMs [32] can form helix bundles in lipids. The hydrophobic side chains in the bundle face the lipid bilayer, while the hydrophilic residues face towards the center; in such a hydrophobic lipid bilayer, they are pierced through by a hydrophilic pore formed by channel proteins to allow charged ions to pass [32]. In contrast, the formation of N-GLYCO and SSs requires enzymatic catalysis, in which the multiplication of these PTMs in a protein occurs sequentially, and each reaction requires a specific local conformation for recognition and release [33,34]. Except for the high-density repeats, most low-occurrence multiplications for SSs and N-glycans do not possess the synergistic aggregational force that exists in TMs. The multiplication of SSs and N-glycans behaved more additively, which could explain the observed length dependence in most low-multiplication proteins. In summary, we observed a similar behavior of SSs and N-glycans that is distinct from TMs in their corresponding proteomes.

2.2. Rate of Modification

Because the occurrence of SSs is based on the presence of cysteine residues, and N-GLYCO mainly takes place in sequons, we further analyzed the utilization rate of cysteines and sequons in the SS proteome and N-glycoproteome, as shown in Figure 2C. Overall, the sequon utilization rate is higher than that of cysteines. A stable modification rate of sequons was observed regardless of N-glycan occurrence in the N-GLYCO proteome. However, in the low SS proteins (1–2 counts/protein), a drop in the cysteine utilization rate was observed. In both the N-GLYCO and SS proteomes, a relatively high utilization rate (>50%) was observed, suggesting that the presence of sequons and cysteines in the protein sequence is not random. Among all amino acids, cysteine has the least abundance in proteins [35], and its presence is often conserved and functionally important [36]. SS formation is a reversible reaction [37], and free cysteines in proteins increase the risk of SS scrambling in their folding [38,39] and function during environmental insults such as heat shock [40] and oxidative stresses [37]. Our observation prompted us to further analyze how these features relate to each other in individual proteins.

2.3. Hierarchical Clustering

Because SSs and N-glycans are common structural features on MPs, we wanted to further explore whether hidden relationships exist between the two PTMs and TMs in individual proteins at the proteome level. In the past, we observed a complementary relationship between TMs and N-glycans from our experimentally obtained N-glycoproteome [41], and we also discovered a complex relationship between SSs and N-glycans [22] in the literature. Here, we wanted to see whether SSs would behave similarly to the N-glycans in the complementary relation to TMs. To study this, we focused on the SS/N-GLYCO proteome (N = 2722).

To evaluate the overall properties in the SS/N-GLYCO proteome, we first performed hierarchical clustering. The results showed three main clusters. The distinct characteristics in Cluster 1 (Figure 3A) from the remaining two clusters are the high TM, low SS, and low N-glycan counts. The remaining two clusters are close to each other, with Cluster 2 (Figure 3B) having relatively high sequon and N-glycan counts and Cluster 3 (Figure 3C) having high cysteine and SS counts. Both Clusters 2 and 3 are low on TMs. Complementarity in TM to the two PTMs on proteins becomes obvious to us, in which TMs are not only complementary to N-GLYCO but also to SSs.

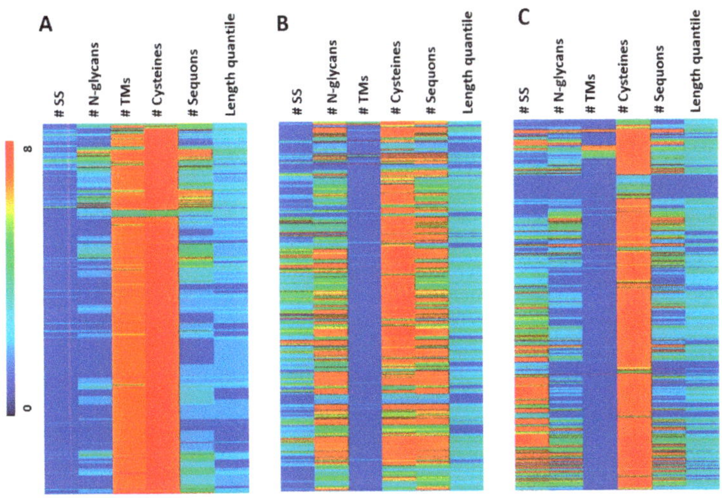

Figure 3. The heatmap of the three clusters of the SS/N-GLYCO proteome based on the counts of SSs, N-glycans, TMs, cysteines, and sequons, as well as the length quantile of the corresponding protein. Length quantile refers to dividing the length of each protein into four equal parts from N-terminus to C-terminus, and their corresponding value is 1–4. (**A**) Heatmap of Cluster 1; (**B**) Heatmap of Cluster 2; (**C**) Heatmap of Cluster 3. Each row represents one protein in the respective proteome, and the color encodes the count values of each protein for a specific property described by each column. The correspondence between different colors and the values is provided as the sidebar on the left.

To further confirm this relationship, we used the SS count as the X-axis and examined the average numbers of N-glycans and TMs in the proteins within each SS count bin. The results are shown in Figure 4A. When the SS occurrence per protein is less than 70, consistent with the analysis so far, the N-glycan occurrence exhibits a tight positive linear correlation with SS frequency, whereas the TMs show a negative exponential correlation with the counts of SSs in proteins. The less obvious trends in high multiplications of SSs can be caused by fewer proteins and less statistical power. When using the number of N-glycans as the X-axis and repeating the same analysis, the results shown in Figure 4B are similar to those of Figure 4A. These results suggest that SSs and N-glycans behave similarly. This observation agrees with our previous findings on experimentally examined individual proteins with both SSs and N-GLYCO [22], in which a close but complex relationship was discovered and was supported by the intertwined biogenesis and processing pathways of both PTMs in the ER [23]. We have reviewed in detail how the potential interacting molecular regulations on SSs and N-glycans in the ER result in complex (can be promoting or inhibiting) relationships of both PTMs in individual membranes and secreted proteins [22]. Here, we observed similar behavior of SSs and N-GLYCO at the proteome level.

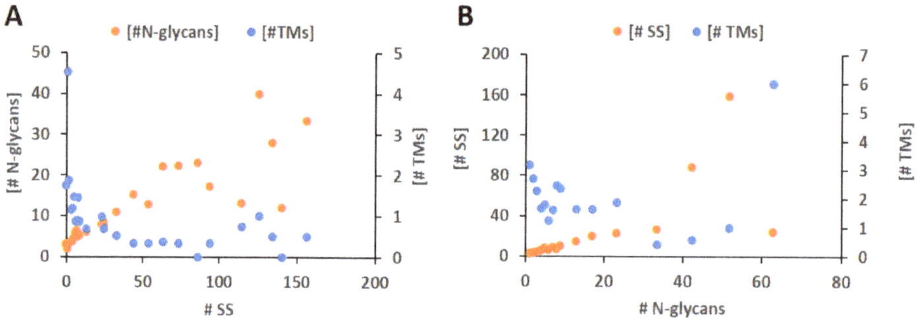

Figure 4. Scatter plot of the numbers of N-glycans, TMs, and SSs in a protein. (**A**) The average numbers of N-glycans and TMs as a function of SS counts in a protein; (**B**) The average numbers of SSs and TMs as a function of N-glycan counts in a protein.

The negative correlation of TMs with either SSs or N-GLYCO in low-SS or low-N-glycan occurrence proteins agrees with the observation of non-length dependence of TMs in MPs, as shown in Figure 2 and the cluster results in Figure 3. This also agreed with our previous discovery based on the experimentally detected N-GLYCO on membrane proteins [41] using our glyco-peptide capture technique [42]. Such complementarity between TMs and PTMs can be understood from their structural functions. Both SSs and N-glycans help form and stabilize desired conformations of proteins, which can be critical in structurally disordered large loops. When TMs are rich, large, and unstructured, external loops are less frequent than when TMs are poor.

This triggered us to further investigate the single-feature carrying proteins that would defy such complementarity. Figure 5A shows the Venn diagram of such comparisons. The smallest section is observed in proteins carrying all features, i.e., 52 entries (2.59% of the total unique entries). The marginal overlap suggests that the complementarity we observed is quite robust with only limited exceptions. To verify whether the remaining entries, particularly those unique to each structural feature proteome, are proteins carrying a large number of complementary structural features, we enlarged the comparison by including the feature occurrence from one to three, as shown in Figure 5B. The size of the section shared by the three features is expanded to 301 entries, which represents an over four-fold increase from 52 entries in Figure 5A. However, the percentage is still marginal, accounting for only 12.1% of the total unique entries. This result suggests that most proteins possess more than three of the other structural feature elements, which explains why MP proteins are challenging to study structurally.

We then shifted the comparison to the higher end of the feature occurrence entries (>10 features/protein), which could also defy the complementarity we observed. The results shown in Figure 5C resemble those of the single-feature comparison shown in Figure 5A. There are only two proteins in the same family shared by all three categories, i.e., NPC intracellular cholesterol transporter 1 (Niemann–Pick C1 protein) and NPC1-like intracellular cholesterol transporter 1 (Niemann–Pick C1-like protein 1), with a cellular localization in endosomes and lysosomes. Both proteins function in transporting low-density lipoproteins and egress cholesterol from endosomes and lysosomes. The two proteins both carry 15 SSs, 19 N-glycans, and 13 TMs. Malfunction of these proteins can cause the over accumulation of cholesterol and glycosphingolipids in the late endosomal/lysosomal compartments. Otherwise, there are only four entries carrying both >10 TMs and >10 N-glycans and 78 entries carrying both >10 N-glycans and >10 SS. No entries carry exclusively both >10 TM and >10 SS. The Venn diagram shown in Figure 5 agrees with the correlation results shown in Figure 4. The small percentage of proteins sharing a high occurrence of the three features (Figure 5C) suggests a certain level of exclusiveness or crowdedness, potentially due to space constraints or avoiding unnecessary redundancy. Accordingly, the large percentage of unique proteins in high-feature-count categories in Figure 5C (TMs at 10.4%, N-GLYCO at 38.2%, and SSs at 52.1%) suggests that MPs of extremely high (>10 counts) structural features are typically high in only one feature. All the proteins in each Venn diagram in Figure 5 are provided in Supplementary Tables S1–S3.

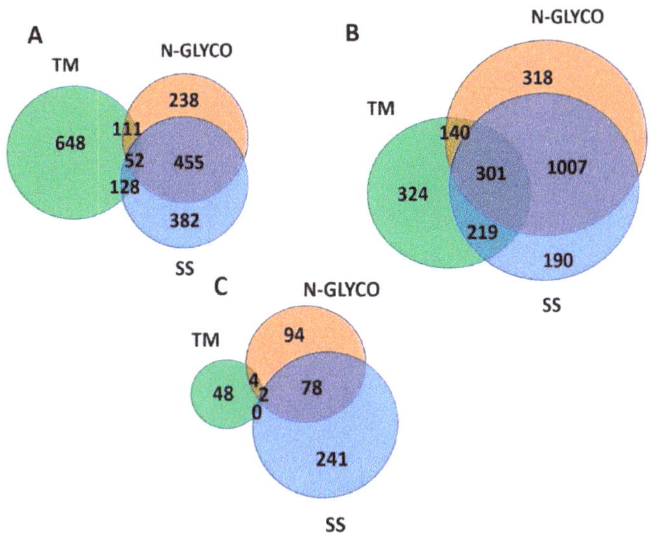

Figure 5. Venn diagrams of proteins carrying various numbers of TMs, N-glycans, and SSs in the SS/N-GLYCO proteome. (**A**) Venn diagram of proteins carrying a single count of either TM, N-glycan, or SS; (**B**) Venn diagram of proteins carrying up to three counts of either TMs, N-glycans, and SSs; (**C**) Venn diagram of proteins carrying more than ten counts of either TMs, N-glycans, and SSs.

2.4. Functional Analysis

The complementary relationship between N-GLYCO, SSs, and TMs revealed in our results inspired us to further disclose the functions carried by these unique classes of proteins in Figure 5. Functional enrichment analysis on the 111 entries having one TM and one N-glycan in Figure 5A revealed proteins in MHC class II, while the 128 entries in Figure 5A having one TM and one SS were enriched by proteins functioning in heparin and heparan sulfate proteoglycan biosynthesis and in the regulation of T cell proliferation. The enrichment analysis was also carried out on the results from the three-feature carrier protein comparisons shown in Figure 5B. The entries shared by one to three TMs and one to three N-glycans are 140 in Figure 5B and enriched by proteins functioning in O-glycosylation, and the entries shared by one to three TMs and one to three SSs exclusively are 219 and are enriched by cell adhesion and cytokine-mediated signal transduction. The commonality between all three categories with features of one to three have 301 entries in Figure 5B and are enriched in immune response proteins. For the largest sharing category, i.e., between one to three N-glycans and one to three SS, there are 1007 entries, which are dominated again by GPCRs. Similarly, in Figure 5A, the section shared by one N-glycan and one SS has 455 entries, the largest overlap section, and is also dominated by GPCRs. The top 10 enriched GO terms from the major sections in Figure 5 are provided in Supplementary Table S4.

The presence of TM domains confined the location of the target protein, and the high number of TMs in low occurrence N-glycan and SS proteins, as we suggested in our previous study, provides stability to the overall protein structure and docking to a special/stable membrane location. Because the membrane is commonly considered a two-dimensional fluidic environment, MPs form close interactions with nearby lipid molecules such that a heterogeneous distribution of lipid molecules in the membrane is observed. Conversely, the phase separation that occurs in the mixed lipid system also encourages the segregation of MPs into different membrane domains to form lipid rafts [43]. The number of TMs is one of the biggest factors in determining the interaction strength of lipids and MPs [30], and the high TMs in low SS and N-GLYCO proteins suggest their special membrane location. Using GPCRs as an example, which have seven TMs and important signaling proteins, they are known to interact closely with lipid components and preferentially localize in lipid rafts and caveolae [44]. Our proteomic level analysis suggests its uniqueness in the entire MPs that is likely contributed to by its seven TMs. A recent discovery suggested that hydrophobic proteins drove phase separation in lipids [45], and GPCRs have been proposed to be one class of these drivers [44]. In addition, the sequon and cysteine presence in high TM proteins, such as GPCRs, is also low, as the utilization ratio is more or less constant across protein populations, as shown in Figure 2C. This result suggests that the high frequency of these PTMs may hinder TM stacking possibly by forming rigid and stable structures in the loop. Through evolution, the potential for these PTMs to occur can be minimized by eliminating their carrier amino acids from the sequence through random mutation and natural selection [46].

2.5. Length Quantile Distribution of Features

To further examine the relative locations of these features in individual proteins, we analyzed the quantile distribution of these features with respect to individual protein length, as shown in Figure 6. Among the four quantiles, the first quantile describes the N-termini, the last quantile describes the C-termini, and the rest describe the middle section of a protein. Figure 6A shows that both SSs and N-glycans follow a similar trend, in which the N-terminus is more utilized than the C-terminus for the two PTMs. Among the two middle quantiles, the second quantile is used more than the third quantile. Because both PTMs are preferentially compartmentalized in the lumen or the ectodomain of the MPs, the dominance in the N-terminus of MPs suggests a dominant N-terminal orientation, which describes type I MPs; vice versa, the C-terminal SS and N-glycan modification suggests the property of type II MPs. The formation of type II MPs mostly requires co-translational inversion [47] and is not as frequent as that of type I MPs. Similar to previous observations, the trend of TM distribution is opposite to those of N-glycans and SSs. TMs exhibit an increase with increasing quantile, which is from the N-terminus to the C-terminus. Consequently, the complementarity between TMs and PTMs in MPs is reflected in the protein length-quantile distribution as well. We also analyzed the cysteine and sequon quantile distribution and their rate of utilization, as shown in Figure 6B. For both cysteines and sequons, the quantile distribution resembled the SS and N-glycan distributions, respectively, suggesting that the utilization rate is not affected by the protein length.

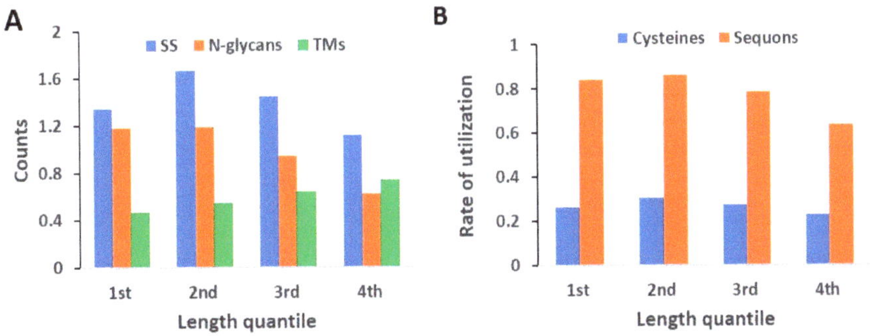

Figure 6. Analysis of feature counts and rate of utilization in length quantile distribution of individual proteins. Length quantile refers to dividing the length of each protein into four equal parts from N-terminus to C-terminus, and named first to fourth quantile. (**A**) Counts of SSs, N-glycans, or TMs as a function of the protein length quantile; (**B**) The rate of utilization of cysteine thiol in SSs and the sequons in N-GLYCO as a function of protein length quantile.

2.6. Conflicts

Conflict analysis was carried out to identify the ectodomain localization of SSs and N-glycans with respect to the MP topology predicted by Deep TMHMM. Compared to an older version of TMHMM 2.0 (364 entries with conflicts), the latest Deep TMHMM prediction (83 entries with conflicts, Supplementary Table S5) has corrected a large number of previous conflicts using the neural network deep learning algorithms, a result suggesting the high accuracy of the current Deep TMHMM prediction. Using the Deep TMHMM results, we found only 1.7% of the total predicted MPs with conflicts, which is within the estimated error rate of Deep TMHMM. This high accuracy rate encouraged us to use the predicted topology to examine the potential location conflicts of sequons and cysteines as well.

Figure 7 summarizes all the results. The protein counts carrying the conflicts are plotted as a function of the conflict occurrence in one protein. The plot includes both the cysteine and sequon conflicts to the Deep TMHMM topology as well as the conflicts of the observed SSs and N-glycans. The SS and N-glycan conflicts are marginal, and most of them only have a single conflict, which are summarized in Supplementary Table S5. What surprised us from the results was the conflict distribution of cysteines and sequons. Sequon conflicts quickly dropped with increasing occurrence, but cysteine conflicts did not, which remained the same. The trend in sequons suggests a nonrandom appearance in the protein sequence because of the agreement with the topology distribution, with single conflict occurring the most. This observation is also congruent with the high utilization rate for N-GLYCO (~85%), as shown in Figure 2B. Experiments using molecular engineering to introduce novel sequons into proteins [46] and clinical analysis of the emergence of novel sequons from genetic mutations [48] have both observed a gain of N-GLYCO that has largely disrupted the normal structure and functions of the target proteins. Collectively, the sequons in proteins do not appear to be random, and most of these sequons have the potential to be N-glycosylated, as Apweiler et al. pointed out 20 years ago when they analyzed the sequons in UniProt [49]. Therefore, we believe that the sequon location defines the dynamic range of this PTM, which could likely trigger local conformational rearrangement due to mostly the observed single conflict. Then, the prevalent cysteine conflicts irrespective of the MP topology suggest that SSs potentially possess larger dynamics than N-GLYCO. Although not all cysteines participate in the SSs, the presence of cysteines does provide the molecular foundation to potentially form SSs. The observed free cysteines can be constrained by local folding [50,51], which could be subject to dynamic perturbations such as oxidative stress and genetic mutations. It is reasonable to speculate that under environmental, developmental, and genetic insults, free cysteines in a protein can increase the possible dynamics of SSs. Collectively, compared to the prevalence of cysteine residues, our results suggest that N-GLYCO (macroheterogeneity) possesses fewer dynamics as SS modifications.

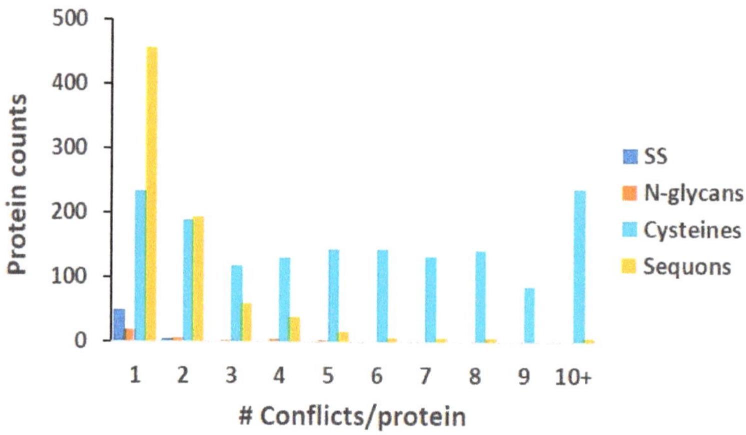

Figure 7. The protein counts as a function of number of conflicts in a single protein. Colors code different types of conflicts. Conflicts were defined as the positions of SSs, N-glycans, or their related cysteines and sequons contradicting the predicted membrane topology of the protein.

2.7. Visualization

Due to the observed complex relationships, we wanted to visualize these structural features in individual proteins. The existing web applications do not include all the information we study. For example, Protter [52], developed by Dr. Wollscheid's lab, does not include SS linkages, and the PTM features displayed on the UniProt website do not have TM information. To meet our needs, we developed a graphical user interface (GUI) using Javascript (https://sfu-sun-lab.github.io/protein-visualizer/, accessed on 1 November 2023) and displayed eight properties for every protein. Unique to our visualization, it includes cysteine and sequon positions. Figure 8 shows examples of our GUI website. Figure 8A shows the main and the top display once a protein is selected. In Figure 8A, a clickable manual sign on the top left corner toggles the display legend panel. In this panel, numerical values of SSs, N-glycans, unoccupied cysteines, and unoccupied sequons are included. A selection window in the top middle of the display enables the selection of the proteins included in our portal. Currently, it can visualize some extreme cases of PTMs as well as all MPs with conflicts. The middle and main parts of the display show the schematic of the selected protein, in which the polypeptide chain is represented by a bar, and the color of the bar encodes the topology, including the beginning and ending of each TM, and the segment orientation. The positions of N-glycans (the bar) and SSs (colored dots) are specified with unique icons. The linkage of SSs is represented using a bracket, and the same pair has matching color in their dots. Such a display allows users to visualize some complex spatial arrangements of SSs, such as disulfide knots, as shown in Figure 8B. The position of unoccupied sequons is denoted using black dots linked to the letter "N" for the asparagine in the sequon, while the unoccupied cysteines are denoted using white open dots.

We designed two zoom features to accommodate MPs with complex structures. First, an expansion ion ". . . " on the top right of the display is clickable, which brings out a scaling bar that allows users to customize the scaling. The entire displayed protein will expand accordingly. This often makes the protein too long to be viewed in the window. To compensate for such inconvenience, a particular section of the protein can be selected using the "Protein Window" by specifying the "start" and "end" positions. An expanded view of the disulfide nots in LRP1 is shown in Figure 8B, which is located below the top main window. Rolling the side bar on the right down allows access to this window. The top of this window has two cells that allow the input of a particular region of the selected protein for display. Using the protein visualizer, we displayed the conflicts (also provided in Supplementary Table S5) between the known compartmentalization of SSs and N-GLYCO to the topology depicted by Deep TMHMM and the extreme cases mentioned in the Results and Discussion.

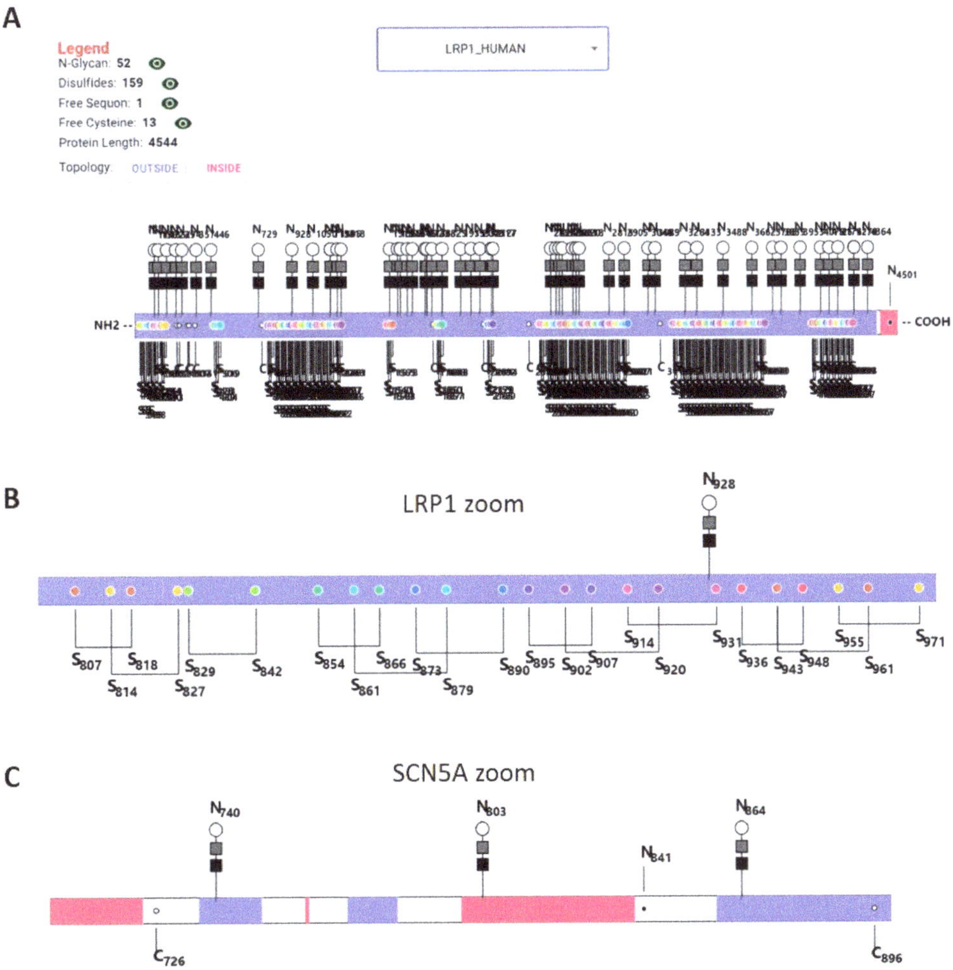

Figure 8. The example displays of the Protein Visualizer. (**A**) The main display of LRP1; (**B**) The zoom display of LRP1 from 800–972 amino acids; (**C**) The zoom display of an N-GLYCO conflict region (N803) in SCN5A between the known compartmentalization of N-glycans to the predicted topology, and the positions of unoccupied sequons and cysteines.

3. Materials and Methods

We accessed the Uniprot human reviewed proteome for protein species, annotation, length, SS, and N-GLYCO information. Proteins with either SSs or N-GLYCO or both features were selected for further analysis and were named the SS proteome, N-GLYCO proteome, and SS/N-GLYCO proteome. The FASTA file of the human reviewed proteome was also used for membrane topology prediction using TMHMM 2.0 and Deep TMHMM.

3.1. Analysis of General Distribution

We first analyzed the distribution of protein length, counts of SSs, N-glycan, and TMs in the entire human proteome. Second, we obtained the position of cysteine residues and sequons in the SS and N-GLYCO proteomes and evaluated their distribution. We then examined the quantile distribution of SSs, N-glycan, TM, cysteine, and sequons in the SS/N-GLYCO proteome. Specifically, we divided the length of each protein into four equal parts (four quantiles, Q1–Q4) from N-terminus to C-terminus, such that the N-terminus falls in Q1 while C-termini falls in Q4. Finally, we assessed the similarity and uniqueness among the obtained distributions.

3.2. Rate of Modification

The rates of sequon occupancy and cysteine utilization were also evaluated in the respective proteomes. These rates were further subjected to correlation analysis for protein length and total number of PTMs per protein as mentioned in the above section.

3.3. Hierarchical Clustering

Nonsupervised hierarchical clustering was carried out in Multiple Experiment Viewer (MeV) run on the local machine to the SS/N-GLYCO proteome for the counts of SSs, N-glycans, TMs, cysteine residues, sequons, and the protein length quantile values. A Pearson correlation with the average linkage clustering method was used to generate hierarchical trees of the proteins. Three main clusters were obtained that accounted for all except one protein.

3.4. Identification and Analysis of Conflicts

We mapped the predicted topology to each SS, N-glycan, cysteine residue, and sequon in every protein. Using the criteria that SSs and N-glycans are located in the extracellular compartment or in the lumen of the secretory pathway, we identified the conflicts. The proteins on the conflict list were further analyzed for their distribution patterns using the method discussed above for similarities and differences.

3.5. Functional Analysis

Special classes of proteins from the abovementioned analyses were assessed for their enriched Gene Ontology terms using the Database for Annotation, Visualization and Integrated Discovery (DAVID) [53]. Specifically, the enrichment in biological processes of Gene Ontology was used to assess their biological functions. The enrichment was determined using the following cutoff parameters: the maximum EASE score (a modified Fisher's exact p-value) of 0.05 and the minimum protein count of 5.

3.6. Visualization Tool

We built an interactive web application named "Protein Visualizer" (https://sfu-sun-lab.github.io/protein-visualizer/, accessed on 1 November 2023) using JavaScript (version 1.5) through React.JS framework to display the special relationship between the reported SSs and N-GLYCO to TMs predicted by Deep TMHMM. A total of 8 features are displayed, including protein length, N-GLYCO sites, SS sites, sites of free sequons and cysteines, TMs, and internal and external loops with respect to the plasma membrane, in which the extracellular domain of the protein is depicted in blue, the intracellular domain is represented in pink, and the TM domain is in white. Users can toggle the visibility of all these features. The application also has two built-in scaling features that allow the expansion of the entire protein or a specific section. Currently, the web portal displays 83 conflicting human PMs and a few representative MPs resolved from the above analyses.

4. Conclusions

Using the existing PTM information in UniProt on the human proteome and the predicted topology of MPs, we analyzed the hidden relations among TMs, SSs, and N-

GLYCO. All three features affect the structure and conformation of MPs and therefore critically influence their functions. All these features have been intensively studied in their own corresponding fields; however, less effort has been focused on their potential relationships. In the past, a serendipitous opportunity led us to discover a complementarity between N-GLYCO and TMs from our experimentally obtained N-glycoproteome [41]. Recently, we further uncovered a complex but closely related interaction between N-GLYCO and SSs in individual proteins by reviewing the published literature [42]. Here, at the human proteome level with the ever-growing accuracy from the computational predictions, we were able to integrate all three structural features together to discover an overall complementarity between TMs and PTMs of SSs and N-GLYCO, in which the TMs do not depend on protein length. On the other hand, the SSs follow an almost identical trend as N-GLYCO in terms of dependence on total protein length, distribution in protein length quantile, and their relationships to TMs, suggesting a similarity between the two PTMs in regulating protein structure and function. With the intriguingly close relationship of enzymes regulating these two PTMs in the ER and the complex relationship of these two PTMs from experimental observations in individual proteins, our analysis at the proteome level suggests that their close relationship not only occurs occasionally in some special proteins but is also more ubiquitously present. Our results encourage more researchers to integrate multiple PTMs and to systematically study proteins carrying them. Furthermore, the tight correlation between sequons and N-GLYCO and consistently less conflict with membrane topology suggest the importance of the sequon position in a protein sequence, which could largely determine the macroheterogeneity of N-GLYCO dynamics. In addition, the wider spread of available cysteine residues and a more constrained presence of sequons from our analysis further hint at a higher dynamic of SSs as opposed to N-GLYCO. Finally, to facilitate future investigation of the hidden relationship between the TM and the two PTMs examined here, we developed a JavaScript-supported "Protein Visualizer" that allows users to visualize these features in detail in individual proteins.

Protein PTMs are an actively studied area. For glycosylation alone, in addition to N-glycosylation, O-glycosylation, in which saccharides are added to the side chains of Ser and Thr, is another common PTM for MPs [54]. However, O-glycosylation is less strict in cellular compartmentalization than N-glycosylation, because O-GlcNAc modification, a special type of O-glycosylation, has been frequently observed in cytosolic and nuclear proteins [55,56]. Interestingly, phosphorylation can compete with O-GlcNAcylation to modify the same side chains of Ser and Thr in certain proteins for cell signaling [57]. How these PTMs contribute to the structure and stability of proteins is another interesting topic that can be further explored in the future. We hope that our efforts will stimulate more studies regarding the correlation between numerous known structural features, and more hidden relationships can be discovered to help better understand functions of MPs and to facilitate future protein engineering and drug design in MPs.

Supplementary Materials: The supporting information can be downloaded at: https://www.mdpi.com/article/10.3390/ijms242216182/s1.

Author Contributions: Conceptualization, B.S.; data analysis, M.D. and A.S.; software development, M.D., A.S., D.P. and S.R.C.; data curation, A.S. and B.S.; writing and editing the manuscript, M.D. and B.S.; visualization, M.D. and B.S.; project administration and funding acquisition, B.S. All authors have read and agreed to the published version of the manuscript.

Funding: This work is supported by the National Sciences and Engineering Research Council of Canada (RGPIN06073), the Canada Foundation of Innovation, the British Columbia Knowledge Development Fund, and the Digital Research Alliance of Canada.

Institutional Review Board Statement: Not applicable.

Informed Consent Statement: Not applicable.

Data Availability Statement: See Supplementary Materials and our website listed in the text.

Conflicts of Interest: The authors declare no conflict of interest.

References

1. Von Heijne, G. The membrane protein universe: What's out there and why bother? *J. Intern. Med.* **2007**, *261*, 543–557. [CrossRef] [PubMed]
2. Arinaminpathy, Y.; Khurana, E.; Engelman, D.M.; Gerstein, M.B. Computational analysis of membrane proteins: The largest class of drug targets. *Drug Discov. Today* **2009**, *14*, 1130–1135. [CrossRef] [PubMed]
3. Von Heijne, G. Membrane-protein topology. *Nat. Rev. Mol. Cell Biol.* **2006**, *7*, 909–918. [CrossRef] [PubMed]
4. White, S.H.; Wimley, W.C. MEMBRANE PROTEIN FOLDING AND STABILITY: Physical Principles. *Annu. Rev. Biophys. Biomol. Struct.* **1999**, *28*, 319–365. [CrossRef]
5. Feng, X.; Barth, P. A topological and conformational stability alphabet for multipass membrane proteins. *Nat. Chem. Biol.* **2016**, *12*, 167–173. [CrossRef] [PubMed]
6. Bill, R.M.; Henderson, P.J.F.; Iwata, S.; Kunji, E.R.S.; Michel, H.; Neutze, R.; Newstead, S.; Poolman, B.; Tate, C.G.; Vogel, H. Overcoming barriers to membrane protein structure determination. *Nat. Biotechnol.* **2011**, *29*, 335–340. [CrossRef] [PubMed]
7. Danmaliki, G.I.; Hwang, P.M. Solution NMR spectroscopy of membrane proteins. *Biochim. Biophys. Acta (BBA)-Biomembr.* **2020**, *1862*, 183356. [CrossRef]
8. McGilvray, P.T.; Anghel, S.A.; Sundaram, A.; Zhong, F.; Trnka, M.J.; Fuller, J.R.; Hu, H.; Burlingame, A.L.; Keenan, R.J. An ER translocon for multi-pass membrane protein biogenesis. *eLife* **2020**, *9*, e56889. [CrossRef]
9. Moss, K.; Helm, A.; Lu, Y.; Bragin, A.; Skach, W.R. Coupled translocation events generate topological heterogeneity at the endoplasmic reticulum membrane. *Mol. Biol. Cell* **1998**, *9*, 2681–2697. [CrossRef]
10. Almeida, J.G.; Preto, A.J.; Koukos, P.I.; Bonvin, A.M.; Moreira, I.S. Membrane proteins structures: A review on computational modeling tools. *Biochim. Biophys. Acta (BBA)-Biomembr.* **2017**, *1859*, 2021–2039. [CrossRef]
11. Kyte, J.; Doolittle, R.F. A simple method for displaying the hydropathic character of a protein. *J. Mol. Biol.* **1982**, *157*, 105–132. [CrossRef]
12. Cserzo, M.; Wallin, E.; Simon, I.; von Heijne, G.; Elofsson, A. Prediction of transmembrane alpha-helices in prokaryotic membrane proteins: The dense alignment surface method. *Protein Eng. Des. Sel.* **1997**, *10*, 673–676. [CrossRef]
13. Yuan, Z.; Mattick, J.S.; Teasdale, R.D. SVMtm: Support vector machines to predict transmembrane segments. *J. Comput. Chem.* **2004**, *25*, 632–636. [CrossRef] [PubMed]
14. Krogh, A.; Larsson, B.; von Heijne, G.; Sonnhammer, E.L. Predicting transmembrane protein topology with a hidden markov model: Application to complete genomes. *J. Mol. Biol.* **2001**, *305*, 567–580. [CrossRef] [PubMed]
15. Rost, B.; Fariselli, P.; Casadio, R. Topology prediction for helical transmembrane proteins at 86% accuracy–Topology prediction at 86% accuracy. *Protein Sci.* **1996**, *5*, 1704–1718. [CrossRef]
16. Krogh, A.; Tsirigos, K.D.; Pedersen, M.D.; Almagro Armenteros, J.J.; Marcatili, P.; Nielsen, H.; Hallgren, J.; Winther, O. Deep TMHMM predicts alpha and beta transmembrane proteins using deep neural networks. *Biorxiv* **2022**, *4*, 487609. [CrossRef]
17. Walsh, C.T.; Garneau-Tsodikova, S.; Gatto, G.J., Jr. Protein Posttranslational Modifications: The Chemistry of Proteome Diversifications. *Angew. Chem. Int. Ed.* **2005**, *44*, 7342–7372. [CrossRef]
18. Dutta, D.; Mandal, C.; Mandal, C. Unusual glycosylation of proteins: Beyond the universal sequon and other amino acids. *Biochim. Biophys. Acta (BBA)-Gen. Subj.* **2017**, *1861*, 3096–3108. [CrossRef] [PubMed]
19. Van Geest, M.; Lolkema, J.S. Membrane topology and insertion of membrane proteins: Search for topogenic signals. *Microbiol. Mol. Biol. Rev.* **2000**, *64*, 13–33. [CrossRef] [PubMed]
20. Braunger, K.; Pfeffer, S.; Shrimal, S.; Gilmore, R.; Berninghausen, O.; Mandon, E.C.; Becker, T.; Förster, F.; Beckmann, R. Structural basis for coupling protein transport and N-glycosylation at the mammalian endoplasmic reticulum. *Science* **2018**, *360*, 215–219. [CrossRef]
21. Mohorko, E.; Glockshuber, R.; Aebi, M. Oligosaccharyltransferase: The central enzyme of N-linked protein glycosylation. *J. Inherit. Metab. Dis.* **2011**, *34*, 869–878. [CrossRef] [PubMed]
22. Bakshi, T.; Pham, D.; Kaur, R.; Sun, B. Hidden Relationships between *N*-Glycosylation and Disulfide Bonds in Individual Proteins. *Int. J. Mol. Sci.* **2022**, *23*, 3742. [CrossRef] [PubMed]
23. Patel, C.; Saad, H.; Shenkman, M.; Lederkremer, G.Z. Oxidoreductases in Glycoprotein Glycosylation, Folding, and ERAD. *Cells* **2020**, *9*, 2138. [CrossRef] [PubMed]
24. Lluis, M.W.; Godfroy, J.I.; Yin, H. Protein engineering methods applied to membrane protein targets. *Protein Eng. Des. Sel.* **2013**, *26*, 91–100. [CrossRef] [PubMed]
25. Sowlati-Hashjin, S.; Gandhi, A.; Garton, M. Dawn of a new era for membrane protein design. *BioDesign Res.* **2022**, *2022*, 9791435. [CrossRef]
26. Vorobieva, A.A. Principles and Methods in Computational Membrane Protein Design. *J. Mol. Biol.* **2021**, *433*, 167154. [CrossRef]
27. Rosenbaum, D.M.; Rasmussen, S.G.; Kobilka, B.K. The structure and function of G-protein-coupled receptors. *Nature* **2009**, *459*, 356–363. [CrossRef]

28. Sriram, K.; Insel, P.A. G Protein-Coupled Receptors as Targets for Approved Drugs: How Many Targets and How Many Drugs? *Mol. Pharmacol.* **2018**, *93*, 251–258. [CrossRef]
29. Tu, L.; Khanna, P.; Deutsch, C. Transmembrane segments form tertiary hairpins in the folding vestibule of the ribosome. *J. Mol. Biol.* **2014**, *426*, 185–198. [CrossRef]
30. MacKenzie, K.R. Folding and stability of α-helical integral membrane proteins. *Chem. Rev.* **2006**, *106*, 1931–1977. [CrossRef]
31. Yogurtcu, O.N.; Johnson, M.E. Cytosolic proteins can exploit membrane localization to trigger functional assembly. *PLOS Comput. Biol.* **2018**, *14*, e1006031. [CrossRef]
32. Jiang, D.; Shi, H.; Tonggu, L.; El-Din, T.M.G.; Lenaeus, M.J.; Zhao, Y.; Yoshioka, C.; Zheng, N.; Catterall, W.A. Structure of the Cardiac Sodium Channel. *Cell* **2020**, *180*, 122–134.e10. [CrossRef] [PubMed]
33. Mamathambika, B.S.; Bardwell, J.C. Disulfide-linked protein folding pathways. *Annu. Rev. Cell Dev. Biol.* **2008**, *24*, 211–235. [CrossRef]
34. Schwarz, F.; Aebi, M. Mechanisms and principles of N-linked protein glycosylation. *Curr. Opin. Struct. Biol.* **2011**, *21*, 576–582. [CrossRef]
35. Eitner, K.; Koch, U.; Gawęda, T.; Marciniak, J. Statistical distribution of amino acid sequences: A proof of Darwinian evolution. *Bioinformatics* **2010**, *26*, 2933–2935. [CrossRef]
36. Brooks, D.J.; Fresco, J.R.; Lesk, A.M.; Singh, M. Evolution of amino acid frequencies in proteins over deep time: Inferred order of introduction of amino acids into the genetic code. *Mol. Biol. Evol.* **2002**, *19*, 1645–1655. [CrossRef] [PubMed]
37. Cremers, C.M.; Jakob, U. Oxidant sensing by reversible disulfide bond formation. *J. Biol. Chem.* **2013**, *288*, 26489–26496. [CrossRef] [PubMed]
38. Liu-Shin, L.P.-Y.; Fung, A.; Malhotra, A.; Ratnaswamy, G. Evidence of disulfide bond scrambling during production of an antibody-drug conjugate. *mAbs* **2018**, *10*, 1190–1199. [CrossRef] [PubMed]
39. Resemann, A.; Liu-Shin, L.; Tremintin, G.; Malhotra, A.; Fung, A.; Wang, F.; Ratnaswamy, G.; Suckau, D. Rapid, automated characterization of disulfide bond scrambling and IgG2 isoform determination. *mAbs* **2018**, *10*, 1200–1213. [CrossRef] [PubMed]
40. Ilbert, M.; Horst, J.; Ahrens, S.; Winter, J.; Graf, P.C.F.; Lilie, H.; Jakob, U. The redox-switch domain of Hsp33 functions as dual stress sensor. *Nat. Struct. Mol. Biol.* **2007**, *14*, 556–563. [CrossRef]
41. Sun, B.; Ma, L.; Yan, X.; Lee, D.; Alexander, V.; Hohmann, L.J.; Lorang, C.; Chandrasena, L.; Tian, Q.; Hood, L. N-Glycoproteome of E14.Tg2a Mouse Embryonic Stem Cells. *PLoS ONE* **2013**, *8*, e55722. [CrossRef]
42. Sun, B.; Ranish, J.A.; Utleg, A.G.; White, J.T.; Yan, X.; Lin, B.; Hood, L. Shotgun Glycopeptide Capture Approach Coupled with Mass Spectrometry for Comprehensive Glycoproteomics. *Mol. Cell. Proteom.* **2007**, *6*, 141–149. [CrossRef] [PubMed]
43. Sezgin, E.; Levental, I.; Mayor, S.; Eggeling, C. The mystery of membrane organization: Composition, regulation and roles of lipid rafts. *Nat. Rev. Mol. Cell Biol.* **2017**, *18*, 361–374. [CrossRef]
44. Nezhady, M.A.M.; Rivera, J.C.; Chemtob, S. Location Bias as Emerging Paradigm in GPCR Biology and Drug Discovery. *iScience* **2020**, *23*, 101643. [CrossRef]
45. Alberti, S. Phase separation in biology. *Curr. Biol.* **2017**, *27*, R1097–R1102. [CrossRef]
46. Medus, M.L.; Gomez, G.E.; Zacchi, L.F.; Couto, P.M.; Labriola, C.A.; Labanda, M.S.; Bielsa, R.C.; Clérico, E.M.; Schulz, B.L.; Caramelo, J.J. N-glycosylation Triggers a Dual Selection Pressure in Eukaryotic Secretory Proteins. *Sci. Rep.* **2017**, *7*, 8788. [CrossRef]
47. Denic, V.; Dötsch, V.; Sinning, I. Endoplasmic reticulum targeting and insertion of tail-anchored membrane proteins by the get pathway. *Cold Spring Harb. Perspect. Biol.* **2013**, *5*, a013334. [CrossRef] [PubMed]
48. Vogt, G.; Chapgier, A.; Yang, K.; Chuzhanova, N.; Feinberg, J.; Fieschi, C.; Boisson-Dupuis, S.; Alcais, A.; Filipe-Santos, O.; Bustamante, J.; et al. Gains of glycosylation comprise an unexpectedly large group of pathogenic mutations. *Nat. Genet.* **2005**, *37*, 692–700. [CrossRef]
49. Apweiler, R.; Hermjakob, H.; Sharon, N. On the frequency of protein glycosylation, as deduced from analysis of the SWISS-PROT database. *Biochim. Biophys. Acta (BBA)-Gen. Subj.* **1999**, *1473*, 4–8. [CrossRef]
50. Tsai, C.-H.; Chan, C.-H.; Chen, B.-J.; Kao, C.-Y.; Liu, H.-L.; Hsu, J.-P. Bioinformatics approaches for disulfide connectivity prediction. *Curr. Protein Pept. Sci.* **2007**, *8*, 243–260. [CrossRef]
51. Tüdős, E.; Mészáros, B.; Fiser, A.; Simon, I. A word of caution about biological inference—Revisiting cysteine covalent state predictions. *FEBS Open Bio* **2014**, *4*, 310–314. [CrossRef]
52. Omasits, U.; Ahrens, C.H.; Müller, S.; Wollscheid, B. Protter: Interactive protein feature visualization and integration with experimental proteomic data. *Bioinformatics* **2014**, *30*, 884–886. [CrossRef]
53. Sherman, B.T.; Hao, M.; Qiu, J.; Jiao, X.; Baseler, M.W.; Lane, H.C.; Imamichi, T.; Chang, W. DAVID: A web server for functional enrichment analysis and functional annotation of gene lists (2021 update). *Nucleic Acids Res.* **2022**, *50*, W216–W221. [CrossRef] [PubMed]
54. Reily, C.; Stewart, T.J.; Renfrow, M.B.; Novak, J. Glycosylation in health and disease. *Nat. Rev. Nephrol.* **2019**, *15*, 346–366. [CrossRef] [PubMed]
55. Bond, M.R.; Hanover, J.A. A little sugar goes a long way: The cell biology of O-GlcNAc. *J. Cell Biol.* **2015**, *208*, 869–880. [CrossRef]

56. Guinez, C.; Morelle, W.; Michalski, J.-C.; Lefebvre, T. O-GlcNAc glycosylation: A signal for the nuclear transport of cytosolic proteins? *Int. J. Biochem. Cell Biol.* **2005**, *37*, 765–774. [CrossRef] [PubMed]
57. Hart, G.W.; Slawson, C.; Ramirez-Correa, G.; Lagerlof, O. Cross talk between o-glcnacylation and phosphorylation: Roles in signaling, transcription, and chronic disease. *Annu. Rev. Biochem.* **2011**, *80*, 825–858. [CrossRef]

International Journal of
Molecular Sciences

MDPI

Review

Structures, Mechanisms, and Physiological Functions of Zinc Transporters in Different Biological Kingdoms

Han Ba Bui [1,2,3] and Kenji Inaba [1,2,3,4,5,*]

1 Institute of Multidisciplinary Research for Advanced Materials, Tohoku University, Sendai 980-8577, Japan; bui.ba.han.e2@tohoku.ac.jp
2 Department of Molecular and Chemical Life Sciences, Graduate School of Life Sciences, Tohoku University, Sendai 980-8577, Japan
3 Medical Institute of Bioregulation, Kyushu University, Fukuoka 812-8582, Japan
4 Department of Chemistry, Graduate School of Science, Tohoku University, Sendai 980-8578, Japan
5 Core Research for Evolutional Science and Technology (CREST), Japan Agency for Medical Research and Development (AMED), Chiyoda-ku, Tokyo 100-0004, Japan
* Correspondence: kenji.inaba.a1@tohoku.ac.jp; Tel.: +81-22-217-5604; Fax: +81-22-217-5605

Abstract: Zinc transporters take up/release zinc ions (Zn^{2+}) across biological membranes and maintain intracellular and intra-organellar Zn^{2+} homeostasis. Since this process requires a series of conformational changes in the transporters, detailed information about the structures of different reaction intermediates is required for a comprehensive understanding of their Zn^{2+} transport mechanisms. Recently, various Zn^{2+} transport systems have been identified in bacteria, yeasts, plants, and humans. Based on structural analyses of human ZnT7, human ZnT8, and bacterial YiiP, we propose updated models explaining their mechanisms of action to ensure efficient Zn^{2+} transport. We place particular focus on the mechanistic roles of the histidine-rich loop shared by several zinc transporters, which facilitates Zn^{2+} recruitment to the transmembrane Zn^{2+}-binding site. This review provides an extensive overview of the structures, mechanisms, and physiological functions of zinc transporters in different biological kingdoms.

Keywords: zinc transporter; ZnT; YiiP; histidine-rich loop; cryo-EM

1. Introduction

Zinc ions (Zn^{2+}), an essential trace element in bacteria, fungi, plants, and animals, including humans [1], serve as a key component in many signal transduction processes and act as an essential cofactor for many proteins and enzymes [2,3]. Zinc deficiency causes several human diseases [4–13]; indeed, zinc supplements have beneficial effects on human health [8,14–19]. However, excessive adsorption of Zn^{2+} leads to disruption of the gastrointestinal flora balance, deficiency of other essential heavy metals, including iron, copper, and manganese, and reduction in immune function [20–23]. Zn^{2+} also plays an important role in the physiology of organisms such as plants and bacteria [24,25]. In plants, zinc deficiency is linked to growth defects and inhibition of flowering [26,27]. Additionally, Zn^{2+} is responsible for the virulence of some bacteria [28]. Since Zn^{2+} is involved in numerous biological events, humans, plants, yeasts, and bacteria have evolved elaborate Zn^{2+} transport systems that respond to Zn^{2+} perturbation.

Failure of the Zn^{2+} transport systems plays a role in diseases such as cancer [29,30], Alzheimer's [31,32], and Parkinson's [33,34], as well as temporary zinc deficiency in newborns [35], perinatal fatal cardiomyopathy [36], risk of febrile seizures [37], Lowe's syndrome [38], disorders of muscle tone with polycythemia [39,40], and chronic liver disease [40]. Therefore, human zinc transporters (ZnTs) are potential targets of drugs and preclinical diagnostic tests. Owing to the important physiological roles, and pharmacological and preclinical diagnostic significance of Zn^{2+} transport systems, a variety of biochemical, structural, physiological, and genetic experiments have been carried out over the past several decades to better understand their functions and mechanisms. The most comprehensively studied bacterial zinc transporter is YiiP, which works in *Escherichia coli* and *Shewanella oneidensis* (EcYiiP and SoYiiP, respectively) [41–49]. These transporters are a convenient model to study the general mechanisms underlying Zn^{2+} transport. The most intensively studied mammalian ZnTs are SLC30A7/ZnT7 [50] and SLC30A8/ZnT8 [51,52]. Our interests in ZnT family members stem mainly from their roles in maintaining Zn^{2+} homeostasis in cellular organelles throughout the body and the fact that their dysfunction causes serious diseases.

As is the case for other membrane transporters, ZnTs undergo conformational conversion to transport Zn^{2+} across biological membranes. To fully understand the mechanism underlying Zn^{2+} transport, high-resolution structures of the transporters have been captured in different states. The first X-ray crystal structure of a zinc transporter (Table 1) was reported for EcYiiP [41,42], followed by the EM structure of SoYiiP [43–46]. More recently, cryo-EM structures of vertebrate ZnTs have been reported (Table 1); these include *Homo sapiens* ZnT7 (HsZnT7) [50], *Homo sapiens* ZnT8 (HsZnT8) [51], and *Xenopus tropicalis* ZnT8 (XtZnT8) [52]. These structures allow us to propose an updated model of ZnTs-mediated Zn^{2+} transport. Of note, our recent structural and biochemical studies on HsZnT7 revealed the role of its cytosolic histidine-rich loop (His-loop) in efficient Zn^{2+} uptake [50]. Thus, we have built on the structural and mechanistic foundations of ZnTs in the biological kingdom, while making significant progress regarding research into other members with Zn^{2+} transport functions.

Table 1. X-ray and cryo-EM structure of zinc transporters (ZnTs).

Proteins	Main Functions	Organisms	States	Conformations (PDB Code)	Ligands	Methods	References
YiiP	Transport Zn^{2+} out of the cytoplasm and into the periplasm	*Escherichia coli*	Homodimer	Outward-facing (2QFI, 3H90)	Zn^{2+}	X-ray diffraction	[41,42]
		Shewanella oneidensis	Homodimer	Inward-facing (3J1Z, 5VRF, 7KZZ [(1)])	Zn^{2+}	Electron microscopy	[44–46]
			Homodimer	Inward-facing occluded (7KZX)	Zn^{2+}		[43]
ZnT7	Transport Zn^{2+} out of the cytoplasm and into the Golgi lumen	*Homo sapiens*	Homodimer	Outward-facing (8J7T)	Apo	Electron microscopy	[50]
			Homodimer	Outward-facing (8J7U)	Zn^{2+}		
			Heterodimer	Inward-facing and outward-facing (8J7V [(2)])	Apo		
			Heterodimer	Inward-facing with Zn^{2+} and outward-facing (8J80 [(3)])	Zn^{2+}, Apo		
			Heterodimer	Inward-facing with Zn^{2+} and outward-facing with Zn^{2+} (8J7W) [(4)]	Zn^{2+}		

Table 1. *Cont.*

Proteins	Main Functions	Organisms	States	Conformations (PDB Code)	Ligands	Methods	References
ZnT8	Transport Zn^{2+} out of the cytoplasm and into the insulin secretory granule	*H. sapiens*	Homodimer	Outward-facing (6XPE)	Zn^{2+}	Electron microscopy	[51]
			Heterodimer	Outward-facing and inward-facing (6XPF)	Apo		
		Xenopus tropicalis	Homodimer	Outward-facing (7Y5G)	Zn^{2+}		[52]
			Homodimer	Outward-facing (7Y5H [(5)])	Apo		

[(1)] This structure was observed in the presence of 0.5 mM EDTA. [(2)] This structure was observed in the absence of Zn^{2+}. [(3)] This structure was observed in the presence of 10 μM Zn^{2+}. [(4)] This structure was observed with addition of 200 and 300 μM Zn^{2+}. [(5)] This structure was observed at low pH.

2. Zn^{2+} Transport Systems in Prokaryotes and Eukaryotes

Prokaryotes and eukaryotes have developed a variety of Zn^{2+} transport systems to promote the uptake or efflux of Zn^{2+} across biological membranes. ZnTs can be divided into three major groups depending on the mode of transport: Uniporters that transport Zn^{2+} alone; symporters that transport Zn^{2+} in the same direction as other ions, such as protons; and antiporters that transport Zn^{2+} and another ion in opposite directions, such that the binding of one is concomitant with the release of the other. In general, uniporters require no external energy input and transport specific molecules along their concentration gradients; they are therefore passive transporters. However, it can also act as an active transporter if the transport process is against the concentration gradient. By contrast, symporters and antiporters use the energy stored in the concentration gradient of another ion, in many cases, a proton, to transport specific molecules against their concentration gradients. In this regard, symporters and antiporters can be regarded as active transporters. In addition, some P-ATPases and ABC transporters transport Zn^{2+} using ATP as an external energy source to overcome the Zn^{2+} concentration gradient.

Zinc transporters (ZnTs) and ZRT- and IRT-related proteins (ZIPs) are the two major Zn^{2+} transport families found universally in bacteria, yeasts, plants, and animals, including humans. ZnTs and ZIPs selectively transport Zn^{2+}, but in opposite directions: ZnTs export Zn^{2+} from the cytoplasm, whereas ZIPs import Zn^{2+} into the cytoplasm. Thus, ZnTs and ZIPs play important roles in maintaining homeostasis of intracellular and intra-organelle Zn^{2+} levels.

While ZntB from *Escherichia coli* (EcZntB) acts as a Zn^{2+}/H^+ symporter [53], many ZnTs function as proton-driven antiporters, exchanging H^+ in the extracellular space or organelle lumens for Zn^{2+} in the cytoplasm [41–55]. By contrast, there is no clear evidence that ZIPs use proton energy flux to transport Zn^{2+} across the membranes. However, recent biochemical studies suggest that, like ZnTs, *Bordetella bronchiseptica* ZIP (BbZIP) may function as a Zn^{2+}/H^+ antiporter [56].

3. ZnTs

ZnTs belong to the cation diffusion facilitator (CDF) family, which can be classified into three groups: Zn-CDFs, Zn/Fe-CDFs, and Mn-CDFs [57,58]. Zn-CDFs consist of Zn^{2+} and Co^{2+} transporters, including ZitB-like, ZnT1-like, and Zrc1-like proteins. The ZitB-like clusters are from *E. coli*. The ZnT1-like clusters include only metazoans. The Zrc1-like cluster includes only fungal CDFs originating from Ascomycetes, Basidiomycetes, and Zygomycetes. Zn/Fe-CDFs are cation-efflux pumps that transport Fe^{2+} or Zn^{2+}, and also Co^{2+}, Cd^{2+}, and Ni^{2+}. Mn-CDFs include metal tolerance proteins (MTPs) from plants.

3.1. Mammalian ZnTs

Ten ZnTs (ZnTs 1–10) have been identified in mammals, including humans [59,60]. All ZnTs are Zn-CDF members, although ZnT10 is more likely a manganese transporter [59–61]. Based on their amino acid sequence similarities, ZnTs are divided into four subgroups: Group 1 includes ZnT5 and ZnT7; group 2 includes ZnT2-ZnT4 and ZnT8; group 3 includes ZnT1 and ZnT10; and group 4 includes ZnT6 and ZnT9 [60]. Most ZnTs form a homodimer composed of the same protomers [50–52], whereas ZnT5 and ZnT6 form a heterodimer including two different protomers [62], and all are located on the plasma or organelle membranes, where they control intracellular and extracellular Zn^{2+} balance [59,63]. Specifically, ZnT7 transports Zn^{2+} into the lumen of the pre-*cis*- and *cis*-Golgi, whereas ZnT5/6 and ZnT4 transport Zn^{2+} into the lumen of the *medial*- and *trans*-Golgi [64]. ZnT7 and ZnT5/6 are responsible for the Golgi-to-ER retrograde transport of the ER chaperone ERp44 [64]. This system is involved in the maturation and activation of some secretory proteins during transport through the early secretory pathway [65].

3.2. Plant ZnTs

Metal tolerance proteins (MTPs) are bivalent cationic transporters in plants that play crucial roles in metal tolerance and homeostasis in metal non-hyperaccumulators (e.g., *Arabidopsis thaliana*) and hyperaccumulators (e.g., *Arabidopsis halleri* and *Noccaea caerulescens*) [66]. MTPs are classified into seven groups based on their amino acid sequence similarities [67]. Thus, plant MTPs are very diverse so as to satisfy the need to absorb or detoxify specific metals. *A. thalaina* has 12 MTPs, while *P. trichocarpa* MTP has up to 22 MTP genes [68]. In *A. thaliana*, AtMTP1 and AtMTP3 ZnTs localized on the vacuole membrane maintain Zn^{2+} homeostasis [69–71]. AtMTP1 and AtMTP3 are involved in the sequestration of excess cytoplasmic Zn^{2+} into vacuoles [71]. Whereas AtMTP1 is more ubiquitously expressed, expression of AtMTP3 is restricted to the root epidermis and cortex [69,72]. Like mammalian ZnT5 and ZnT6, AtMTP5 and AtMTP12 form a heterodimer at the Golgi membrane and transport Zn^{2+} into the Golgi lumen [73].

3.3. Yeast ZnTs

Our understanding of ZnTs in yeast derives primarily from *Saccharomyces cerevisiae*. In *S. cerevisiae*, vacuolar ZnTs ZRC1 and COT1 act as Zn^{2+}/H^+ antiporters and regulate Zn^{2+} homeostasis by transporting and storing Zn^{2+} in the vacuole [74,75]. ScZRC1 senses Zn^{2+} availability in the cytosol, possibly through the histidine-repeat motifs, and transports Zn^{2+} from the cytosol to the vacuole when cytosolic Zn^{2+} is abundant, thereby conferring resistance to Zn^{2+} toxicity [76,77].

S. cerevisiae also possesses Msc2 and Zrg17, which transport Zn^{2+} from the nucleus and ER to the cytoplasm [78]. ScMsc2 and ScZrg17 interact physically to form a heterodimer and likely serve to maintain the Zn^{2+} levels in the ER of Zn^{2+}-adequate cells [79–81]. *Schizosaccharomyces pombe* also has a zinc transporter, called ZHF1, which maintains Zn^{2+} homeostasis in the ER and nucleus and sequesters Cd^{2+} into the ER [82]. The structures of yeast ZnTs have not yet been reported. While ScZRC1, ScCOT1, and ScZrg17 are predicted to have six transmembrane (TM) helices, ScMsc2 is presumed to contain up to 16 TM helices.

3.4. Bacterial ZnTs

Bacterial ZnTs YiiP, ZitB, and CzcD have been functionally characterized. Insight into the structural features and Zn^{2+} transport mechanisms of bacterial ZnTs comes primarily from YiiP. YiiP was first identified in *Escherichia coli* [83]. In vitro, YiiP also binds Hg^{2+}, Co^{2+}, Ni^{2+}, Mn^{2+}, Ca^{2+}, and Mg^{2+} but is unlikely to transport them efficiently [84]. Like mammalian ZnTs, YiiP functions as a Zn^{2+}/H^+ antiporter [43,48].

Other ZnTs have been identified recently in bacteria. ZitB conducts Zn^{2+} efflux across the cytoplasmic membrane, thereby reducing Zn^{2+} accumulation in the cytoplasm and rendering bacteria more resistant to Zn^{2+} [85]. By contrast, ZntA, a Zn^{2+}-transporting

P-ATPase, is required for growth at more toxic concentrations [85]. CzcD is a Cd^{2+}, Co^{2+}, and Zn^{2+}/H^{+}-K^{+} antiporter involved in maintaining intracellular divalent cation and potassium homeostasis through active efflux of Zn^{2+}, Cd^{2+}, and Co^{2+} in exchange for K^{+} and protons [86].

4. Structural Basis of ZnTs

In general, ZnTs adopt inward-facing (IF) or outward-facing (OF) forms and undergo conversion between these two forms during Zn^{2+} transport (Figure 1A). The IF form creates an open cavity for Zn^{2+} recruitment from the cytoplasm to the TM metal-binding site. Conversely, in the OF form, an open cavity is formed on the extracellular or organelle luminal side to release Zn^{2+} (Figure 1A). Unlike ZnTs, ZIPs are supposed to operate by a mechanism named elevator-type transport. Membrane transporters using this mechanism commonly consist of moving and fixed domains. Switching between the outward- and inward-facing forms involves the sliding of the entire moving domain through the bilayer as a rigid body. The substrate-binding site translocates some distance across the bilayer during transport, along with the moving domain [87].

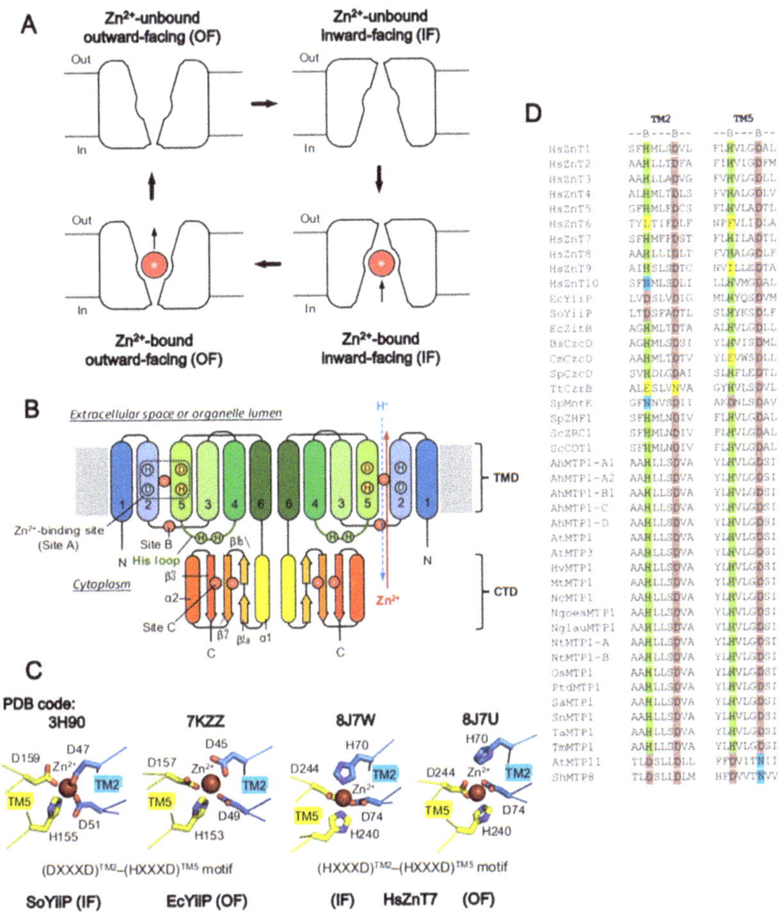

Figure 1. Molecular architecture of zinc transporters (ZnTs). (A) Two conformations of ZnTs, inward-facing (IF) and outward-facing (OF), with the metal-binding site open to the cytoplasmic side (In) and extracellular or organelle luminal side (Out), respectively. ZnTs convert between these two forms

during the Zn^{2+} transport cycle. (**B**) Typical topology model of a ZnT homodimer in which two transmembrane domains (TMDs) and two C-terminal domains (CTDs) tightly contact each other. The Zn^{2+}-binding sites are labeled as site A in the TMD, site B at the interface of the TMD and CTD, and site C in the CTD. Red and blue arrows indicate the directions of Zn^{2+} and H^+ transports, respectively. For clarity, the arrows are shown only in a right protomer. (**C**) Zn^{2+} coordination structures of bacterial YiiP (PDB code: 3H90 for SoYiiP in the IF form and 7KZZ for EcYiiP in the OF form) and human ZnT7 (PDB code: 8J7W for the IF form and 8J7U for OF form). Red spheres indicate bound Zn^{2+}. "A" in the red sphere indicates Zn^{2+} bound to site A. (**D**) Sequence alignment of Zn^{2+} coordinating residues at site A. Light green and light pink highlight conserved His and Asp residues, respectively. Yellow highlights loss of the conserved His and Asp residues. Cyan highlights conserved Asn residues required for Mn^{2+} transport.

The first X-ray crystal structure of ZnTs was solved for EcYiiP in a Zn^{2+}-bound OF state [41,42]. Subsequently, cryo-EM single-particle analysis identified the structure of a Zn^{2+}-bound IF state for SoYiiP [44,46] (Table 1). More recently, the cryo-EM structures of vertebrate ZnTs human ZnT7 (HsZnT7) [50], human ZnT8 (HsZnT8) [51], and Xenopus ZnT8 (XtZnT8) [52] were reported (Table 1), revealing that these ZnTs can adopt both OF-OF homodimeric and IF-OF heterodimeric conformations [50–52]. The presence of the homodimeric and heterodimeric conformations may suggest that the two protomers work independently during Zn^{2+} transport [50–52], although more detailed studies are required to draw a firm conclusion.

All Zn-CDFs share common structural folds. The structural core consists of six TM helices (TM1-TM6) in the TM domain (TMD), and a cytosolic domain (CTD) with a ferredoxin-like fold having an $\alpha\beta\beta\alpha\beta$ secondary structure topology (Figure 1B). The Zn^{2+}-binding site is formed by an HXXXD motif located on TM2 and TM5, individually, near the center of the TMD (Figure 1C). In some Zn-CDFs, the His (H) and Asp (D) residues on the motifs are replaced by Asp (D) and Asn (N), respectively (Figure 1D). Some eukaryotic Zn-CDFs possess a histidine-rich loop (His-loop) flanked by TM4 and TM5. The length of the His loop, and the number and distribution of His residues in the loop, vary among the Zn-CDFs (Figure 2).

Figure 2. Amino acid sequence of the histidine-rich loop (His-loop) of ZnTs. Some ZnTs have no His-loop (**A**), while others have a short (**B**) or exceptionally long His-loops (**C**). Histidine residues are highlighted in light green. The C-terminal part of TM4 and the N-terminal part of TM5 are shown in

orange. Bold numbers on the right denote the number of histidine residues contained in the His-loop. Accession numbers are as follows: *Escherichia coli* YiiP (EcYiiP, UniProt code: P69380), *Shewanella oneidensis* YiiP (SoYiiP, UniProt code: Q8E919), *Streptococcus pneumoniae* MntE (SpMntE, UniProt code: Q8DP19S), *Homo sapiens* ZnT6 (HsZnT6, UniProt code: Q6NXT4), *Homo sapiens* ZnT9 (HsZnT9, UniProt code: Q6PML9), *Homo sapiens* ZnT10 (HsZnT10, UniProt code: Q6XR72), *Arabidopsis thaliana* MTP5 (AtMTP5, UniProt code: Q6ICY4), *Arabidopsis thaliana* MTP6 (AtMTP6, UniProt code: Q8L725), *Arabidopsis thaliana* MTP11 (AtMTP11, UniProt code: O80632), *Homo sapiens* ZnT1 (HsZnT1, UniProt code: Q9Y6M5), *Homo sapiens* ZnT2 (HsZnT2, UniProt code: Q9BRI3), *Homo sapiens* ZnT3 (HsZnT3, UniProt code: Q99726), *Homo sapiens* ZnT4 (HsZnT4, UniProt code: O14863), *Homo sapiens* ZnT8 (HsZnT8, UniProt code: Q8IWU4), *Homo sapiens* ZnT5 (HsZnT5, UniProt code: Q8TAD4), *Homo sapiens* ZnT7 (HsZnT7, UniProt code: Q8NEW0), *Arabidopsis thaliana* MTP1 (AtMTP1, UniProt code: Q9ZT63), *Oryza sativa* MTP1 (OsMTP1, UniProt code: Q688R1), *Schizosaccharomyces pombe* ZHF1 (SpZHF1, UniProt code: O13918), *Saccharomyces cerevisiae* ZRC1 (ScZRC1, UniProt code: P20107), *Saccharomyces cerevisiae* COT1 (ScCOT1, UniProt code: P32798), *Hordeum vulgare* MTP1 (HvMTP1, UniProt code: A0JJL9), and *Noccaea goesingensis* MTP1 (NgMTP1, UniProt code: Q6Q4F7).

As aforementioned, all ZnTs belong to the SLCA30 family, allowing us to compare the conformational details between them. ZnTs show different TM helix arrangements in both the IF and OF forms. In the IF form of bacterial YiiP and HsZnT8, TM4 and TM5 largely swing to the outside on the cytosolic side (relative to their positions in the OF form), using their luminal ends as pivot points (Figure 3A,B). By contrast, TM2 moves slightly toward the center [51]. Notably, in the IF form of HsZnT7, TM5 kinks at the middle, and its cytosolic half is largely bent toward the outside, resulting in the very open cytosolic cavity (Figure 3C) [50]. In this form, part of the His loop is integrated into the cytosolic cavity, and His164 in this loop is directly coordinated to Zn^{2+} along with His70, Asp74, and Asp244 (Figure 3D) [50]. The detailed mechanistic and functional roles of the His-loop are discussed later.

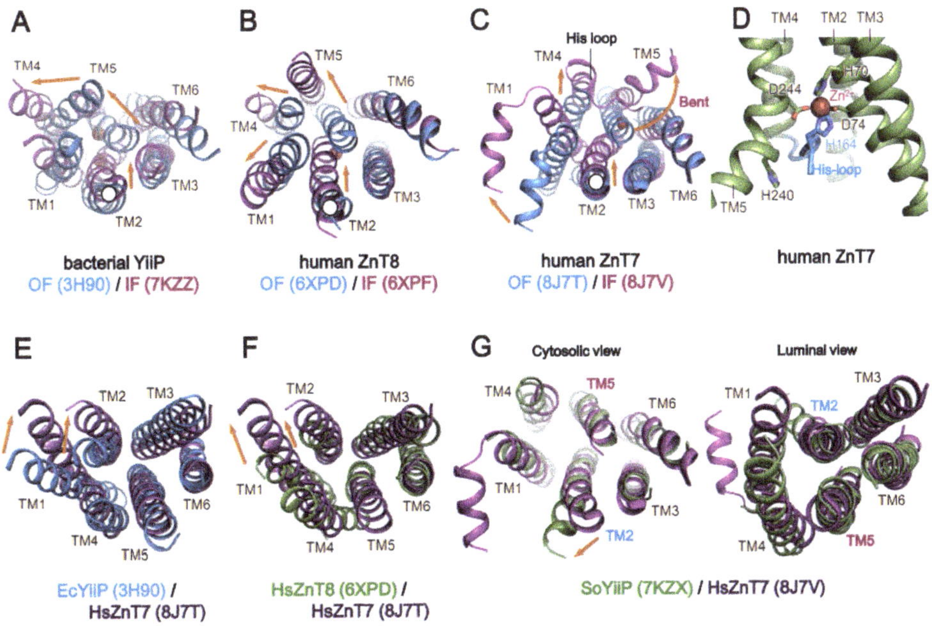

Figure 3. Rearrangement of the transmembrane helix of ZnTs during transition between the OF and IF forms. (**A–C**) Superimpositions of the OF (blue) and IF (magenta) forms of bacterial YiiP (PDB codes:

3H90 and 7KZZ, respectively), human ZnT8 (PDB codes: 6XPD and 6XPF, respectively), and human ZnT7 (PDB codes: 8J7T and 8J7V, respectively), viewed from the cytosolic side. (**D**) Integration of the His-loop of HsZnT7 into the widely open cytosolic cavity, viewed from the side. (**E**) Superimposition of the OF forms of EcYiiP (blue, PDB code: 3H90) and HsZnT7 (violet, PDB code: 8J7T), viewed from the luminal side. (**F**) Superimposition of the OF forms of HsZnT8 (green, PDB code: 6XPD) and HsZnT7 (violet, PDB code: 8J7T), viewed from the luminal side. (**G**) Superimposition of the Zn^{2+}-unbound IF states of HsZnT7 (violet, PDB code: 8J7V) and SoYiiP (green, PDB code: 7KZX), viewed from the cytosolic side. The cytosolic domains and TM helix loops are omitted for clarity. Orange arrows indicate the movement of TM helices during the conversion from the OF to the IF state. TM, transmembrane.

There are also significant differences in the OF forms of ZnTs with a known structure. The OF form of HsZnT7 has a wider cavity at the luminal side than that of HsZnT8 and EcYiiP due to the more "outside" positions of TM1 and TM2 (Figure 3E,F). An additional conformation has been identified for Zn^{2+}-unbound HsZnT7, in which TM5 packs tightly with TM2, TM3, and TM6 on both the cytoplasmic and luminal sides (Figure 3G) [50]. The TM helix arrangement in this state is highly superimposable to that of a previously reported occluded state of Zn^{2+}-unbound SoYiiP (Figure 3G) [44]. Although the physiological relevance of the Zn^{2+}-unbound occluded state remains unclear, this state may be formed after the release of Zn^{2+} to the luminal or periplasmic side and before complete conversion to the IF state with a widely open cytosolic cavity. Hereafter, we refer to this state as the "IF resting state" in this review article.

5. Zn^{2+}-Binding Sites on ZnTs

Zn^{2+}-Binding Sites and Metal Ion Selectivity

Metal specificity is an important functional feature of all metal transporters. X-ray crystallographic and cryo-EM analyses revealed one to three Zn^{2+}-binding sites (site A, site B, and site C) in ZnTs. All ZnTs commonly possess a Zn^{2+}-binding site in the TMD (site A). Zn^{2+} binds transiently to site A before effluxing to the other side, indicating that site A is located on the Zn^{2+}-translocation pathway. Site A contains highly conserved Zn^{2+}-binding motifs formed by the BXXXB motif on TM2 and TM5, in which B is His (H), Asp (D), Asn (N), or Glu (E), and X can be any residues (Figure 1C,D). There are, however, some differences in the sequence of site A (Figure 1D). While human, plant, and yeast ZnTs conserve the $(HXXXD)^{TM2}$-$(HXXXD)^{TM5}$ motif, different amino acid sequences are seen in the motifs of bacterial ZnTs; for instance, $(DXXXD)^{TM2}$-$(HXXXD)^{TM5}$ in EcYiiP and SoYiiP, and $(HXXXD)^{TM2}$-$(HXXXD)^{TM5}$ in EcZitB and CzcD (Figure 1D). Notably, mammalian ZnT6 loses the BXXXB motif in the TMD, and does not therefore have Zn^{2+} transport activity [62]. Mammalian ZnT10 harboring $(NXXXD)^{TM2}$-$(HXXXD)^{TM5}$ transports Mn^{2+} rather than Zn^{2+} (Figure 1D) [88–90]. AtMTP11 and ShMTP8 harbor $(DXXXD)^{TM2}$-$(DXXXN)^{TM5}$ and transport both Mn^{2+} and Cu^{2+} (Figure 1D) [66]. Thus, not all ZnTs transport Zn^{2+} exclusively.

Site B is less conserved at the TMD-CTD interface (Figures 1B and 4A). In bacterial YiiP, this site is located in the TM2-TM3 loop, which contains a DHH motif (Figure 4B) [42–44]. In HsZnT8, this site is constituted by His residues from the TM2-TM3 loop in the TMD and from the α2-β3 loop in the CTD, and it has low affinity for Zn^{2+} [51]. While the functional role of site B is unclear in HsZnT8, bacterial YiiP participates directly in Zn^{2+} transport [43,44]. Since site B is positioned near the cytosolic Zn^{2+} entry gate, mutation of two His residues at site B reduces the Zn^{2+} transport activity of bacterial YiiP [43]; thus, site B is thought to trap Zn^{2+} in the cytosol, thereby facilitating Zn^{2+} transport [43,51]. Although the structure of AtMTP1 has not been solved, homology modeling based on the crystal structure of EcYiiP suggests that it lacks site B (Figure 4B). However, mutations of some residues in the TM2-TM3 loop impair the Zn^{2+} transport activity of AtMTP1, indicating that the TM2-TM3 loop itself is essential for AtMTP1 function [91].

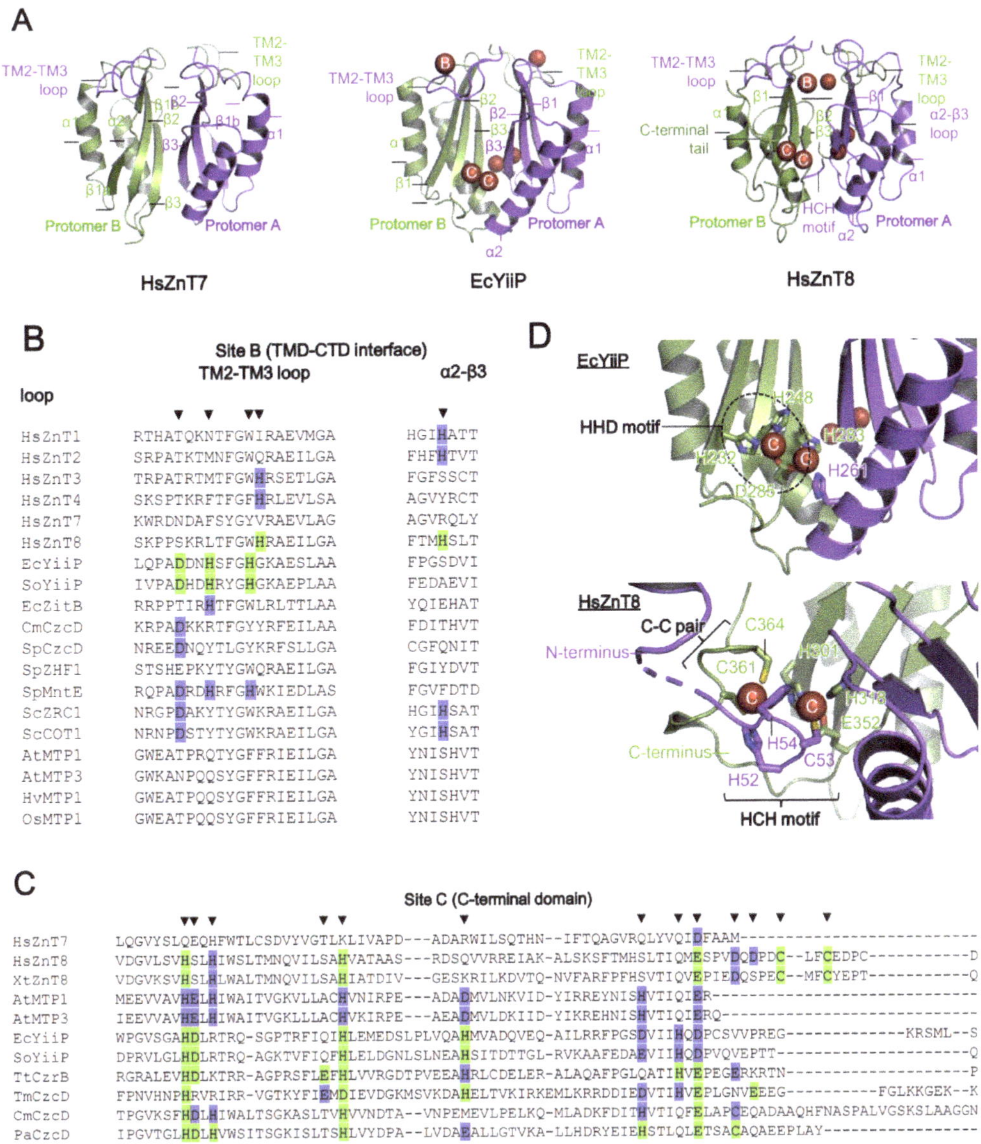

Figure 4. Zn^{2+}-binding sites in the C-terminal domain of ZnTs. (**A**) The C-terminal domain (CTD) of dimeric HsZnT7 (**left**), EcYiiP (**middle**), and HsZnT8 (**right**). Red spheres indicate Zn^{2+} ions. "B" and "C" in the red spheres indicate Zn^{2+} bound to site B and site C, respectively. Note that HsZnT7 contains no Zn^{2+} binding sites in the CTD. (**B**) Sequence alignment of representative ZnT-family members from human, bacteria, and plants around site B, located at the TMD-CTD interface. Zn^{2+}-coordinating residues confirmed by structure analysis are highlighted in light green and marked by black down-pointing triangles. Light violet highlights His and Asp residues that are predicted to be coordinated with Zn^{2+}. (**C**) Sequence alignment of representative ZnT-family members around site C in the CTD. Zn^{2+}-coordinating residues identified by structure analyses are highlighted in light green

and marked by black down-pointing triangles. Light violet highlights residues to be involved in Zn^{2+} coordination. Note that HsZnT7 lacks site B and site C. TM, transmembrane. (**D**) Closed-up views of Zn^{2+}-binding sites around the CTD (site C) in EcYiiP (upper) and HsZnT8 (lower). Bound Zn^{2+} ions are represented by red spheres. Black circle indicates the HHD motif located at site C of EcYiiP. "C" in the red spheres indicate Zn^{2+} bound to site C. Accession numbers are as follows: *Homo sapiens* ZnT1 (HsZnT1, UniProt code: Q9Y6M5), *Homo sapiens* ZnT2 (HsZnT2, UniProt code: Q9BRI3), *Homo sapiens* ZnT3 (HsZnT3, UniProt code: Q99726), *Homo sapiens* ZnT4 (HsZnT4, UniProt code: O14863), *Homo sapiens* ZnT7 (HsZnT7, UniProt code: Q8NEW0), *Homo sapiens* ZnT8 (HsZnT8, UniProt code: Q8IWU4), *Escherichia coli* YiiP (EcYiiP, UniProt code: P69380), *Shewanella oneidensis* YiiP (SoYiiP, UniProt code: Q8E919), *Escherichia coli* ZitB (EcZitB, UniProt code: P75757), *Cupriavidus metallidurans* CzcD (CmCzcD, UniProt code: P13512), *Streptococcus pneumoniae* CzcD (SpCzcD, UniProt code: A0A0B7LW62), *Schizosaccharomyces pombe* ZHF1 (SpZHF1, UniProt code: O13918), *Streptococcus pneumoniae* MntE (SpMntE, UniProt code: Q8DP19S), *Saccharomyces cerevisiae* ZRC1 (ScZRC1, UniProt code: P20107), *Saccharomyces cerevisiae* COT1 (ScCOT1, UniProt code: P32798), *Arabidopsis thaliana* MTP1 (AtMTP1, UniProt code: Q9ZT63), *Arabidopsis thaliana* MTP3 (AtMTP3, UniProt code: Q9LXS1), *Hordeum vulgare* MTP1 (HvMTP1, UniProt code: A0JJL9), *Oryza sativa* MTP1 (OsMTP1, UniProt code: Q688R1), *Thermus thermophilus* CzrB (TtCzrB, UniProt code: Q8VLX7), *Thermotoga maritima CzcD* (TmCzcD, UniProt code: Q9WZX9), and *Pseudomonas aeruginosa CzcD* (PaCzcD, UniProt code: Q9I6A3).

Site C is located at the dimer interface between two CTDs (Figures 1B and 4A). In bacterial YiiP, site C forms a binuclear Zn^{2+} complex composed of the $(HHD)_2$ motif and stabilizes its dimeric conformation (Figure 4C,D) [42,43]. In HsZnT8, the exact location of site C differs from that in bacterial YiiP. In HsZnT8, two Zn^{2+} ions are coordinated by an HCH motif from the N-terminal loop, and a Cys-Cys pair from the C-terminal tail, thereby forming a tetrahedral complex (Figure 4D) [51]. The HCH motifs seal off site C and bury the Zn^{2+} ions inside the protein (Figure 4D). This motif is highly conserved among the ZnT8 orthologues but not among bacterial ZnTs [51]. The N-terminal truncation that accompanies the loss of the HCH motif reduces the Zn^{2+} uptake activity of HsZnT8 significantly [51]. AtMTP1 also possesses site C within the CTD (Figure 4C), and the lack of this domain results in loss of function [91]. Thus, Zn^{2+}-mediated dimerization via the CTD seems likely to be essential for Zn^{2+} transport by HsZnT8, bacterial YiiP, and plant MTP1.

By contrast, no Zn^{2+} ions have been identified at either site B or site C of HsZnT7, although its cryo-EM structures were determined in the presence of high Zn^{2+} concentrations (10, 200, or 300 μM) of Zn^{2+} (Figure 4). Consistent with this, residues required for Zn^{2+} binding are not conserved at site B or site C of HsZnT7 (Figure 4C). Presumably, other elements contribute to the dimerization of the CTD and the Zn^{2+} transport activity of HsZnT7. Indeed, dimerization of the CTD is mediated by residues within four β-strands and TM2-TM3 loops located at the dimer interface (Figure 4A) [50].

6. Mechanism of Zn^{2+} Transport by Human ZnT7 and Bacterial YiiP

6.1. Zn^{2+} Transport by HsZnT7

HsZnT7 transports Zn^{2+} from the cytoplasm to the TM Zn^{2+}-binding site (site A), and then to the Golgi lumen. The cryo-EM structures of HsZnT7 in multiple states have helped to paint a full picture of Zn^{2+} transport mediated by this transporter (Figure 5) [50]. In the absence of Zn^{2+}, TM5 packs against TM2, TM3, and TM6 at both the cytosolic and luminal sides, forming an "IF resting" state (Figures 3G and 5A). In this state, Zn^{2+} uptake to site A seems to be blocked due to the closed Zn^{2+} entry gate (Figure 5A–C(i)). In the presence of Zn^{2+}, however, the cytosolic cavity opens by bending the N-terminal half of TM5, and the His-loop is integrated into the cavity to coordinate with Zn^{2+} at site A (Figure 5A–C(ii)). In the next step, TM5 returns to a straight conformation concomitant with pulling His164 out of site A. Consequently, His240 on TM5 coordinates with Zn^{2+} instead of His164 (Figure 5A–C(iii)). Upon conversion to the OF state, the His70 side chain moves away from Zn^{2+} slightly, while Asp74, His240, and Asp244 remain coordinated with Zn^{2+}

(Figure 5B(iv)). Simultaneously (or shortly thereafter), protons presumably enter the Zn^{2+} transport channel from the weakly acidic Golgi lumen to protonate His70 and His240. As a result, His70 flips out of site A, and Zn^{2+} is released to the Golgi lumen (Figure 5A–C(v)). In this context, it is notable that while a His residue is highly conserved among human ZnTs, it is substituted by an Asp residue in bacterial YiiP (Figure 1C,D). Although it is not fully elucidated how many protons are involved in the release of Zn^{2+}, the above mechanistic model suggests that HsZnT7 exchanges one Zn^{2+} for two protons per cycle.

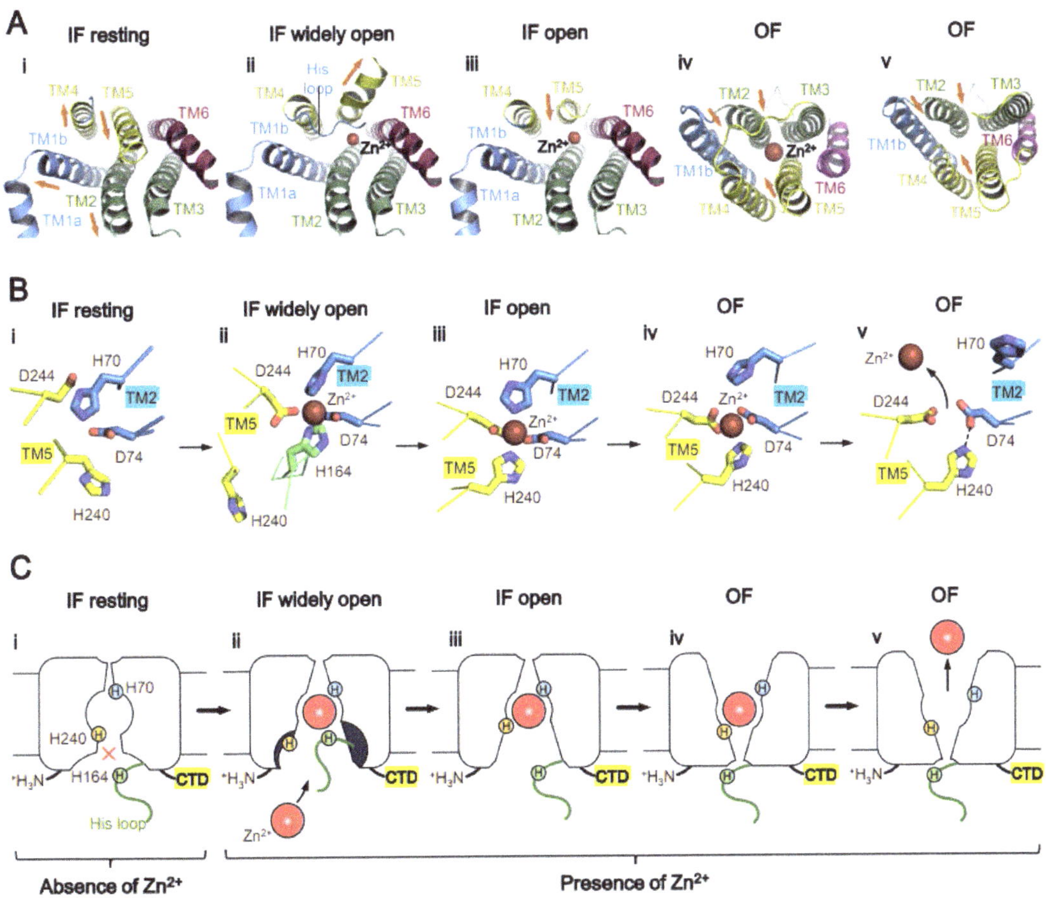

Figure 5. Zn^{2+} transport mechanism used by human ZnT7. (**A**) TM helix rearrangement during the transition from the IF state (**i–iii**) to the OF state (**iv**,**v**), viewed from the cytosolic side (**i–iii**) and from the luminal side (**iv**,**v**). (**B**) Zn^{2+}-coordination structure at site A. Note that His240 (TM5) and His70 (TM2) undergo striking positional shifts during the Zn^{2+} transport cycle. His164 (His-loop) is coordinated directly with Zn^{2+} in the "IF widely open" state and is subsequently replaced by His240 (TM5). His70 moves away from site A to facilitate Zn^{2+} release to the luminal side. (**C**) Simplified cartoon showing the Zn^{2+} transport cycle of human ZnT7. States **i–v** indicate intermediates generated during the transition from the IF state (**i–iii**) to OF states (**iv**,**v**).

6.2. Zn^{2+} Transport by Bacterial YiiP

Recent biochemical, structural, and computational studies provide deep insight into the mechanism of Zn^{2+} transport mediated by bacterial YiiP [43,44]. Initially, Zn^{2+} is recruited to site B, formed by the TM2-TM3 loop, and then transferred to site A of another protomer (Figure 6) [43,44]. In the Zn^{2+}-free IF state, the TM2-TM3 loop is disordered to allow Zn^{2+} to approach site B. After binding Zn^{2+} site B, the TM2-M3 loop forms an ordered conformation [43]. Importantly, the higher affinity of Zn^{2+} for site A than site B explains sequential Zn^{2+} transfer from site A to site B [43]. At this step, the negatively charged cytoplasmic cavity seems advantageous for directing Zn^{2+} to site A [43,92]. In the OF state, the TM2-TM3 loop in a Zn^{2+}-free state interacts with the CTD, forming an extended conformation [43]. Zn^{2+} is released to the periplasm and is facilitated by the protonation of two residues in site A, or potentially three residues at a lower pH [43].

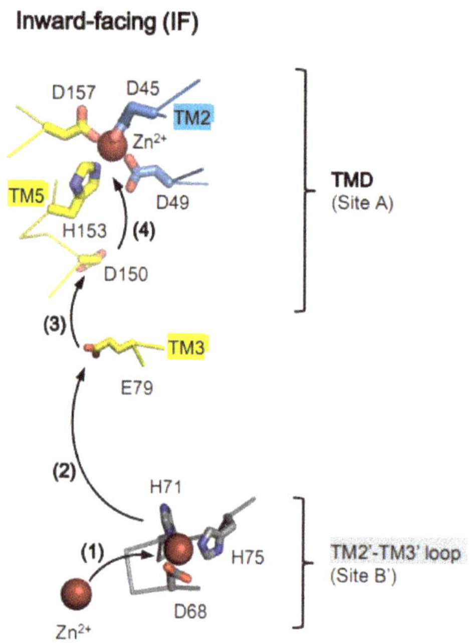

Figure 6. Mechanism of Zn^{2+} recruitment and transfer to site A by bacterial YiiP. Initially, Zn^{2+} is recruited to site B (step (**1**)) located at the TMD-CTD interface, and then transferred to site A via steps (**2**–**4**), sequentially.

MD simulations using EcYiiP and SoYiiP as a model metal transporter demonstrate that Zn^{2+} binds transiently to Asp150 (Asp152 in SoYiiP) on TM5, and to Glu79 (Glu81 in SoYiiP) on TM3, before reaching site A (Figure 6) [92]. In the next step, His153 (His155 in SoYiiP), located at site A, plays a key role in guiding Zn^{2+} to site A. In support of this, His153 can adopt two conformations to receive Zn^{2+} from Asp150, after which it forms a tetrahedral Zn^{2+} complex with Asp45, Asp49, and Asp157 (Figure 6) [49]. Release of Zn^{2+} can be triggered by protonation of site A residues [43,44]. His153 is likely deprotonated again when it faces the neutral-pH cytoplasm to receive Zn^{2+} from site B via Asp150 [49,51, 54,92,93].

6.3. *Role of the Proton Gradient in ZnT-Mediated Zn^{2+} Transport*

Lactose permease (LacY) utilizes a proton motive force for the efficient transport of lactose [94]. Presumably, ZnTs also uses the proton motive force to transport Zn^{2+} from the cytoplasm to the extracellular space or the organelle lumen [55]. Given that the concentration of labile Zn^{2+} in the Golgi is much higher than in the cytosol [64,95,96], and that the Golgi is a weakly acidic organelle [97], it makes sense that Golgi-resident ZnTs utilize the energy provided by the Golgi-to-cytosol proton flow to transport Zn^{2+} to the luminal side. As mentioned above, protonation of His70 and His240 is critical for the

IF-to-OF conversion and subsequent release of Zn^{2+}. While HsZnT7 exchanges one Zn^{2+} for (possibly) two H^+ per cycle, EcZitB [47] and EcYiiP [48] exchange one Zn^{2+} for one H^+. SoYiiP conducts the exchange at a ratio of 1:2–3, depending on the external pH [43]. Human ZnT2 is thought to operate at a ratio of 1:2 [54].

7. Structural and Functional Roles of the Histidine-Rich Loop (His-Loop)

The His-loop is a unique sequence located at the cytoplasmic side between TM4 and TM5 (Figures 1B and 2). The length of this loop, as well as the number and arrangement of His residues, can differ (Figure 2). Most bacterial ZnTs (e.g., YiiP and MntE), several mammalian ZnTs (e.g., ZnTs 6, 9, and 10), and plant MTPs (e.g., AtMTPs 5, 6, and 11) possess a short TM4-TM5 loop without any His residues. Mammalian ZnTs such as ZnTs 2-4 and ZnT8 are also characterized by a short His-loop. Of note, several mammalian ZnTs (e.g., ZnTs 1, 5, and 7), plant MTP1s (e.g., AtMTP1, NgMTP1, OsMTP1, and HvMTP1), and yeast ZnTs (e.g., SpZHF1, ScZRC1, and ScCOT1) have an exceptionally long His-loop (Figure 2). Despite having a relatively long TM4-TM5 loop, ZnT10 has no His residues in the loop (Figure 2). Such variations in the His-loop may underlie the diverse functions and different mechanisms of Zn^{2+} transport mediated by ZnTs. Indeed, the His-loop also functions as a Zn^{2+}-buffering pocket [98], a Zn^{2+} selectivity filter [99], a Zn^{2+} sensor [98,100,101], or a Zn^{2+}-fishing loop [50].

As described above, cryo-EM analysis of HsZnT7 revealed that a short segment of the His-loop inserts into the cytoplasmic cavity, where His164 contained in the His-loop coordinates directly with Zn^{2+} at site A, while His240 on TM5 is kicked out from site A due to bending of the N-terminal half of TM5 (Figure 5A,B(ii)) [50]. Thus, the His-loop appears to contribute to efficient Zn^{2+} uptake to site A, but its task seems transient. In other words, HsZnT7 may use the His-loop as a "Zn^{2+} fishing loop", and His residues in the loop function as a "fishing hook" that captures Zn^{2+} and transfers it to site A. Consistent with this, the deletion of the His-loop, or the mutations of all His residues in the loop to Ser, abolishes the Zn^{2+} transport activity of HsZnT7 [50]. Similarly, deletion of the entire His-loop segment in *Noccaea caerulescens* MTP1 and AtMTP1 results in substantial loss of Zn^{2+} transport activity [98,102] although deletion of the first half of the His-loop increases the activity of AtMTP1 significantly [98].

The 63-residue His-loop of HsZnT7 contains 21 His residues and can be divided into two halves: The first half contains seven His residues close to the C-terminus of TM4, and the second half contains 14 His residues close to the N-terminus of TM5 (Figure 2) [50]. The His-loop of HsZnT7 binds Zn^{2+} with a K_d value of 12 μM; although each half can bind one Zn^{2+}, the first and second halves appear to bind Zn^{2+} co-operatively [50]. AtMTP1 also contains an exceptionally long His-loop containing 25 His residues and can also be divided into two halves: The first half contains 18 His residues close to the C-terminus of TM4, while the second half contains seven His residues close to the N-terminus of TM5 (Figure 2) [98]. The His-loop of AtMTP1 can bind four Zn^{2+} ions, with a K_d value of 25 μM [101]. The lack of either half leads to impairment of Zn^{2+}-binding by AtMTP1, suggesting that the two halves of the His-loop work co-operatively to form multiple Zn^{2+}-binding sites [101]. However, stable Zn^{2+} binding may lead to slower Zn^{2+} release from AtMTP1, which likely explains the less efficient Zn^{2+} transport by AtMTP1 WT than by the mutant lacking the first half [98,101]. Consistent with this, the His-loop of AtMTP1 adopts a β-strand conformation to suppress its Zn^{2+} transport activity at high concentrations of Zn^{2+}, implying a role in sensing excessive Zn^{2+} levels in the cytosol [101].

8. Zn^{2+} Transport by Other Zn^{2+} Transport Systems

8.1. ZIPs

ZIPs play a crucial role in mediating metal import across the plasma membrane as well as export across organelle membranes. ZIPs are classified into four subfamilies based on phylogenetic relationships [102]. In humans, 14 ZIPs have been identified. Despite their relevance to human diseases, the structural and mechanistic details of mammalian ZIPs remain poorly understood. Recent studies on *Bordetella bronchiseptica* ZIP (BbZIP) have shed light on some structural and mechanistic features [56,103,104]. Crystal structures of BbZIP revealed an IF monomer, with binuclear metal-binding sites potentially occupied by Zn^{2+} or Cd^{2+} [104], while cryo-EM studies revealed an IF homodimer [56]. BbZIP is an elevator-type transporter in which a four-TM helix bundle with bound metal ions slides as a rigid body against the dimeric domain composed of the other static TM helices to exert an alternating access mechanism, and its Zn^{2+} uptake is pH-dependent [56,105–107]. However, it is unclear whether ZIPs transfer Zn^{2+} through a symport or antiport mechanism [56,105]. The presence of a His-rich loop in many mammalian ZIPs, including BbZIP, adds complexity to the mechanisms underlying their regulated Zn^{2+} transport [108,109]. Plants possess more ZIPs [110], but detailed structural and mechanistic characterization has not been reported for any of them.

8.2. ZntB

ZntB belongs to the prokaryotic CorA family [111,112]. *E. coli* and *P. aeruginosa* ZntB transport multiple metal ions, including Zn^{2+}, Co^{2+}, Ni^{2+}, and Cd^{2+} [53,113]. ZntB forms homogeneous pentagonal structures that span the biological membrane, similar to other members of the CorA family. Each protomer consists of two TM helices and a large cytoplasmic domain [53,113]. ZnTB-mediated Zn^{2+} transport is stimulated by a pH gradient across the biological membrane. ZntB transports protons and Zn^{2+} together in the same direction, thereby working as a symporter [53]. In contrast to bacterial YiiP and human ZnTs, ZntB does not seem to use an alternating access mechanism mediated by conformational transitions between IF and OF forms [53]. The cryo-EM structure of PaZntB is similar to that of EcZntB and *T. maritima* CorA, but with different helical arrangements [53,113,114]. PaZntB is capable of transporting Zn^{2+}, Mg^{2+}, Cd^{2+}, Ni^{2+}, and Co^{2+}. In contrast to other ZntBs, PaZntB is unlikely to use the proton motive force [53,113]. Thus, different mechanisms may operate among ZntBs from various bacterial species [53, 111,113,115].

8.3. P-Type ATPases with Zn^{2+} Transport Activity

Zn^{2+}-transporting P-type ATPases, which belong to class IB (P_{IB-2}-ATPases), are active zinc transporters. While no P-type ATPases with Zn^{2+} transport activity have been identified in mammals, several have been identified in bacteria and plants. In plants, the P_{IB-2}-ATPases, also known as the heavy metal ATPases (HMAs), are thought to play an important role in the transport of transition metals. In *A. thaliana*, AtHMA2 and AtHMA4 are zinc transporters that increase Zn^{2+} levels in roots, stems, and leaves [66,116]. AtHMA1 functions to detoxify Zn^{2+} in the chloroplast, while AtHMA3 is thought to supply Zn^{2+} to the ER and vacuole. Little is known about the structural features and Zn^{2+} transport mechanisms of HMAs [66]. In bacteria, ZntA is an active zinc transporter that is crucial for cellular detoxification and sub-cellular redistribution of Zn^{2+} [117]. ZntA couples ATP hydrolysis with Zn^{2+} transport via the "Post-Albers" cycle, during which at least four primary intermediate states (E1, E1P, E2P, and E2) accumulate [118–121]. Structurally, P_{IB-2}-ATPases possess four domains: A TMD, a nucleotide-binding domain (N domain), an actuator/dephosphorylation domain (A domain), and a phosphorylation domain (P domain) [118,120,122]. In addition, P_{IB-2}-ATPases contain metal-binding domains (MDBs) in the N-terminal region to promote metal recruitment to the TM metal-binding site (114, 116, 118). To date, only the crystal structure of *Shigella sonnei* ZntA has been reported [117], and the cryo-EM structures of other P_{IB-2}-ATPases are currently unknown.

8.4. Zn^{2+} Transport by ATP-Binding Cassette Transporters

ATP-binding cassette (ABC) transporters are a ubiquitous superfamily of integrated membrane proteins that transport various substrates across biological membranes by utilizing ATP as an external energy source [123]. In bacteria, ZnuABC is a specialist zinc transporter. It comprises three protein subunits: A periplasmic Zn^{2+}-binding protein (ZnuA), an integrated membrane protein that transports Zn^{2+} across the cytoplasmic membrane (ZnuB), and an ATPase protein in charge of coupling Zn^{2+} transport to ATP hydrolysis (ZnuC) [124,125]. In most Gram-negative bacteria, ZnuABC expression is regulated by a Zur (zinc uptake regulatory) protein induced by Zn^{2+} acquisition [126]. The structure of ZnuABC and its mechanism of Zn^{2+} transport coupled to ATP hydrolysis are only poorly understood, although crystal structures have been solved for EcZnuA [124,127,128], and *Salmonella enterica* ZnuA [129].

9. Conclusions

As described above, while some members of the Zn-CDF family share structural and mechanistic features, there are considerable variations among them. It is widely accepted that ZnTs commonly operate as dimers; however, detailed structural analyses conducted so far show that whereas bacterial YiiPs adopt OF-OF and IF-IF homodimers, human ZnTs exist as either OF-OF homodimers or IF-OF heterodimers. No IF-IF homodimers have been observed for mammalian ZnTs, likely due to their conformational instability. Indeed, modes of the TM helix rearrangement between the IF and OF states among ZnTs seem to differ significantly. Of particular note, HsZnT7 undergoes marked bending of TM5 to allow efficient Zn^{2+} uptake. Concomitantly, a part of the long cytoplasmic His-loop is integrated into the negatively charged cytosolic cavity to facilitate acqusition and efficient transfer of Zn^{2+} to the TM Zn^{2+}-binding site. In this state, His164 in the His-loop is coordinated directly with Zn^{2+} in the TMD. Subsequently, His240 (TM5) repaces His164, followed by the conversion from the IF to the OF state. In this context, the essential role of His164 needs to be further explored by conducting additional mutational and structural studies. It will also be interesting to see if another His residue in the His-loop can replace His164 for efficient Zn^{2+} recruitment when His164 is deleted. Recently, a de novo heterozygous variant of *SLC30A7*, His164Ser, was found in Joubert syndrome patients [130]. Although no *SLC30A7* variants have yet been shown to cause human phenotypes or diseases, *SLC30A7* is a candidate gene associated with Joubert syndrome [130].

Undoubtedly, multiple ZnTs work co-operatively to maintain Zn^{2+} homeostasis in cells. Indeed, disruption of Zn^{2+} homeostasis leads to disruption of protein homeostasis [64,131] and dysfunction of essential enzymes, transcription factors, and other biomolecules [132–135], eventually causing many fatal diseases [29–40]. In this regard, the mechanisms of Zn^{2+} homeostasis mediated by various kinds of ZnTs need to be understood more comprehensively. Zinc biology is an important field of research that encompasses biochemical, structural, computational, physiological, and medical approaches.

Author Contributions: Conceptualization: H.B.B. and K.I.; literature collection and analysis: H.B.B.; writing—original draft: H.B.B. and K.I.; writing—review and editing: H.B.B. and K.I. All authors have read and agreed to the published version of the manuscript.

Funding: This work was supported by funding from AMED-CREST (21gm1410006h0001 to K.I.), JSPS KAKENHI (18H03978, 21H04758, and 21H05247 to K.I), Canon Medical Systems Corporation (to K.I.), and the Basis for Supporting Innovative Drug Discovery and Life Science Research (BINDS) from the Japan Agency for Medical Research and Development (AMED) under grant numbers JP19am0101115 (Support No. 1025), JP19am0101078 (Support No. 2293), and JP21am0101079 (Support No. 2343).

Institutional Review Board Statement: Not applicable.

Informed Consent Statement: Not applicable.

Data Availability Statement: Not applicable.

Conflicts of Interest: The authors declare no conflicts of interest with respect to the contents of this article.

References

1. Skalny, A.V.; Aschner, M.; Tinkov, A.A. Zinc. *Adv. Food Nutr. Res.* **2021**, *96*, 251–310. [CrossRef]

2. Zaher, D.M.; El-Gamal, M.I.; Omar, H.A.; Aljareh, S.N.; Al-Shamma, S.A.; Ali, A.J.; Zaib, S.; Iqbal, J. Recent advances with alkaline phosphatase isoenzymes and their inhibitors. *Arch. Pharm.* **2020**, *353*, e2000011. [CrossRef]
3. Xiong, L.; Peng, M.; Zhao, M.; Liang, Z. Truncated expression of a Carboxypeptidase A from bovine improves its enzymatic properties and detoxification efficiency of Ochratoxin A. *Toxins* **2020**, *12*, 680. [CrossRef]
4. Luan, R.; Ding, D.; Xue, Q.; Li, H.; Wang, Y.; Yang, J. Protective role of zinc in the pathogenesis of respiratory diseases. *Eur. J. Clin. Nutr.* **2023**, *77*, 427–435. [CrossRef]
5. Begum, F.; Me, H.M.; Christov, M. The role of zinc in cardiovascular disease. *Cardiol. Rev.* **2022**, *30*, 100–108. [CrossRef]
6. Gembillo, G.; Visconti, L.; Giuffrida, A.E.; Labbozzetta, V.; Peritore, L.; Lipari, A.; Calabrese, V.; Piccoli, G.B.; Torreggiani, M.; Siligato, R.; et al. Role of zinc in diabetic kidney disease. *Nutrients* **2022**, *14*, 1353. [CrossRef]
7. Li, J.; Cao, D.; Huang, Y.; Chen, B.; Chen, Z.; Wang, R.; Dong, Q.; Wei, Q.; Liu, L. Zinc intakes and health outcomes: An umbrella review. *Front. Nutr.* **2022**, *9*, 798078. [CrossRef]
8. Li, Z.; Liu, Y.; Wei, R.; Yong, V.W.; Xue, M. The important role of zinc in neurological diseases. *Biomolecules* **2022**, *13*, 28. [CrossRef]
9. Garner, T.B.; Hester, J.M.; Carothers, A.; Diaz, F.J. Role of zinc in female reproduction. *Biol. Reprod.* **2021**, *104*, 976–994. [CrossRef] [PubMed]
10. Banupriya, N.; Bhat, B.V.; Sridhar, M.G. Role of zinc in neonatal sepsis. *Indian J. Pediatr.* **2021**, *88*, 696–702. [CrossRef] [PubMed]
11. Tamura, Y. The role of zinc homeostasis in the prevention of diabetes mellitus and cardiovascular diseases. *J. Atheroscler. Thromb.* **2021**, *28*, 1109–1122. [CrossRef] [PubMed]
12. Allouche-Fitoussi, D.; Breitbart, H. The role of zinc in male fertility. *Int. J. Mol. Sci.* **2020**, *21*, 7796. [CrossRef]
13. Michalczyk, K.; Cymbaluk-Płoska, A. The role of zinc and copper in gynecological malignancies. *Nutrients* **2020**, *12*, 3732. [CrossRef] [PubMed]
14. Granero, R.; Pardo-Garrido, A.; Carpio-Toro, I.L.; Ramírez-Coronel, A.A.; Martínez-Suárez, P.C.; Reivan-Ortiz, G.G. The role of iron and zinc in the treatment of ADHD among children and adolescents: A systematic review of randomized clinical trials. *Nutrients* **2021**, *13*, 4059. [CrossRef]
15. Moshtagh, M.; Amiri, R. Role of zinc supplementation in the improvement of acute respiratory infections among Iranian children: A systematic review. *Tanaffos* **2020**, *19*, 1–9.
16. Wang, M.X.; Win, S.S.; Pang, J. Zinc supplementation reduces common cold duration among healthy adults: A systematic review of randomized controlled trials with micronutrients supplementation. *Am. J. Trop. Med. Hyg.* **2020**, *103*, 86–99. [CrossRef]
17. Korkmaz-Icöz, S.; Atmanli, A.; Radovits, T.; Li, S.; Hegedüs, P.; Ruppert, M.; Brlecic, P.; Yoshikawa, Y.; Yasui, H.; Karck, M.; et al. Administration of zinc complex of acetylsalicylic acid after the onset of myocardial injury protects the heart by upregulation of antioxidant enzymes. *J. Physiol. Sci.* **2016**, *66*, 113–125. [CrossRef]
18. Barnett, J.B.; Dao, M.C.; Hamer, D.H.; Kandel, R.; Brandeis, G.; Wu, D.; Dallal, G.E.; Jacques, P.F.; Schreiber, R.; Kong, E.; et al. Effect of zinc supplementation on serum zinc concentration and T cell proliferation in nursing home elderly: A randomized, double-blind, placebo-controlled trial. *Am. J. Clin. Nutr.* **2016**, *103*, 942–951. [CrossRef]
19. Cope, E.C.; Levenson, C.W. Role of zinc in the development and treatment of mood disorders. *Curr. Opin. Clin. Nutr. Metab. Care* **2010**, *13*, 685–689. [CrossRef]
20. Plum, L.M.; Rink, L.; Haase, H. The essential toxin: Impact of zinc on human health. *Int. J. Environ. Res. Public Health* **2010**, *7*, 1342–1365. [CrossRef] [PubMed]
21. Fox, M.R.S. Zinc Excess. In *Zinc in Human Biology. ILSI Human Nutrition Reviews*; Mills, C.F., Ed.; Springer: London, UK, 1989; pp. 365–370. [CrossRef]
22. Brown, M.A.; Thom, J.V.; Orth, G.L.; Cova, P.; Juarez, J. Food poisoning involving zinc contamination. *Arch. Environ. Health* **1964**, *8*, 657–660. [CrossRef]
23. Skalny, A.V.; Aschner, M.; Lei, X.G.; Gritsenko, V.A.; Santamaria, A.; Alekseenko, S.I.; Prakash, N.T.; Chang, J.S.; Sizova, E.A.; Chao, J.C.J.; et al. Gut Microbiota as a Mediator of Essential and Toxic Effects of Zinc in the Intestines and Other Tissues. *Int. J. Mol. Sci.* **2021**, *22*, 13074. [CrossRef] [PubMed]
24. Hamzah-Saleem, M.; Usman, K.; Rizwan, M.; Al-Jabri, H.; Alsafran, M. Functions and strategies for enhancing zinc availability in plants for sustainable agriculture. *Front. Plant Sci.* **2022**, *13*, 1033092. [CrossRef] [PubMed]
25. Choudhury, R.; Srivastava, S. Zinc resistance mechanisms in bacteria. *Curr. Sci.* **2001**, *81*, 768–775.
26. Pandey, N.; Pathak, G.C.; Sharma, C.P. Zinc is critically required for pollen function and fertilisation in lentil. *J. Trace Elem. Med. Biol.* **2006**, *20*, 89–96. [CrossRef]
27. Wissuwa, M.; Ismail, A.M.; Yanagihara, S. Effects of zinc deficiency on rice growth and genetic factors contributing to tolerance. *Plant Physiol.* **2006**, *142*, 731–741. [CrossRef]
28. Corbett, D.; Wang, J.; Schuler, S.; Lopez-Castejon, G.; Glenn, S.; Brough, D.; Andrew, P.W.; Cavet, J.S.; Roberts, I.S. Two zinc uptake systems contribute to the full virulence of Listeria monocytogenes during growth in vitro and in vivo. *Infect. Immun.* **2012**, *80*, 14–21. [CrossRef] [PubMed]
29. Hara, T.; Yoshigai, E.; Ohashi, T.; Fukada, T. Zinc transporters as potential therapeutic targets: An updated review. *J. Pharmacol. Sci.* **2022**, *148*, 221–228. [CrossRef]

Zhu, B.; Huo, R.; Zhi, Q.; Zhan, M.; Chen, X.; Hua, Z.C. Increased expression of zinc transporter ZIP4, ZIP11, ZnT1, and ZnT6 -licts poor prognosis in pancreatic cancer. *J. Trace Elem. Med. Biol.* **2021**, *65*, 126734. [CrossRef]

-ton, S.; Bush, A.I. The essential elements of Alzheimer's disease. *J. Biol. Chem.* **2021**, *296*, 100105. [CrossRef]

32. Xu, Y.; Xiao, G.; Liu, L.; Lang, M. Zinc transporters in Alzheimer's disease. *Mol. Brain* **2019**, *12*, 106. [CrossRef] [PubMed]
33. Sikora, J.; Ouagazzal, A.M. Synaptic zinc: An emerging player in Parkinson's disease. *Int. J. Mol. Sci.* **2021**, *22*, 4724. [CrossRef] [PubMed]
34. Davis, D.N.; Strong, M.D.; Chambers, E.; Hart, M.D.; Bettaieb, A.; Clarke, S.L.; Smith, B.J.; Stoecker, B.J.; Lucas, E.A.; Lin, D.; et al. A role for zinc transporter gene SLC39A12 in the nervous system and beyond. *Gene* **2021**, *799*, 145824. [CrossRef] [PubMed]
35. Chowanadisai, W.; Lönnerdal, B.; Kelleher, S.L. Identification of a mutation in SLC30A2 (ZnT-2) in women with low milk zinc concentration that results in transient neonatal zinc deficiency. *J. Biol. Chem.* **2006**, *281*, 39699–39707. [CrossRef] [PubMed]
36. Lieberwirth, J.K.; Joset, P.; Heinze, A.; Hentschel, J.; Stein, A.; Iannaccone, A.; Steindl, K.; Kuechler, A.; Abou-Jamra, R. Bi-allelic loss of function variants in SLC30A5 as cause of perinatal lethal cardiomyopathy. *Eur. J. Hum. Genet.* **2021**, *29*, 808–815. [CrossRef] [PubMed]
37. Hildebrand, M.S.; Phillips, A.M.; Mullen, S.A.; Adlard, P.A.; Hardies, K.; Damiano, J.A.; Wimmer, V.; Bellows, S.T.; McMahon, J.M.; Burgess, R.; et al. Loss of synaptic Zn^{2+} transporter function increases risk of febrile seizures. *Sci. Rep.* **2015**, *5*, 17816. [CrossRef] [PubMed]
38. Pérez, Y.; Shorer, Z.; Liani-Leibson, K.; Chabosseau, P.; Kadir, R.; Volodarsky, M.; Halperin, D.; Barber-Zucker, S.; Shalev, H.; Schreiber, R.; et al. SLC30A9 mutation affecting intracellular zinc homeostasis causes a novel cerebro-renal syndrome. *Brain* **2017**, *140*, 928–939. [CrossRef] [PubMed]
39. Tavasoli, A.; Arjmandi-Rafsanjani, K.; Hemmati, S.; Mojbafan, M.; Zarei, E.; Hosseini, S. A case of dystonia with polycythemia and hypermanganesemia caused by SLC30A10 mutation: A treatable inborn error of manganese metabolism. *BMC Pediatr.* **2019**, *19*, 229. [CrossRef]
40. Quadri, M.; Federico, A.; Zhao, T.; Breedveld, G.J.; Battisti, C.; Delnooz, C.; Severijnen, L.A.; Di Toro Mammarella, L.; Mignarri, A.; Monti, L.; et al. Mutations in SLC30A10 cause parkinsonism and dystonia with hypermanganesemia, polycythemia, and chronic liver disease. *Am. J. Hum. Genet.* **2012**, *90*, 467–477. [CrossRef]
41. Lu, M.; Fu, D. Structure of the zinc transporter YiiP. *Science* **2007**, *317*, 1746–1748. [CrossRef]
42. Lu, M.; Chai, J.; Fu, D. Structural basis for autoregulation of the zinc transporter YiiP. *Nat. Struct. Mol. Biol.* **2009**, *16*, 1063–1067. [CrossRef]
43. Hussein, A.; Fan, S.; Lopez-Redondo, M.L.; Kenney, I.; Zhang, X.; Beckstein, O.; Stokes, D.L. Energy coupling and stoichiometry of Zn^{2+}/H^+ antiport by the cation diffusion facilitator YiiP. *eLife* **2023**, *12*, RP87167. [CrossRef]
44. Lopez-Redondo, M.L.; Fan, S.; Koide, A.; Koide, S.; Beckstein, O.; Stokes, D.L. Zinc binding alters the conformational dynamics and drives the transport cycle of the cation diffusion facilitator YiiP. *J. Gen. Physiol.* **2021**, *153*, e202112873. [CrossRef] [PubMed]
45. Lopez-Redondo, M.L.; Coudray, N.; Zhang, Z.; Alexopoulos, J.; Stokes, D.L. Structural basis for the alternating access mechanism of the cation diffusion facilitator YiiP. *Proc. Natl. Acad. Sci. USA* **2018**, *115*, 3042–3047. [CrossRef] [PubMed]
46. Coudray, N.; Valvo, S.; Hu, M.; Lasala, R.; Kim, C.; Vink, M.; Zhou, M.; Provasi, D.; Filizola, M.; Tao, J.; et al. Inward-facing conformation of the zinc transporter YiiP revealed by cryoelectron microscopy. *Proc. Natl. Acad. Sci. USA* **2013**, *110*, 2140–2145. [CrossRef] [PubMed]
47. Chao, Y.; Fu, D. Kinetic study of the antiport mechanism of an *Escherichia coli* zinc transporter, ZitB. *J. Biol. Chem.* **2004**, *279*, 12043–12050. [CrossRef] [PubMed]
48. Chao, Y.; Fu, D. Thermodynamic studies of the mechanism of metal binding to the *Escherichia coli* zinc transporter YiiP. *J. Biol. Chem.* **2004**, *279*, 17173–17180. [CrossRef] [PubMed]
49. Sharma, G.; Merz, K.M. Mechanism of zinc transport through the zinc transporter YiiP. *J. Chem. Theory Comput.* **2022**, *18*, 2556–2568. [CrossRef] [PubMed]
50. Bui, H.B.; Watanabe, S.; Nomura, N.; Liu, K.; Uemura, T.; Inoue, M.; Tsutsumi, A.; Fujita, H.; Kinoshita, K.; Kato, Y.; et al. Cryo-EM structures of human zinc transporter ZnT7 reveal the mechanism of Zn^{2+} uptake into the Golgi apparatus. *Nat. Commun.* **2023**, *14*, 4770. [CrossRef]
51. Xue, J.; Xie, T.; Zeng, W.; Jiang, Y.; Bai, X.C. Cryo-EM structures of human ZnT8 in both outward- and inward-facing conformations. *eLife* **2020**, *9*, e58823. [CrossRef]
52. Zhang, S.; Fu, C.; Luo, Y.; Xie, Q.; Xu, T.; Sun, Z.; Su, Z.; Zhou, X. Cryo-EM structure of a eukaryotic zinc transporter at a low pH suggests its Zn^{2+}-releasing mechanism. *J. Struct. Biol.* **2023**, *215*, 107926. [CrossRef] [PubMed]
53. Gati, C.; Stetsenko, A.; Slotboom, D.J.; Scheres, S.H.W.; Guskov, A. The structural basis of proton driven zinc transport by ZntB. *Nat. Commun.* **2017**, *8*, 1313. [CrossRef] [PubMed]
54. Golan, Y.; Alhadeff, R.; Warshel, A.; Assaraf, Y.G. ZnT2 is an electroneutral proton-coupled vesicular antiporter displaying an apparent stoichiometry of two protons per zinc ion. *PLoS Comput. Biol.* **2019**, *15*, e1006882. [CrossRef] [PubMed]
55. Cotrim, C.A.; Jarrott, R.J.; Martin, J.L.; Drew, D. A structural overview of the zinc transporters in the cation diffusion facilitator family. *Acta Crystallogr. D Struct. Biol.* **2019**, *75*, 357–367. [CrossRef] [PubMed]
56. Pang, C.; Chai, J.; Zhu, P.; Shanklin, J.; Liu, Q. Structural mechanism of intracellular autoregulation of zinc uptake in ZIP transporters. *Nat. Commun.* **2023**, *14*, 3404. [CrossRef]
57. Montanini, B.; Blaudez, D.; Jeandroz, S.; Sanders, D.; Chalot, M. Phylogenetic and functional analysis of the Cation Diffusion Facilitator (CDF) family: Improved signature and prediction of substrate specificity. *BMC Genom.* **2007**, *8*, 107. [CrossRef] [PubMed]
58. Gaither, L.A.; Eide, D.J. Eukaryotic zinc transporters and their regulation. *Biometals* **2001**, *14*, 251–270. [CrossRef] [PubMed]

59. Kambe, T.; Taylor, K.M.; Fu, D. Zinc transporters and their functional integration in mammalian cells. *J. Biol. Chem.* **2021**, *296*, 100320. [CrossRef]
60. Kambe, T.; Tsuji, T.; Hashimoto, A.; Itsumura, N. The physiological, biochemical, and molecular roles of zinc transporters in zinc homeostasis and metabolism. *Physiol. Rev.* **2015**, *95*, 749–784. [CrossRef]
61. Hara, T.; Takeda, T.A.; Takagishi, T.; Fukue, K.; Kambe, T.; Fukada, T. Physiological roles of zinc transporters: Molecular and genetic importance in zinc homeostasis. *J. Physiol. Sci.* **2017**, *67*, 283–301. [CrossRef]
62. Fukunaka, A.; Suzuki, T.; Kurokawa, Y.; Yamazaki, T.; Fujiwara, N.; Ishihara, K.; Migaki, H.; Okumura, K.; Masuda, S.; Yamaguchi-Iwai, Y.; et al. Demonstration and characterization of the heterodimerization of ZnT5 and ZnT6 in the early secretory pathway. *J. Biol. Chem.* **2009**, *284*, 30798–30806. [CrossRef]
63. Huang, L.; Tepaamorndech, S. The SLC30 family of zinc transporters: A review of current understanding of their biological and pathophysiological roles. *Mol. Asp. Med.* **2013**, *34*, 548–560. [CrossRef]
64. Amagai, Y.; Yamada, M.; Kowada, T.; Watanabe, T.; Du, Y.; Liu, R.; Naramoto, S.; Watanabe, S.; Kyozuka, J.; Anelli, T.; et al. Zinc homeostasis governed by Golgi-resident ZnT family members regulates ERp44-mediated proteostasis at the ER-Golgi interface. *Nat. Commun.* **2023**, *14*, 2683. [CrossRef]
65. Kambe, T.; Wagatsuma, T. Metalation and activation of Zn^{2+} enzymes via early secretory pathway-resident ZNT proteins. *Biophys. Rev.* **2023**, *4*, 041302. [CrossRef]
66. Ricachenevsky, F.K.; Menguer, P.K.; Sperotto, R.A.; Williams, L.E.; Fett, J.P. Roles of plant metal tolerance proteins (MTP) in metal storage and potential use in biofortification strategies. *Front. Plant Sci.* **2013**, *4*, 144. [CrossRef]
67. Gustin, J.L.; Zanis, M.J.; Salt, D.E. Structure and evolution of the plant cation diffusion facilitator family of ion transporters. *BMC Evol. Biol.* **2011**, *11*, 76. [CrossRef]
68. Gao, Y.; Yang, F.; Liu, J.; Xie, W.; Zhang, L.; Chen, Z.; Peng, Z.; Ou, Y.; Yao, Y. Genome-wide identification of metal tolerance protein genes in *Populus trichocarpa* and their roles in response to various heavy metal stresses. *Int. J. Mol. Sci.* **2020**, *21*, 1680. [CrossRef]
69. Arrivault, S.; Senger, T.; Krämer, U. The Arabidopsis metal tolerance protein AtMTP3 maintains metal homeostasis by mediating Zn exclusion from the shoot under Fe deficiency and Zn oversupply. *Plant J.* **2006**, *46*, 861–879. [CrossRef] [PubMed]
70. Desbrosses-Fonrouge, A.G.; Voigt, K.; Schröder, A.; Arrivault, S.; Thomine, S.; Krämer, U. *Arabidopsis thaliana* MTP1 is a Zn transporter in the vacuolar membrane which mediates Zn detoxification and drives leaf Zn accumulation. *FEBS Lett.* **2005**, *579*, 4165–4174. [CrossRef] [PubMed]
71. Kobae, Y.; Uemura, T.; Sato, M.H.; Ohnishi, M.; Mimura, T.; Nakagawa, T.; Maeshima, M. Zinc transporter of *Arabidopsis thaliana* AtMTP1 is localized to vacuolar membranes and implicated in zinc homeostasis. *Plant Cell Physiol.* **2004**, *45*, 1749–1758. [CrossRef] [PubMed]
72. Sinclair, S.A.; Krämer, U. The zinc homeostasis network of land plants. *Biochim. Biophys. Acta* **2012**, *1823*, 1553–1567. [CrossRef] [PubMed]
73. Fujiwara, T.; Kawachi, M.; Sato, Y.; Mori, H.; Kutsuna, N.; Hasezawa, S.; Maeshima, M. A high molecular mass zinc transporter MTP12 forms a functional heteromeric complex with MTP5 in the Golgi in *Arabidopsis thaliana*. *FEBS J.* **2015**, *282*, 1965–1979. [CrossRef]
74. Miyabe, S.; Izawa, S.; Inoue, Y. The Zrc1 is involved in zinc transport system between vacuole and cytosol in *Saccharomyces cerevisiae*. *Biochem. Biophys. Res. Commun.* **2001**, *282*, 79–83. [CrossRef]
75. MacDiarmid, C.W.; Milanick, M.A.; Eide, D.J. Biochemical properties of vacuolar zinc transport systems of *Saccharomyces cerevisiae*. *J. Biol. Chem.* **2002**, *277*, 39187–39194. [CrossRef]
76. MacDiarmid, C.W.; Milanick, M.A.; Eide, D.J. Induction of the ZRC1 metal tolerance gene in zinc-limited yeast confers resistance to zinc shock. *J. Biol. Chem.* **2003**, *278*, 15065–15072. [CrossRef] [PubMed]
77. Lin, H.; Kumánovics, A.; Nelson, J.M.; Warner, D.E.; Ward, D.M.; Kaplan, J. A single amino acid change in the yeast vacuolar metal transporters ZRC1 and COT1 alters their substrate specificity. *J. Biol. Chem.* **2008**, *283*, 33865–33873. [CrossRef] [PubMed]
78. Li, L.; Kaplan, J. The yeast gene MSC2, a member of the cation diffusion facilitator family, affects the cellular distribution of zinc. *J. Biol. Chem.* **2001**, *276*, 5036–5043. [CrossRef]
79. Wu, Y.H.; Taggart, J.; Song, P.X.; MacDiarmid, C.; Eide, D.J. An MSC2 promoter-lacZ fusion gene reveals zinc-responsive changes in sites of transcription initiation that occur across the yeast genome. *PLoS ONE* **2016**, *11*, e0163256. [CrossRef]
80. Ellis, C.D.; Macdiarmid, C.W.; Eide, D.J. Heteromeric protein complexes mediate zinc transport into the secretory pathway of eukaryotic cells. *J. Biol. Chem.* **2005**, *280*, 28811–28818. [CrossRef]
81. Ellis, C.D.; Wang, F.; MacDiarmid, C.W.; Clark, S.; Lyons, T.; Eide, D.J. Zinc and the Msc2 zinc transporter protein are required for endoplasmic reticulum function. *J. Cell Biol.* **2004**, *166*, 325–335. [CrossRef]
82. Clemens, S.; Bloss, T.; Vess, C.; Neumann, D.; Nies, D.H.; Zur Nieden, U. A transporter in the endoplasmic reticulum of *Schizosaccharomyces pombe* cells mediates zinc storage and differentially affects transition metal tolerance. *J. Biol. Chem.* **2002**, *277*, 18215–18221. [CrossRef] [PubMed]

Grass, G.; Otto, M.; Fricke, B.; Haney, C.J.; Rensing, C.; Nies, D.H.; Munkelt, D. FieF (YiiP) from *Escherichia coli* mediates decreased ...ular accumulation of iron and relieves iron stress. *Arch. Microbiol.* **2005**, *183*, 9–18. [CrossRef] [PubMed]

...u, D. Selective metal binding to a membrane-embedded aspartate in the *Escherichia coli* metal transporter YiiP (FieF). *J.* ...05, *280*, 33716–33724. [CrossRef] [PubMed]

85. Grass, G.; Fan, B.; Rosen, B.P.; Franke, S.; Nies, D.H.; Rensing, C. ZitB (YbgR), a member of the cation diffusion facilitator family, is an additional zinc transporter in *Escherichia coli*. *J. Bacteriol.* **2001**, *183*, 4664–4667. [CrossRef] [PubMed]
86. Guffanti, A.A.; Wei, Y.; Rood, S.V.; Krulwich, T.A. An antiport mechanism for a member of the cation diffusion facilitator family: Divalent cations efflux in exchange for K^+ and H^+. *Mol. Microbiol.* **2002**, *45*, 145–153. [CrossRef] [PubMed]
87. Garaeva, A.A.; Slotboom, D.J. Elevator-type mechanisms of membrane transport. *Biochem. Soc. Trans.* **2020**, *48*, 1227–1241. [CrossRef] [PubMed]
88. Nishito, Y.; Tsuji, N.; Fujishiro, H.; Takeda, T.A.; Yamazaki, T.; Teranishi, F.; Okazaki, F.; Matsunaga, A.; Tuschl, K.; Rao, R.; et al. Direct comparison of manganese detoxification/efflux proteins and molecular characterization of ZnT10 protein as a manganese transporter. *J. Biol. Chem.* **2016**, *291*, 14773–14787. [CrossRef]
89. Zogzas, C.E.; Aschner, M.; Mukhopadhyay, S. Structural elements in the transmembrane and cytoplasmic domains of the metal transporter SLC30A10 are required for its manganese efflux activity. *J. Biol. Chem.* **2016**, *291*, 15940–15957. [CrossRef]
90. Leyva-Illades, D.; Chen, P.; Zogzas, C.E.; Hutchens, S.; Mercado, J.M.; Swaim, C.D.; Morrisett, R.A.; Bowman, A.B.; Aschner, M.; Mukhopadhyay, S. SLC30A10 is a cell surface-localized manganese efflux transporter, and parkinsonism-causing mutations block its intracellular trafficking and efflux activity. *J. Neurosci.* **2014**, *34*, 14079–14095. [CrossRef]
91. Kawachi, M.; Kobae, Y.; Kogawa, S.; Mimura, T.; Krämer, U.; Maeshima, M. Amino acid screening based on structural modeling identifies critical residues for the function, ion selectivity and structure of Arabidopsis MTP1. *FEBS J.* **2012**, *279*, 2339–2356. [CrossRef]
92. Sala, D.; Giachetti, A.; Rosato, A. An atomistic view of the YiiP structural changes upon zinc(II) binding. *Biochim. Biophys. Acta Gen. Subj.* **2019**, *1863*, 1560–1567. [CrossRef] [PubMed]
93. Gupta, S.; Chai, J.; Cheng, J.; D'Mello, R.; Chance, M.R.; Fu, D. Visualizing the kinetic power stroke that drives proton-coupled zinc(II) transport. *Nature* **2014**, *512*, 101–104. [CrossRef] [PubMed]
94. Kaback, H.R. A chemiosmotic mechanism of symport. *Proc. Natl. Acad. Sci. USA* **2015**, *112*, 1259–1264. [CrossRef] [PubMed]
95. Liu, R.; Kowada, T.; Du, Y.; Amagai, Y.; Matsui, T.; Inaba, K.; Mizukami, S. Organelle-level labile Zn^{2+} mapping based on targetable fluorescent sensors. *ACS Sens.* **2022**, *7*, 748–757. [CrossRef] [PubMed]
96. Kowada, T.; Watanabe, T.; Amagai, Y.; Liu, R.; Yamada, M.; Takahashi, H.; Matsui, T.; Inaba, K.; Mizukami, S. Quantitative imaging of labile Zn^{2+} in the Golgi apparatus using a localizable small-molecule fluorescent probe. *Cell Chem. Biol.* **2020**, *27*, 1521–1531.e8. [CrossRef]
97. Maeda, Y.; Ide, T.; Koike, M.; Uchiyama, Y.; Kinoshita, T. GPHR is a novel anion channel critical for acidification and functions of the Golgi apparatus. *Nat. Cell Biol.* **2008**, *10*, 1135–1145. [CrossRef]
98. Kawachi, M.; Kobae, Y.; Mimura, T.; Maeshima, M. Deletion of a histidine-rich loop of AtMTP1, a vacuolar Zn^{2+}/H^+ antiporter of *Arabidopsis thaliana*, stimulates the transport activity. *J. Biol. Chem.* **2008**, *283*, 8374–8383. [CrossRef]
99. Podar, D.; Scherer, J.; Noordally, Z.; Herzyk, P.; Nies, D.; Sanders, D. Metal selectivity determinants in a family of transition metal transporters. *J. Biol. Chem.* **2012**, *287*, 3185–3196. [CrossRef]
100. Tanaka, N.; Fujiwara, T.; Tomioka, R.; Krämer, U.; Kawachi, M.; Maeshima, M. Characterization of the histidine-rich loop of Arabidopsis vacuolar membrane zinc transporter AtMTP1 as a sensor of zinc level in the cytosol. *Plant Cell Physiol.* **2015**, *56*, 510–519. [CrossRef]
101. Tanaka, N.; Kawachi, M.; Fujiwara, T.; Maeshima, M. Zinc-binding and structural properties of the histidine-rich loop of *Arabidopsis thaliana* vacuolar membrane zinc transporter MTP1. *FEBS Open Bio* **2013**, *3*, 218–224. [CrossRef] [PubMed]
102. Jeong, J.; Eide, D.J. The SLC39 family of zinc transporters. *Mol. Asp. Med.* **2013**, *34*, 612–619. [CrossRef]
103. Zhang, Y.; Jiang, Y.; Gao, K.; Sui, D.; Yu, P.; Su, M.; Wei, G.W.; Hu, J. Structural insights into the elevator-type transport mechanism of a bacterial ZIP metal transporter. *Nat. Commun.* **2023**, *14*, 385. [CrossRef]
104. Zhang, T.; Liu, J.; Fellner, M.; Zhang, C.; Sui, D.; Hu, J. Crystal structures of a ZIP zinc transporter reveal a binuclear metal center in the transport pathway. *Sci. Adv.* **2017**, *3*, e1700344. [CrossRef]
105. Hoch, E.; Levy, M.; Hershfinkel, M.; Sekler, I. Elucidating the H^+ coupled Zn^{2+} transport mechanism of ZIP4: Implications in acrodermatitis enteropathica. *Int. J. Mol. Sci.* **2020**, *21*, 734. [CrossRef] [PubMed]
106. Franz, M.C.; Pujol-Giménez, J.; Montalbetti, N.; Fernandez-Tenorio, M.; DeGrado, T.R.; Niggli, E.; Romero, M.F.; Hediger, M.A. Reassessment of the transport mechanism of the human zinc transporter SLC39A2. *Biochemistry* **2018**, *57*, 3976–3986. [CrossRef] [PubMed]
107. Lin, W.; Chai, J.; Love, J.; Fu, D. Selective electrodiffusion of zinc ions in a Zrt-, Irt-like protein, ZIPB. *J. Biol. Chem.* **2010**, *285*, 39013–39020. [CrossRef]
108. Zhang, T.; Sui, D.; Zhang, C.; Cole, L.; Hu, J. Asymmetric functions of a binuclear metal center within the transport pathway of a human zinc transporter ZIP4. *FASEB J.* **2020**, *34*, 237–247. [CrossRef]
109. Mao, X.; Kim, B.E.; Wang, F.; Eide, D.J.; Petris, M.J. A histidine-rich cluster mediates the ubiquitination and degradation of the human zinc transporter, hZIP4, and protects against zinc cytotoxicity. *J. Biol. Chem.* **2007**, *282*, 6992–7000. [CrossRef] [PubMed]
110. Ajeesh-Krishna, T.P.; Maharajan, T.; Victor-Roch, G.; Ignacimuthu, S.; Antony Ceasar, S. Structure, function, regulation and phylogenetic relationship of ZIP family transporters of plants. *Front. Plant Sci.* **2020**, *11*, 662. [CrossRef]
111. Tan, K.; Sather, A.; Robertson, J.L.; Moy, S.; Roux, B.; Joachimiak, A. Structure and electrostatic property of cytoplasmic domain of ZntB transporter. *Protein Sci.* **2009**, *18*, 2043–2052. [CrossRef]

112. Worlock, A.J.; Smith, R.L. ZntB is a novel Zn^{2+} transporter in *Salmonella enterica* serovar Typhimurium. *J. Bacteriol.* **2002**, *184*, 4369–4373. [CrossRef]
113. Stetsenko, A.; Stehantsev, P.; Dranenko, N.O.; Gelfand, M.S.; Guskov, A. Structural and biochemical characterization of a novel ZntB (CmaX) transporter protein from *Pseudomonas aeruginosa*. *Int. J. Biol. Macromol.* **2021**, *184*, 760–767. [CrossRef]
114. Pfoh, R.; Li, A.; Chakrabarti, N.; Payandeh, J.; Pomès, R.; Pai, E.F. Structural asymmetry in the magnesium channel CorA points to sequential allosteric regulation. *Proc. Natl. Acad. Sci. USA* **2012**, *109*, 18809–18814. [CrossRef]
115. Wang, D.; Hurst, T.K.; Thompson, R.B.; Fierke, C.A. Genetically encoded ratiometric biosensors to measure intracellular exchangeable zinc in *Escherichia coli*. *J. Biomed. Opt.* **2011**, *16*, 087011. [CrossRef]
116. Mills, R.F.; Francini, A.; Ferreira da Rocha, P.S.; Baccarini, P.J.; Aylett, M.; Krijger, G.C.; Williams, L.E. The plant P1B-type ATPase AtHMA4 transports Zn and Cd and plays a role in detoxification of transition metals supplied at elevated levels. *FEBS Lett.* **2005**, *579*, 783–791. [CrossRef]
117. Wang, K.; Sitsel, O.; Meloni, G.; Autzen, H.E.; Andersson, M.; Klymchuk, T.; Nielsen, A.M.; Rees, D.C.; Nissen, P.; Gourdon, P. Structure and mechanism of Zn^{2+}-transporting P-type ATPases. *Nature* **2014**, *514*, 518–522. [CrossRef] [PubMed]
118. Dyla, M.; Kjærgaard, M.; Poulsen, H.; Nissen, P. Structure and mechanism of P-type ATPase ion pumps. *Annu. Rev. Biochem.* **2020**, *89*, 583–603. [CrossRef]
119. Raimunda, D.; Subramanian, P.; Stemmler, T.; Argüello, J.M. A tetrahedral coordination of zinc during transmembrane transport by P-type Zn^{2+}-ATPases. *Biochim. Biophys. Acta* **2012**, *1818*, 1374–1377. [CrossRef]
120. Kühlbrandt, W. Biology, structure and mechanism of P-type ATPases. *Nat. Rev. Mol. Cell Biol.* **2004**, *5*, 282–295. [CrossRef] [PubMed]
121. Albers, R.W. Biochemical aspects of active transport. *Annu. Rev. Biochem.* **1967**, *36*, 727–756. [CrossRef] [PubMed]
122. Argüello, J.M.; Eren, E.; González-Guerrero, M. The structure and function of heavy metal transport P1B-ATPases. *Biometals* **2007**, *20*, 233–248. [CrossRef] [PubMed]
123. Rees, D.C.; Johnson, E.; Lewinson, O. ABC transporters: The power to change. *Nat. Rev. Mol. Cell Biol.* **2009**, *10*, 218–227. [CrossRef]
124. Yatsunyk, L.A.; Easton, J.A.; Kim, L.R.; Sugarbaker, S.A.; Bennett, B.; Breece, R.M.; Vorontsov, I.I.; Tierney, D.L.; Crowder, M.W.; Rosenzweig, A.C. Structure and metal binding properties of ZnuA, a periplasmic zinc transporter from *Escherichia coli*. *J. Biol. Inorg. Chem.* **2008**, *13*, 271–288. [CrossRef]
125. Patzer, S.I.; Hantke, K. The ZnuABC high-affinity zinc uptake system and its regulator Zur in *Escherichia coli*. *Mol. Microbiol.* **1998**, *28*, 1199–1210. [CrossRef] [PubMed]
126. Gabbianelli, R.; Scotti, R.; Ammendola, S.; Petrarca, P.; Nicolini, L.; Battistoni, A. Role of ZnuABC and ZinT in *Escherichia coli* O157:H7 zinc acquisition and interaction with epithelial cells. *BMC Microbiol.* **2001**, *11*, 36. [CrossRef]
127. Li, H.; Jogl, G. Crystal structure of the zinc-binding transport protein ZnuA from *Escherichia coli* reveals an unexpected variation in metal coordination. *J. Mol. Biol.* **2007**, *368*, 1358–1366. [CrossRef] [PubMed]
128. Chandra, B.R.; Yogavel, M.; Sharma, A. Structural analysis of ABC-family periplasmic zinc binding protein provides new insights into mechanism of ligand uptake and release. *J. Mol. Biol.* **2007**, *367*, 970–982. [CrossRef]
129. Ilari, A.; Alaleona, F.; Petrarca, P.; Battistoni, A.; Chiancone, E. The X-ray structure of the zinc transporter ZnuA from *Salmonella enterica* discloses a unique triad of zinc-coordinating histidines. *J. Mol. Biol.* **2011**, *409*, 630–641. [CrossRef]
130. Penon-Portmann, M.; Eldomery, M.K.; Potocki, L.; Marafi, D.; Posey, J.E.; Coban-Akdemir, Z.; Harel, T.; Grochowski, C.M.; Loucks, H.; Devine, W.P.; et al. De novo heterozygous variants in *SLC30A7* are a candidate cause for Joubert syndrome. *Am. J. Med. Genet. A* **2022**, *188*, 2360–2366. [CrossRef]
131. Watanabe, S.; Amagai, Y.; Sannino, S.; Tempio, T.; Anelli, T.; Harayama, M.; Masui, S.; Sorrentino, I.; Yamada, M.; Sitia, R.; et al. Zinc regulates ERp44-dependent protein quality control in the early secretory pathway. *Nat. Commun.* **2009**, *10*, 603. [CrossRef]
132. Fan, Y.G.; Wu, T.Y.; Zhao, L.X.; Jia, R.J.; Ren, H.; Hou, W.J.; Wang, Z.Y. From zinc homeostasis to disease progression: Unveiling the neurodegenerative puzzle. *Pharmacol. Res.* **2023**, *18*, 107039. [CrossRef] [PubMed]
133. Cheng, Y.; Chen, H. Aberrance of zinc metalloenzymes-induced human diseases and its potential mechanisms. *Nutrients* **2021**, *13*, 4456. [CrossRef] [PubMed]
134. Kambe, T.; Matsunaga, M.; Takeda, T.A. Understanding the contribution of zinc transporters in the function of the early secretory pathway. *Int. J. Mol. Sci.* **2017**, *18*, 2179. [CrossRef] [PubMed]
135. Murakami, M.; Hirano, T. Intracellular zinc homeostasis and zinc signaling. *Cancer Sci.* **2008**, *99*, 1515–1522. [CrossRef]

MDPI AG
Grosspeteranlage 5
4052 Basel
Switzerland
Tel.: +41 61 683 77 34

International Journal of Molecular Sciences Editorial Office
E-mail: ijms@mdpi.com
www.mdpi.com/journal/ijms

www.ingramcontent.com/pod-product-compliance
Lightning Source LLC
Chambersburg PA
CBHW061450210926
51850CB00008B/1673